Oil and Gas Field Development Techniques

DRILLING

FROM THE SAME PUBLISHER

Publications in English

- Drilling Mud and Cement Slurry Rheology Manual.
- Hydraulic Downhole Drilling Motors.
  W. TIRASPOLSKY
- Directional Drilling and Deviation Control Technology.
- Dictionary of Drilling and Boreholes. English-French, French-English.
  M. MOUREAU and G. BRACE
- Drilling Data Handbook.
  G. GALBOLDE and J.-P. NGUYEN
- Basics of Reservoirs Engineering.
  Oil and Gas Field Development Techniques.
  R. COSSÉ
- Cementing Technology and Procedures.
  Edited by J. LECOURTIER and M.V. CARTALOS
- Cement Evaluation Logging Handbook.
  D. ROUILLAC

INSTITUT FRANÇAIS DU PÉTROLE

Ecole nationale supérieure **du pétrole et des moteurs**

Oil and Gas Field Development Techniques

DRILLING

Jean-Paul NGUYEN
IDN graduate Engineer
Senior Engineer at
Institut Français du Pétrole

Translated from the French
by Barbara Brown Balvet

1996

ÉDITIONS TECHNIP 27 RUE GINOUX 75737 PARIS CEDEX 15

Translation (reviewed Edition) of
"Le Forage", J.-P. Nguyen

ISBN 2-7108-0689-4

Printed in France
by Imprimerie Chirat, 42540 Saint-Just-la-Pendue

FOREWORD

Oil and gas field development and production include a host of specialty fields that can be divided up under four main headings:

- *reservoir engineering,*
- *drilling,*
- *downhole production,*
- *surface production.*

The four aspects are interdependent and oil and gas-related engineering, construction and operations entail the input of a great many operating and service company specialists, as well as equipment manufacturers and vendors. However, these professionals and technicians often lack a comprehensive grasp of all the specialized technology utilized in the process of producing oil and gas.

The aim of the four volumes presented under the title:

OIL AND GAS FIELD DEVELOPMENT TECHNOLOGY

is to provide the basics of the technology and constraints involved in each of the four activities listed above in a condensed form, thereby allowing better interaction among the men and women who work in the different skill categories.

Furthermore, this introduction to oil and gas field technology is also designed to help specialists in other fields, e.g. information processing, law, economy, research, manufacturing, etc. It aims to enables them to situate their work in the context of the other skill categories and in this way make it easier for them to integrate their efforts in the professional fabric.

The four volumes partly recapitulate the contents of the training seminars organized by *ENSPM-Formation Industrie* to meet perceived needs, and also use some of the lecture material given to student engineers attending the *École Nationale Supérieure du Pétrole et des Moteurs (ENSPM).*

A. Leblond
Former Head of Drilling, Production and Reservoir Engineering
at ENSPM-Formation Industrie

CONTENTS

Chapter 3
DOWNHOLE EQUIPMENT

Chapter 4
THE DRILLING RIG

Chapter 5
DRILLING FLUIDS

Chapter 6
WELLHEADS

Chapter 7
CASING AND CEMENTING OPERATIONS

Chapter 8
MEASUREMENTS AND DRILLING

Chapter 9
PRINCIPLES OF KICK CONTROL

Chapter 10
DIRECTIONAL DRILLING

Chapter 11
FISHING JOBS

Chapter 12
THE DRILL STEM TEST (DST)

Chapter 13
DRILLING OFFSHORE

Chapter 1

INTRODUCTION

In 1859, oil came spurting out of the ground for the first time from a well 69.5 feet deep in Titusville, Pennsylvania. Colonel Drake had just gone down in oil prospecting history. But although this event initiated industrial oil well drilling, a large number of wells had been drilled long before to produce water, brine and even naphtha for caulking boats, and for lighting and medicinal purposes.

All boreholes in the olden days, including Drake's, were drilled using the cable system (**Fig. 1.1**).

A massive bit with an edge similar to a sculptor's chisel was attached to the end of a heavy rod (drill collar), which in turn hung from a walking beam. It was dropped free fall onto the rock which it pounded into slivers. The walking beam was actuated by man or animal power long ago, then equipped with a steam engine in the 19th century. But whatever the means of driving the system, the bottom of the hole still had to be cleared of cuttings periodically. The borehole was filled with water and the mud made from the water mixed with broken bits of rock was bailed out with a cylindrical tool. The tool had a valve-like end that was run in open and pulled out closed by means of the drawworks. The deepest well drilled using this method reached 2250 m in 1918. The cable system is still sometimes used today for shallow water wells.

It was at the beginning of the 20th century that Antony Lucas showed the whole world how effective rotary drilling was with the discovery of Spindeltop field (Texas). He combined the use of a rotating bit and continuous mud injection. Since then, rotary drilling has been used worldwide and upgraded by technical developments.

1.1 THE PRINCIPLE OF ROTARY DRILLING

The rotary method uses tricone-type toothed bits or one-piece bits such as diamond or PDC bits. While the bit is being rotated, a force is applied to it by a weight. The advantage is that a fluid can be pumped continuously through the bit, which is crushing the rock formation, and carry cuttings up out of the hole to the surface with the rising fluid flow.

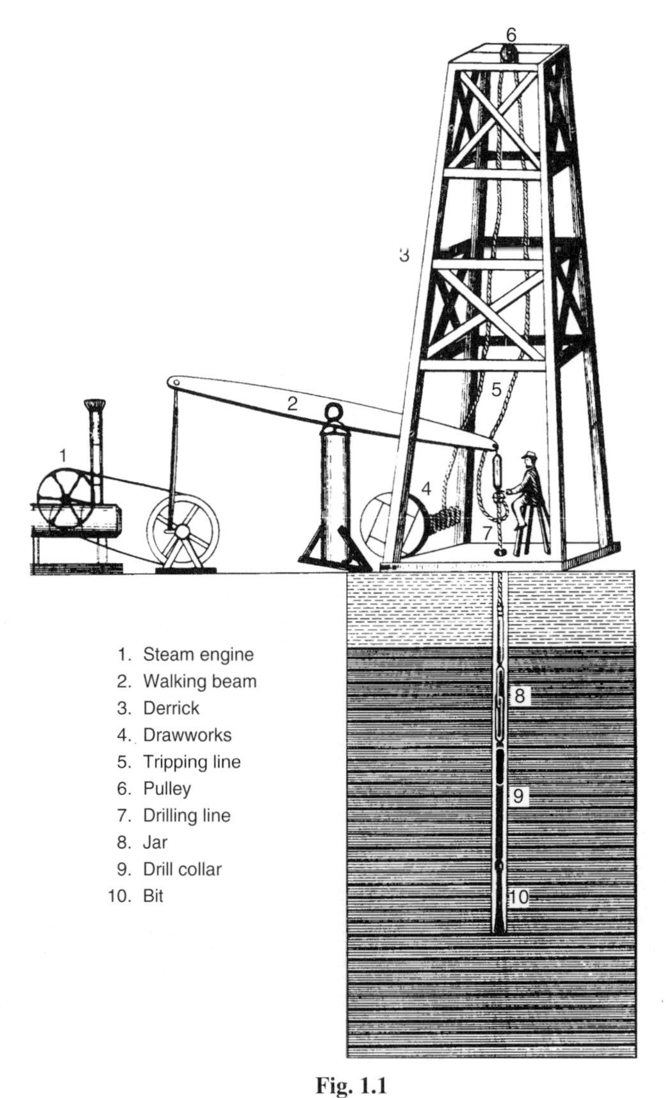

Fig. 1.1

Cable-tool drilling (*Source: "Le pétrole: prospection et production", Esso Standard SAF, Training Department*).

The rotary drilling rig is the apparatus required to fulfill the following three functions (**Figs. 1.2 and 1.3**):

- put weight on the bit,
- rotate the bit,
- circulate a fluid.

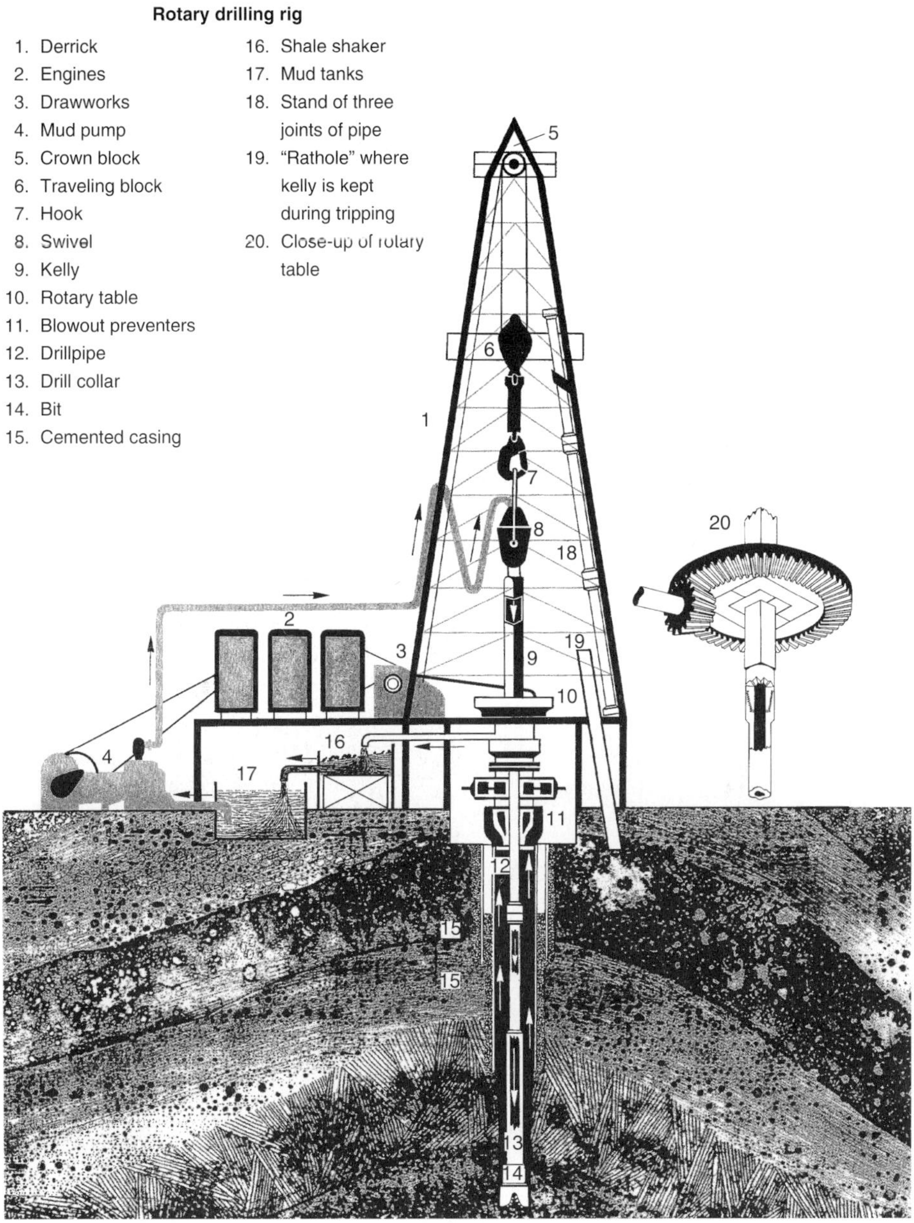

Rotary drilling rig

1. Derrick
2. Engines
3. Drawworks
4. Mud pump
5. Crown block
6. Traveling block
7. Hook
8. Swivel
9. Kelly
10. Rotary table
11. Blowout preventers
12. Drillpipe
13. Drill collar
14. Bit
15. Cemented casing
16. Shale shaker
17. Mud tanks
18. Stand of three joints of pipe
19. "Rathole" where kelly is kept during tripping
20. Close-up of rotary table

Fig. 1.2

Rotary drilling rig (*Source: "Le pétrole: prospection et production", Esso Standard SAF, Training Department*).

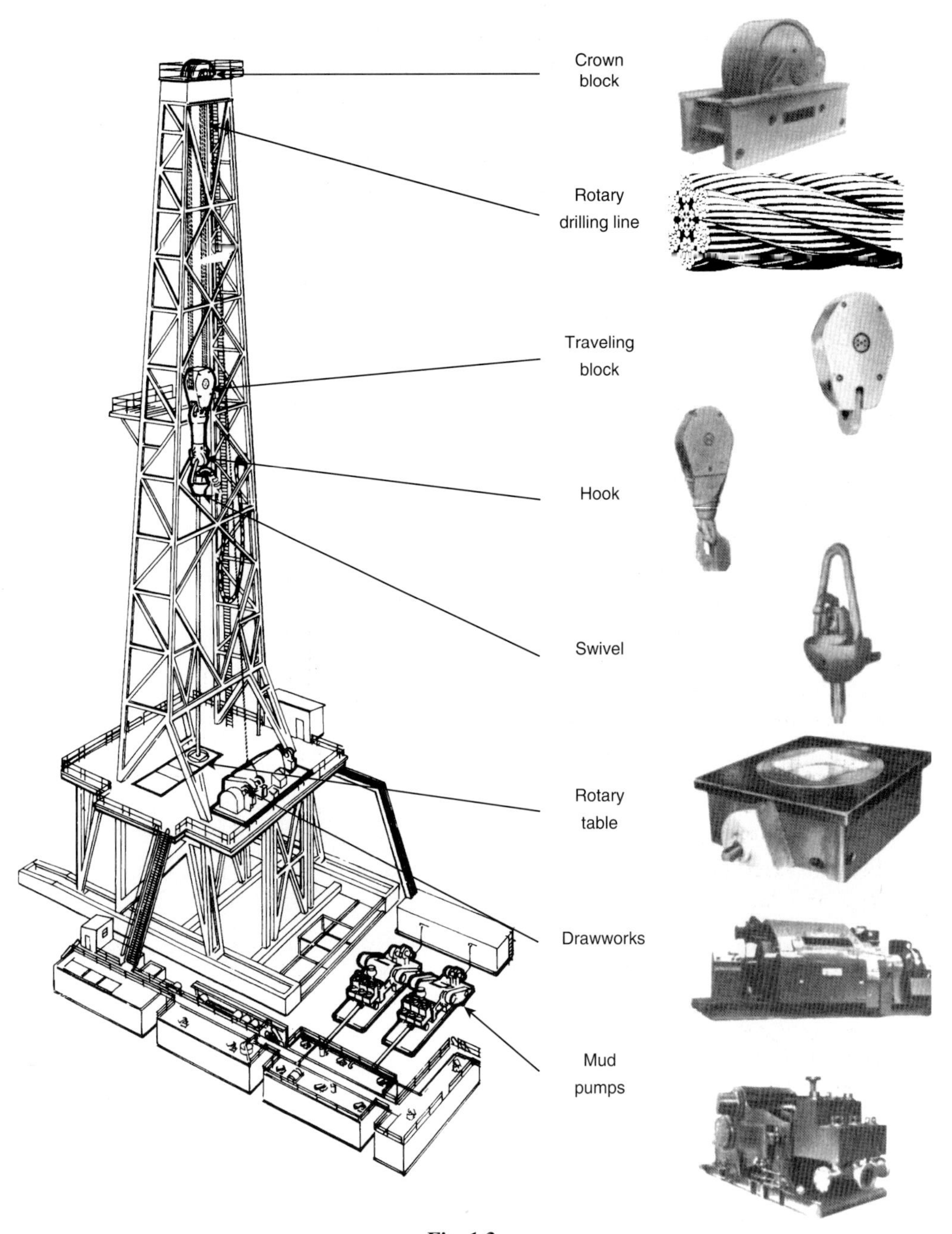

Fig. 1.3
Close-up of a drilling rig (*Source: National Oil Well*).

It is the drill collars, screwed onto the bottom of the drillpipe assembly just above the bit, that provide the necessary weight. Drill collars, along with drillpipe and bit all make up the drill string, which is rotated by the rotary table and the kelly. The drill string component parts are hollow down the middle so that the drilling fluid can be circulated down to the bit. A fluid-tight rotary joint, the swivel, is located at the top of the kelly and provides a connection between the mud pump discharge line and the inside of the drill string. A hoisting system is required to support the weight of the drill string, lower it into the hole and pull it out. This is the function of the derrick, the hook and the drawworks.

The drilling rig is complete with facilities to treat the drilling fluid when it gets back to the surface, a storage area for tubular goods, shelters and offices on the site.

In addition, when a well is being drilled it is regularly cased. It is lined with steel pipe, or casing, which is lowered into the hole under its own weight in smaller and smaller diameters as the hole gets deeper. The first length of pipe is run in as soon as the bit has drilled the surface formation and is then cemented in the hole. A casing housing is connected to the top of the surface casing. All the following lengths of pipe are hung on the casing housing and cemented at their base to the walls of the hole (**Fig. 1.4**).

After the first drilling phase is cased, drilling will be resumed with a bit with a diameter smaller than the inside diameter of the casing string that was run in and cemented. The deeper the borehole gets and the more casings are set in the well, the smaller the diameter of the bit must be.

The casing housing also serves to hold the safety equipment, such as blowout preventers.

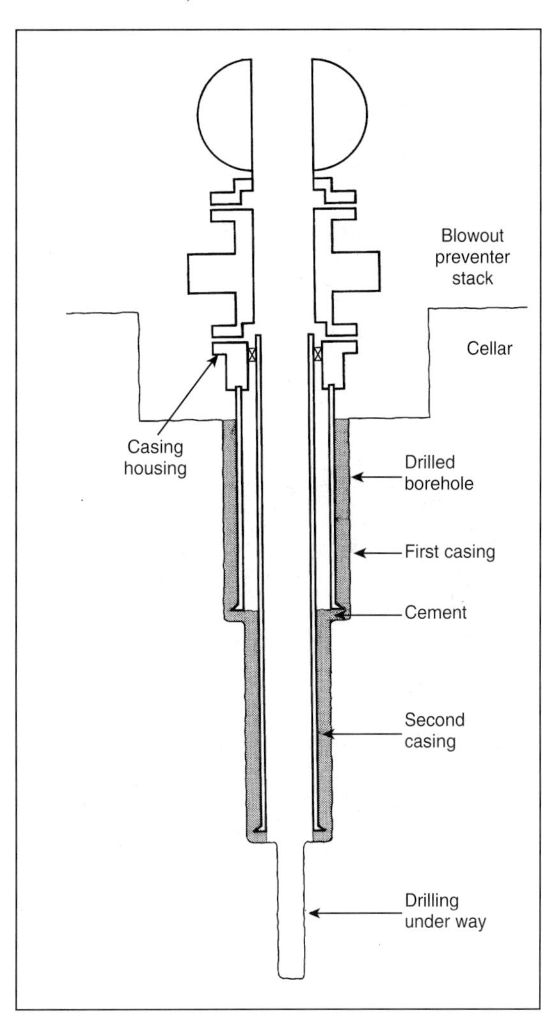

Fig. 1.4
Simplified cross-section of a borehole.

1.2 THE MAJOR OPERATIONS

1.2.1 Drilling (Fig. 1.5)

Though drilling is the basic operation, it is the one that requires the fewest number of people. The driller operates the drawworks alone. The rotary table rotates and drives the drilling bit by means of the drill string and the kelly. The main control device is the brake lever. The driller controls and regulates the downward movement of the hook by putting on the brake. According to the principle of rotary drilling, bits are used at constant weight. The weight of everything that is suspended from the hook is constant and the driller knows this information by reading the weight on the hook when the bit is off bottom (**Fig. 1.6**).

Fig. 1.5
The driller's control console
(*Source: Driller console, Martin Decker*).

The weight applied on the bit is the difference between the weight on the hook off bottom and on bottom.

This is the difference that the driller reads on the weight indicator (commonly called Martin Decker). He must keep it constant by lowering the kelly at the same speed as the rate of penetration of the bit.

The other two parameters, rotation and mud flow rate, are generally preset. The driller checks and adjusts the values depending on the program and mainly sees to it that the pump discharge pressure complies with the program and stays that way.

1.2.2 Adding drillpipe

When the bit has drilled the equivalent of a length of pipe (30 ft.), the drill string must be lengthened by screwing a new joint of drillpipe onto the bottom of the kelly. The sequence is illustrated in **Figs. 1.7a, 1.7b, 1.7c and 1.7d:**

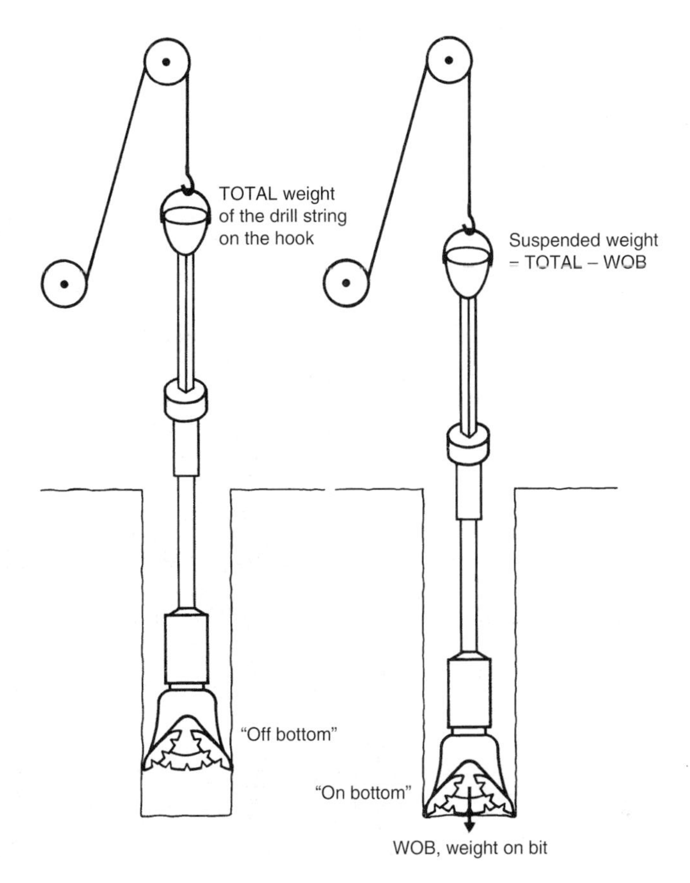

Fig. 1.6
Checking the weight on the bit.

- During drilling, the crew places a joint of pipe in a sheath, called the mousehole, located near the rotary table.
- The driller engages the drawworks to hoist the drill string to the first length of drillpipe under the kelly. The crew puts the slips in place and the kelly can be unscrewed since the drill string is supported by the rotary table. Mud circulation has of course been stopped. In **Fig. 1.7b**, the crew screws the kelly to the box end of the length of drillpipe in the mousehole. Pipe is screwed and made up in the mousehole.
- The driller hoists the kelly and drillpipe with the drawworks (**Fig. 1.7c**). Once the new joint of pipe has been screwed and made up on the drill string, the driller resumes drilling fluid circulation.
- The crew places the kelly bushing back in the rotary table and drilling can be resumed (**Fig. 1.7d**).

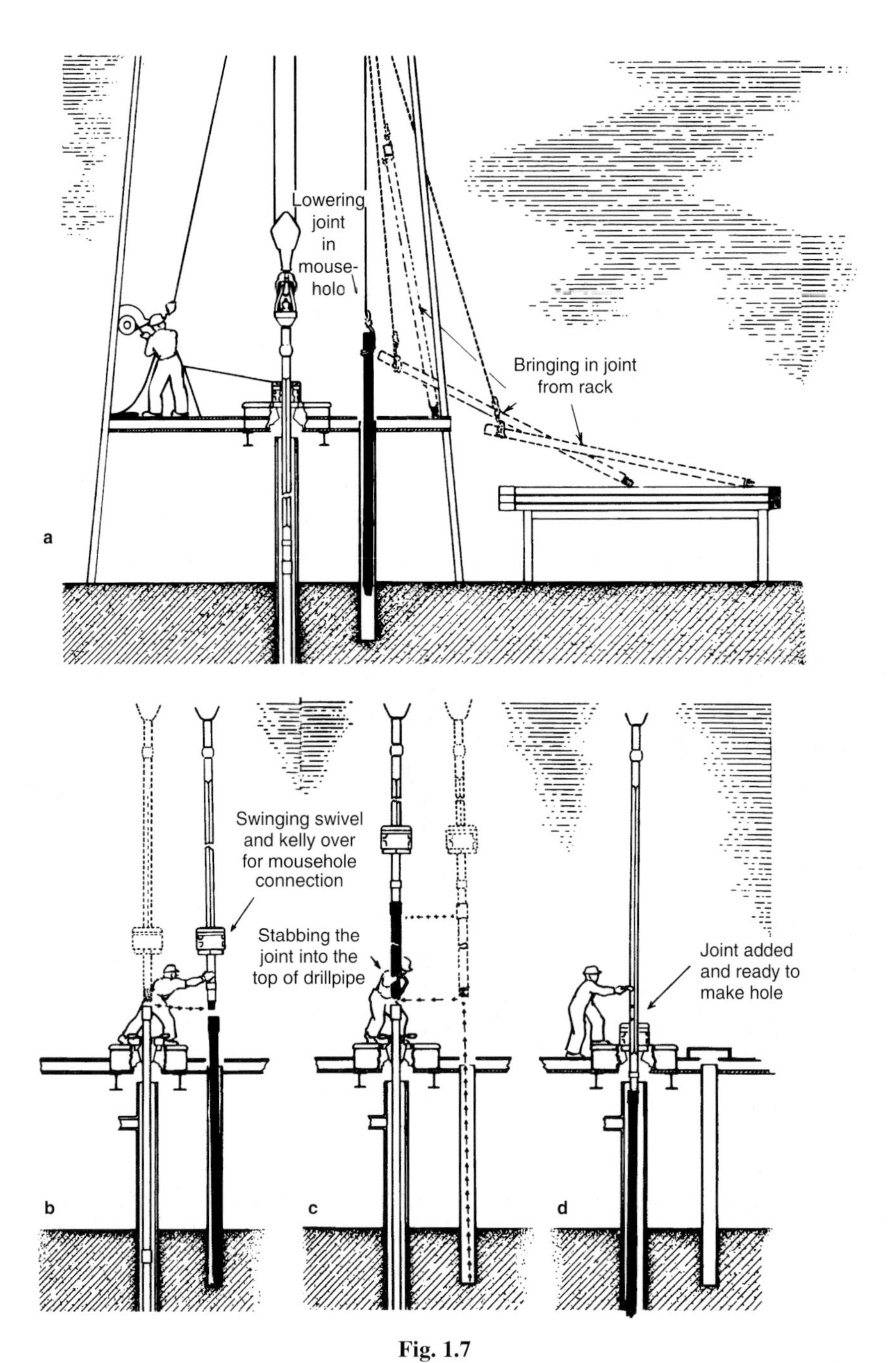

Fig. 1.7

Adding drillpipe (*Source: A primer of oil well drilling, Petex*).

1.2.3 The round trip

When the bit is worn or when total borehole depth has been reached, all of the drill string must be pulled out of the hole to change bits or run in casing pipe.

The first step is to disconnect the swivel from the hook and place the kelly and the swivel —which is still connected to the mud pumps by the hose— in a sheath called the rathole (**Fig. 1.8a**).

The crew latches the elevator under the tool joint of the first length of drillpipe and the driller hoists the drill string to a height of three joints of pipe with the drawworks.

The fourth length of pipe is clamped in the rotary table by the slips and the connection is unscrewed with the tongs (**Fig. 1.8b**). A stand of three lengths of pipe is then hanging from the elevator. The crew pushes the lower end of the stand so that it rests on the setback. Then the derrickman, who stands on the monkey board, unlatches the elevator, holds the stand and places the upper end of the stand in the pipe rack (**Figs. 1.8c and 1.9**).

The operation continues down to the drill collars, that are also stacked vertically in threes.

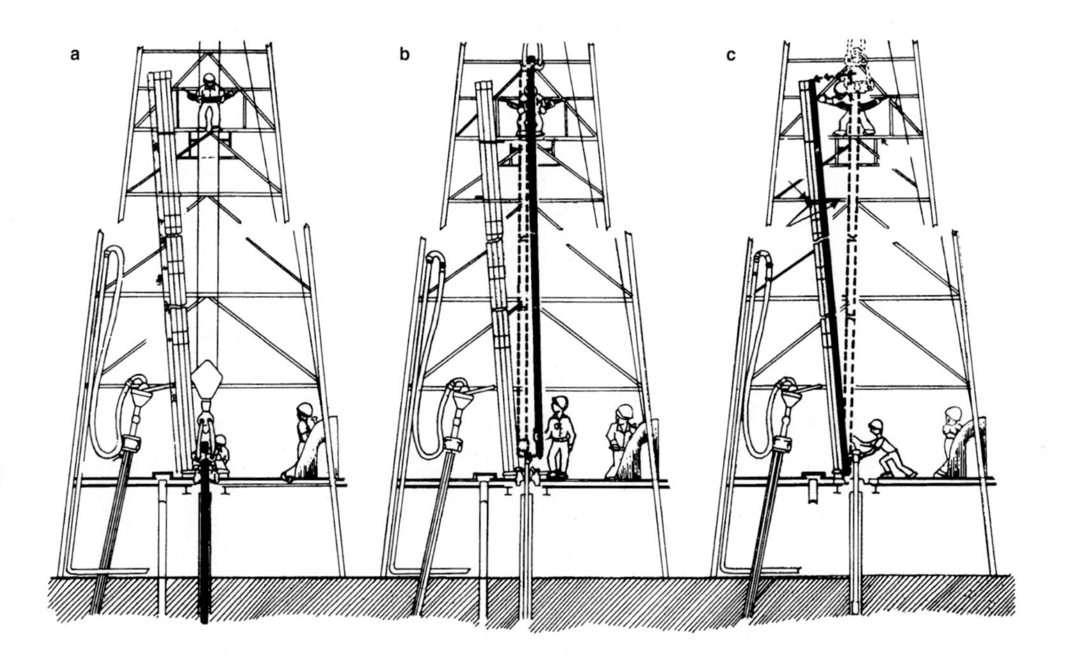

Fig. 1.8

Round trip.

a. Latching the elevator; **b.** Pulling out a stand (thribble); **c.** Stacking the stand on the setback (*Source: A primer of oil well drilling, Petex*).

Monkey board seen from rig floor.

Racking board.

Fig. 1.9

Drilling mast with stacked drill string (*Source: Reynolds Aluminum Drill Pipe*).

The stand length depends on how high the derrick is. The largest rigs handle stands in threes, lightweight rigs in twos and the smallest ones can manage only singles. The running in, or tripping in, operation is carried out in the same way.

During a round trip both rotation and circulation are at a standstill. If either is needed, the kelly is taken out of the rathole and screwed back onto the drill string.

1.2.4 Casing

Once the borehole has been drilled to the depth planned for the current phase, the casing pipe is run into the well. The operation is hazardous because of the narrow clearance between the casing and the borehole, and since it is almost impossible to rotate the casing string. Casing pipe is run in singly joint by joint. Once it is run in, normal circulation (i.e. down the inside of the pipe) is used to pump cement into the annulus between the casing and the borehole wall.

1.2.5 Installing the wellhead

When casing has been run into the well and cemented, a variety of hanging and sealing equipment must be installed on top of the well.

The operations are done manually when wellheads are above ground or above the surface of the sea offshore.

Wellhead equipment also accommodates the blowout preventers (BOP) that have a high-pressure system called kill line and choke line.

A series of pressure tests on the casing, hangers and BOP finalizes the installation. If everything complies with safety requirements, the following drilling phase can then commence.

1.2.6 Completion

This is the final operation just after running in the last casing string (production casing). The production equipment is run into the well: packer, tubing, safety valve, etc. The connection between the producing formation and the well must often be enhanced by drilling, perforations, acidizing, fracturing, etc.

Though these operations are often performed by drillers, the techniques involved come under the heading of downhole production which is dealt with in another book.

The purpose of the following chapters is to describe as completely as possible the materials, equipment and operating techniques that are involved during drilling. This book aims to serve as an introduction to an operation that is more complex than it seems. The goal is to help the reader who is not a driller, but whose activity is related to drilling, understand the basics.

Chapter 2

DESIGNING AN OIL WELL

INTRODUCTION

An oil or gas well is a borehole drilled in the subsoil to reach a hydrocarbon deposit. The primary objective may be exploration (prospecting for a deposit) or development (bringing a deposit on stream). A well fulfills the following functions:

- gain access to the underground deposit,
- make a connection with the producing formations,
- allow the hydrocarbon effluent to reach the earth's surface safely and effectively,
- support equipment on the surface to control production and allow maintenance (wireline and workover operations, etc.)

2.1 WELL ARCHITECTURE

2.1.1 Data required to construct a well

There is no way or technique to detect the presence of hydrocarbons in the subsoil with any certainty from the surface. Drilling is done to confirm the existence and nature of the hydrocarbons that may be contained in the reservoir rock.

The depth of an oil well can vary from a few hundred meters to 10,000 meters to reach its objective. On the average, a well goes down 2000 meters; deep is beyond 3500 meters and extra deep is beyond 4500 meters.

It is therefore an expensive venture and like any industrial investment it must be studied and planned before it is carried out.

The first study phase entails defining the well profile and its general architecture.

Working out the profile and the methods to reach the objective is the result of the efforts of several people in different departments and divisions. The well architecture is designed afterward in several stages under the responsibility of the drilling department. The stages can be summarized by the documents that are drawn up, generally:

- the well prognosis,
- the well program,
- the drilling and casing program.

2.1.1.1 The well prognosis

This is the basic document established as soon as the possibility of drilling a well arises. It defines:

- the well site and location (coordinates, altitude or water depth),
- the drilling objective(s) (type, relative size, estimated depth(s)).

It includes basic data on:

- geology, geophysics,
- information on nearby wells if there are any,
- constraints and deadlines to be complied with.

The well prognosis is a working document that varies depending on the well, which can be for exploration or development.

The document must indicate the decisions to be made in due time since there are relatively long lead times, such as for preparing the site and choosing the rig. It also initiates studies to confirm the feasibility of the project, which in turn lead to a positive or negative technical and economic decision on drilling the well.

If the well is for exploration purposes, there is automatically a certain lack of precision as to the probable depth where formations will change and even as to the nature of the formations. The geologist can supply estimates of probable mud losses, presence of abnormally high-pressure fluids and stability of formations. If there are any wells nearby, all the information they have provided will carefully be collected.

Establishing an exploratory drilling program therefore requires gathering the most information possible beforehand and knowing the degree of reliability of the data.

For development drilling, the experience of the first wells drilled on the structure quickly leads to a master program, since the problem is the same for all the wells.

The drilling objective is perfectly clear for a development well. In exploratory drilling, the geologist must be able to specify a main objective and the whole program will be constructed around it. Secondary objectives will be taken into consideration whenever possible.

2.1.1.2 The well program

This new document is established when the decision to drill has been made. It is similar to contract terms and conditions and covers all of the operations to be carried out. It is used to determine the budget of the venture.

It is the result of collaboration among the various relevant departments:

- exploration,
- reservoir engineering,
- drilling,
- production.

The main chapters in the well program are:
- the geographical location,
- the purpose of the well and petroleum objectives (this is a recapitulation of the base data, along with the objective(s) defined in the well prognosis that have been modified or supplemented since then by studies and research),
- the geological context,
- the geophysical context,
- the drilling and casing program,
- the logging program,
- the coring and testing program,
- the sample-taking and analysis program.

The document can also define requirements for:
- manpower,
- equipment (rig, etc.),
- services,
- expendables.

2.1.2 The drilling and casing program

2.1.2.1 Background basics

Establishing a drilling and casing program is of prime importance in preparing a drilling venture.

The actual aim is to choose construction characteristics so that the borehole can reach its objective as economically as possible while complying with a number of specifications.

The characteristics to be chosen are as follows:
- respective diameters of bits and pipe strings,
- number of strings and setting depth,
- height of cement between casing and borehole walls.

A good drilling and casing program is a decisive factor in the success of drilling, in operational safety and in the final cost price of the well.

An inadequately developed program may mean that proposed objectives can not be reached or that the venture is a total failure. In contrast, an overloaded program adds unnecessarily to the cost price of exploration. It will reduce the number of wells that can be drilled with a given budget, thereby reducing the odds of hitting pay sands.

A drilling and casing program is generally set up as follows:
- The diameter of the last string is determined.
- The depths of casing shoes are chosen along with the diameters of the various strings and drilling phases. The different diameters are deduced on the basis of the last phase near the objective, moving up progressively to the surface.

The drilling and casing program will be worked out in a totally different way depending on whether it is for an exploratory well or a production well.

An exploratory well is designed to go through formations that geologists know little about, especially as to the difficulties that can be encountered: lost circulation, swelling shales, aquifers, fluids under high pressures, etc. As a result, a certain safety coefficient must be adopted when the drilling and casing program is being set up.

In contrast, a development well goes through known formations to reach a clearly determined objective. Additionally, it is due to keep producing for a long time.

Before the methods for establishing the program are discussed, a few basic principles need to be recalled.

As a borehole gets deeper, the drilling fluid provides a plastering effect to consolidate the walls among its other functions. The mechanical instability of some formations, in particular fairly loose, often unconsolidated surface layers, requires setting one or even two casings. Otherwise there will be problems of caving or narrowing down in the hole. The aim is to avoid extra reaming, drill string sticking and unsuccessful fishing that may sometimes cause the hole to be abandoned.

In addition, the column of mud must be heavy enough to keep formations from flowing and prevent unwanted fluid influxes (water, gas, oil) without fracturing the formations, however. Successive changes in the type of formations and in the internal pressure in the rock, also called pore pressure, can make the two conditions incompatible. This is when setting extra casing, called intermediate casing, becomes necessary.

Besides these considerations, it would not be possible to set only one type of pipe capable of withstanding all the stresses, internal and external pressures, pipe yield stresses, sour gases, etc. found in deep, high-pressure, or deviated wells. All these reasons are why several casings are set from the surface down to the objective.

The casings isolate the formations from one another when they need it by means of the sheath of cement.

2.1.2.2 Casing and tubing pipe technology

Casing and tubing are tubular goods that are made to *American Petroleum Institute* (*API*) 5 CT standards. They are characterized by:

- the dimensions of the pipe body,
- the grade of steel,
- the dimensions and style of the connection.

A. *Dimensions of the pipe*

• The nominal diameter of a casing is the outside diameter of the pipe body. There are fourteen sizes in inches:

4 1/2 – 5 – 5 1/2 – 6 5/8 – 7 – 7 5/8 – 8 5/8 – 9 5/8 – 10 3/4 – 11 3/4 – 13 3/8 – 16 – 18 5/8 – 20.

The larger diameters are classified as Line Pipe.

There are ten tubing sizes:

1.050 – 1.315 – 1.660 – 1.900 – 2.063 – 2 3/8 – 2 7/8 – 3 1/2 – 4 – 4 1/2.

• Lengths are defined by the range.

Type of pipe	Range 1	Range 2	Range 3
Casing	16 to 25 ft	25 to 34 ft	34 to 48 ft
Tubing		25 to 34 ft	34 to 48 ft

The length of a pipe is measured between the rim of the box end and the nose of the pin thread.

• The nominal weight in pounds/foot in fact defines the thickness of the pipe by means of a formula that is found in API 5 CT. For example, a 9 5/8" casing is available in the following nominal weights:

Nominal weight (lb/ft)	32.30	36	40	43.50	47	53.50	58.40	61.10	71.80
Thickness (mm)	7.90	8.90	10	11.10	12	13.80	15.10	15.90	19.10

The inside diameter decreases when the nominal weight increases.

• Drift (or diameter of the mandrel) is the diameter of a gage that must be able to pass through the pipe. This is a point that must systematically be checked on the well site before the string is run in.

B. Dimensions and style of connections

There are all sorts of connections differentiated by their length, thickness and thread. The two main types of thread are:

- round thread (API round) (**Fig. 2.1**),
- buttress thread (**Fig. 2.2**).

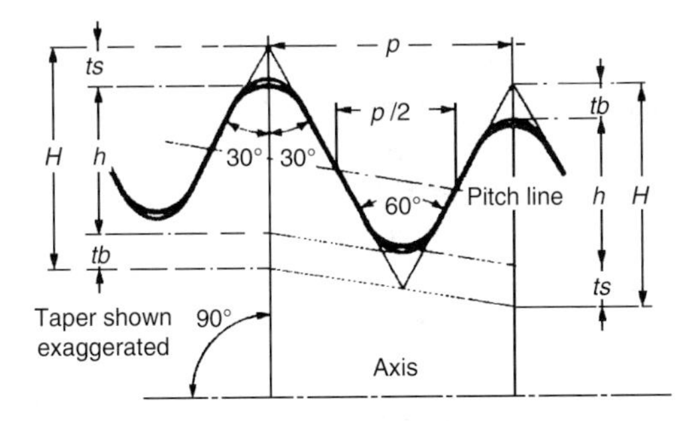

Taper: 6.25%

8 threads/in, p = 3.175 mm

$H = 0.866\ p$ = 2.750 mm

$h = 0.626\ p - 0.178$ = 1.810 mm

$tb = 0.120\ p + 0.051$ = 0.432 mm

$ts = 0.120\ p + 0.127$ = 0.508 mm

Fig. 2.1

API Round thread *(Source: Drilling Data Handbook, Editions Technip, Paris, 1989).*

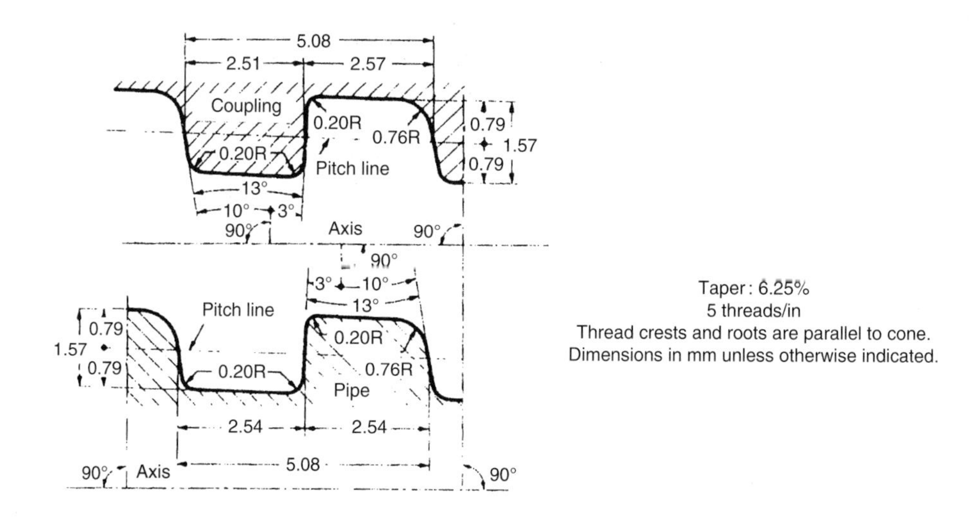

Fig. 2.2

Buttress thread *(Source: Drilling Data Handbook, Editions Technip, Paris, 1989)*.

There are three main styles of connections:

- Conventional (**Fig. 2.3a**) where the seal is ensured by a grease and metallic additive based compound packed down into the root of the threads.
- VAM (**Fig. 2.3b**) is a connection that uses the buttress thread but has a metal/metal seal provided by a shoulder shape. This type of connection is used when greater seal efficiency is required, e.g. for high pressures and gas wells.
- Extreme line (**Fig. 2.3c**) is a connection without a coupling. The thread is directly machined in one end of the pipe itself, usually thickened by forging. It screws together with a shouldered connection.

C. *Grade of steel*

The yield strength as defined by API is actually the stress that has already produced permanent strain on a test coupon. Elongation is usually 0.5%.

The grade is expressed by a letter followed by the minimum yield strength in thousands of psi.

API considers three types of steel for casing pipe.

Normal steels (Table 2.1)

The pipe must be seamless or electrically welded with no addition of extra metal. It must undergo heat treatment (standardization).

The base product must be manufactured of Martin steel, in an electric furnace or oxygen basic converter. Maximum phosphate and sulfur content is 0.040% and 0.060% respectively.

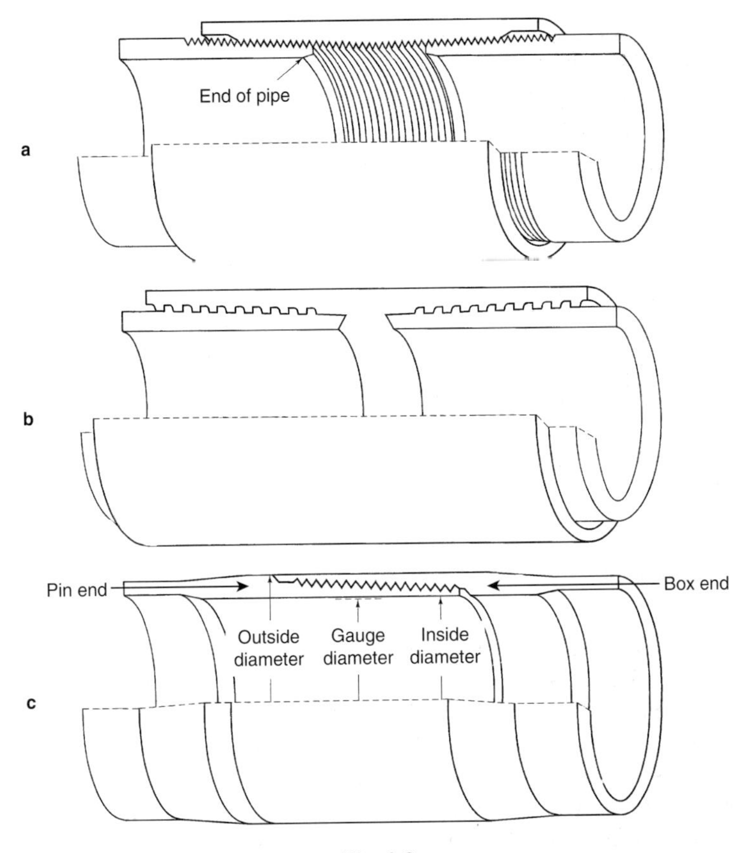

Fig. 2.3

Pipe connections. **a.** API round; **b.** VAM; **c.** Extreme line.
(*Source: Drilling Data Handbook, Editions Technip, Paris, 1989*).

TABLE 2.1

Normal steels

Grade	Yield strength				Tensile strength	
	Minimum		Maximum		Minimum	
	psi	MPa	psi	MPa	psi	MPa
H40	40,000	276	80,000	552	60,000	414
J55	55,000	379	80,000	552	75,000	517
K55	55,000	379	80,000	552	95,000	665
N80	80,000	552	110,000	758	100,000	689

High-strength steels (Table 2.2)

The pipe must be seamless. The base product and maximum phosphorus and sulfur content are the same as for normal steels.

TABLE 2.2

High-strength steels

Grade	Yield strength				Tensile strength	
	Minimum		Maximum		Minimum	
	psi	MPa	psi	MPa	psi	MPa
P105	105,000	724	135,000	931	120,000	827
P110	110,000	758	140,000	965	125,000	827
Q125	125,000	862	155,000	1069	135,000	931
V150	150,000	1034			160,000	1104

Steels with a limited yield strength range (Table 2.3)

The pipe must be seamless or electrically welded with no addition of extra metal. It must undergo heat treatment (standardization + tempering or quenching + tempering).

Besides carbon (0.50%), these steels may contain manganese, molybdenum, chromium, nickel or copper. They have a phosphorus, sulfur and silicon content lower than 0.040%, 0.060% and 0.35% respectively.

In fact, in the category with a yield strength of 75,000 to 95,000 psi there are a number of quite different special steels. To differentiate between them, the different types of corrosion must first be discussed.

TABLE 2.3

Steels with a limited yield strength range

Grade	Yield strength				Tensile strength		Hardness
	Minimum		Maximum		Minimum		Maximum
	psi	MPa	psi	MPa	psi	MPa	HRC
C75	75,000	517	90,000	620	95,000	655	22
L80	80,000	552	95,000	655	95,000	655	23
C90	90,000	620	105,000	724	100,000	690	25.4
C95	95,000	655	110,000	758	105,000	723	

H_2S embrittlement

The presence of H_2S causes cracks to form under stresses lower than the theoretical strength limits of materials (SSC = sulfide stress cracking). Embrittlement results from the action of high stress along with free hydrogen produced on the surface of the metal by a reaction between iron and H_2S. The hydrogen atoms tend to lock the steel strain mechanisms.

The mechanism depends on temperature and yield strength as illustrated on the graph in **Fig. 2.4**.

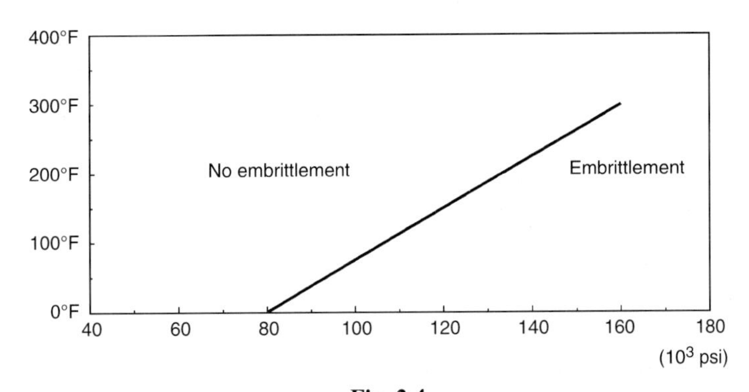

Fig. 2.4
Influence of H_2S versus temperature.

The graph shows that there is embrittlement for a yield strength of over 80,000 psi and that embrittlement is at a maximum at low temperature. It is no longer a problem from 150°C onward whatever the type of steel.

The best steels are those with a fine-grained homogeneous structure, with the lowest content of nonmetallic inclusions (sulfur, phosphorus). The steels can be modified by quenching (enhanced by manganese, chromium or nickel) and tempering (molybdenum and vanadium retarders).

For a given grade, the lower the steel's mechanical properties, the less sensitive it will be to embrittlement. The yield strength range will therefore be reduced to 10,000 or 15,000 psi.

The hardness threshold below which the risk of embrittlement is theoretically averted is 22 HRC. This is true only for steels with a yield strength of less than 95,000 psi.

The grades in **Table 2.3** for H_2S duty are manufactured by Vallourec, for example, under the names VH and VHS (**Table 2.4**).

TABLE 2.4

Grade	Yield strength		Tensile strength	Hardness
	Minimum	Maximum	Minimum	Max HRC
L80 VH	80,000	95,000	95,000	23
C90 VHS	90,000	105,000	100,000	24
C95 VH	95,000	110,000	105,000	27

CO_2 corrosion

Produced waters from fields containing CO_2 are chloride waters acidified by carbonic acid. The Cl^- ion content ranges from a few ppm to several moles per liter. The pH can go from 2.5 to 6.5.

Aluminum and copper alloys could be used in corrosion control but are excluded because they lack the right mechanical properties. The only possibility is steels of the following types: 9% Cr + 1% Mo, 13% Cr, or nickel-chromium-molybdenum alloys with a small percentage of iron.

The grades in **Table 2.3** for CO_2 corrosion duty are manufactured by Vallourec, for instance, under the names:

- C 75 VC, L 80 VC, N 80 VC for 13% Cr steels.
- L 80 VCM, C 95 VCM for 9% Cr + 1% Mo steels.

The letters VC and VCM are replaced by:

- VS 22 for 22% Cr – 5.5% Ni – 3% Mo,
- VS 28 for 27% Cr – 31% Ni – 3.5% Mo,
- VS 42 N for 42% Ni – 22% Cr – 3% Mo,

and there are grades 80 VS 22, 95 VS 22 or 110 VS 42 N.

2.1.2.3 The different types of strings and their functions (Fig. 2.5)

A. *Conductor pipe*

a. *Onshore well*

Here it is a string of very lightweight pipe (rolled sheet metal) set in the subsoil at a depth of around a dozen meters. The purpose is to channel the mud when drilling is started up and to keep the loose surface formations from being washed away. It is usually run in and cemented by the service companies that prepare the well site and is connected to a mud return line.

b. *Offshore well drilled from a stationary rig*

Here the conductor pipe must go down through the water head to the bottom. It is generally driven down into the bottom to refusal. It is made up of components 26", 30" or 36" in diameter with a wall thickness of 1" that are welded one after the other as the string is driven in.

c. *Offshore well drilled from a floating rig*

Here the series of operations to start up a well is as follows:

- positioning on location,
- a temporary base plate with guidelines is lowered,
- drilling begins in 36" with lost circulation,

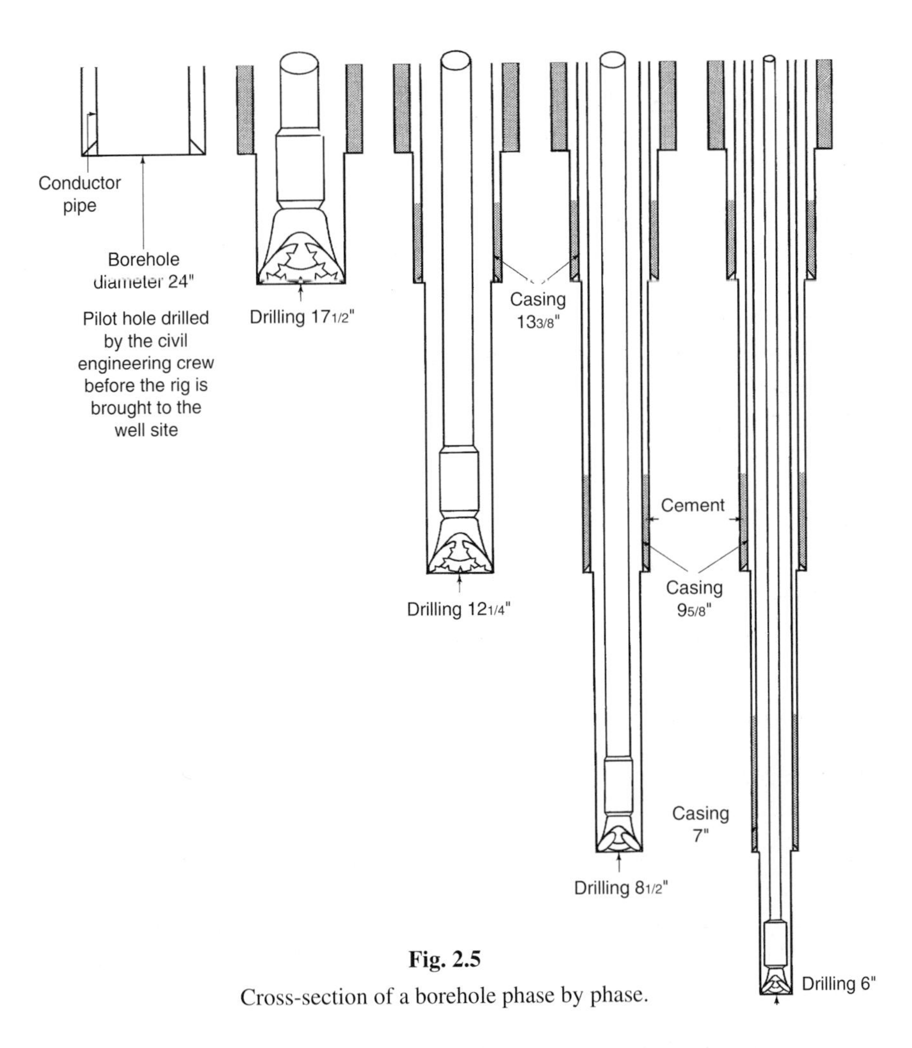

Fig. 2.5

Cross-section of a borehole phase by phase.

- a 30” conductor pipe is lowered and cemented, and on the upper part there is a guide structure to accommodate:
  - casing hangers,
  - the blowout preventer stack with the riser going up to the surface.

B. Surface casing

This is the first real string of pipe run in and cemented in the borehole with routine oil industry procedure. Its purpose is to secure formations near the surface which quite often have a tendency to cave in. The casing also serves to seal off surface fresh water to keep it from being polluted by drilling mud, and to provide an anchoring point for blowout preventers and for hangers for further casing strings.

To fulfill all these functions, the surface casing is cemented right up to the surface. In the event this is impossible (due to cement losses, insufficient cement, etc.), further cementing through the annular space is needed in order to give mechanical stability to the whole well architecture. The aim is for this string to be solidly set and fluid-proof over its complete height.

The length of the string may vary, according to local conditions, from several dozen to several hundred meters.

C. *Intermediate casing string*

This string is designed to allow drilling to continue in the following circumstances:

- An uncased portion of the hole which is a source of trouble during drilling or tripping (walls may cave in, a key-seat may be formed, etc.). Progress in adapting drilling mud to the formation and improvement in drilling methods mean that fewer strings are run in for this reason.
- Even so, in difficult cases, the drilling mud can not provide sufficient protection for drilling to be able to continue under normal safety conditions. Some clayey formations are very prone to hydration, they are a constant hazard, despite the use of calcium mud (swelling shales, caving and sloughing walls). Poorly consolidated shales and sands deteriorate gradually as time goes by with the mechanical action of mud circulation and drill string movement in the hole. Once these troublesome formations have been completely drilled through, a good precaution is to case them with a cemented string.
- The need to isolate formations containing high-pressure fluids. In fact, the formations are drilled with heavy mud, which has two drawbacks: oil shows are hard to spot and mud may be lost into underlying normal-pressure reservoirs. It is recommended to case these formations with a cemented intermediate string to avoid these drawbacks.
- The need to isolate formations containing fluids under low pressure. Here the problem is the reverse of the preceding one. The formations are usually drilled with partial or total mud losses and it is advisable to cover them up with a casing string after drilling through them. This prevents costly mud losses and allows leeway to weight up the mud after casing the hole, in the event formations with high-pressure fluids are encountered. The same is true when the drilling program requires the use of salt saturated mud to drill a salt formation. This mud's specific gravity needs to be higher than 1.23 and underlying formations may not be able to withstand it. They therefore have to be cased.

If none of the reasons mentioned above applies, it may still be advisable to set a protection casing string if the borehole is due to be drilled very deep.

Before the protection casing is set, borehole safety is secured by blowout preventers connected to the surface casing. Whatever the quality of the blowout preventers, the well as a whole can not withstand pressure above the formation breakdown pressure of the weakest uncased portion. Since the surface string seat is usually not very deep, it would be dangerous to approach very deep reservoirs with inadequate safety. For example, a reservoir at 4000 m can contain a fluid under a pressure of 450 bar. If the surface casing is 200 m deep, the formation at this depth can withstand a maximum of 48 bar given a geostatic gradient of 2.4 bar/10 m.

If a kick begins to occur, there can be a pressure of 50 bar at the wellhead. It can cause the formations around the casing seat to break down, in turn causing a crater and an uncontrollable blowout.

Sometimes two intermediate casing strings have to be run in for ultradeep drilling. The intermediate string is cemented over a height of several hundred meters above the shoe.

If need be, staged cementing can be done to isolate a shallower formation without having to cement the height from the casing seat on up (see Chapter 7, Cementing).

D. Production string

In development drilling, this string is indispensable to provide full protection for the pay zone and allow implementation of production equipment.

In exploratory drilling, if oil or gas shows warrant more extensive testing than a simple open-hole test, a string is run in that temporarily has the same functions as a production string. If the well is abandoned, most of it can be retrieved.

The production string is set either at the top of the pay zone or goes right through it. It is cemented in the same way as the intermediate string.

The surface and intermediate strings are commonly termed protection casing in contrast with production or test strings. Protection strings are designed to make continued drilling technically possible while production or test strings are needed to produce a reservoir.

E. Liner

All the strings described above cover the borehole walls from the bottom to the surface and are hung inside the preceding casing string.

Liners are hung by a mechanical system, a liner hanger, in the lower part of the preceding string. The advantages are savings on pipe, less weight and a larger inside diameter between the surface and the top of the liner. But there are also a number of drawbacks: casing and cementing operations are more complex, the preceding string must be sized at the same bursting pressure as the liner, and the preceding string may be seriously weakened by the wear and tear of drillpipe rotating when the well is deepened.

2.1.2.4 Required clearance (Fig. 2.6)

A borehole is a telescope-like construction since each casing set in the hole reduces the diameter that could be drilled later on. Due consideration must be given to two types of clearance. First is the clearance between the inside of a casing string and the bit that could be used to continue drilling. And second is the indispensable clearance between the borehole walls and the outside of the casing string.

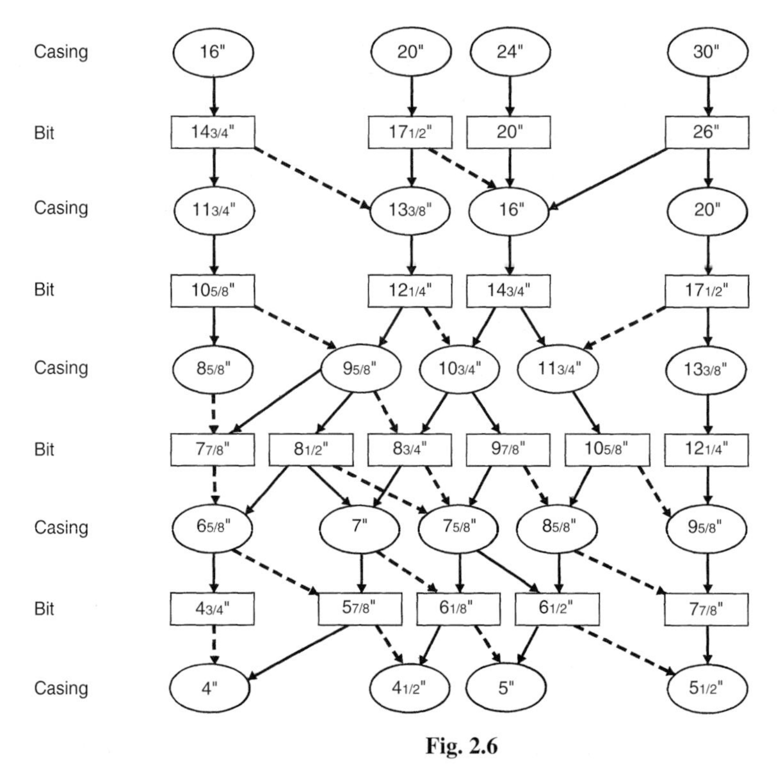

Fig. 2.6

Diagram: selecting bit diameters according to casing diameters
(Source: Drilling Engineering, Neals Adams, Penwell Books).

A. Bit/inside of casing clearance

This is the difference between the smallest inside diameter of the casing string and the diameter of the bit that will pass through it.

The thickest-walled pipe has the smallest inside diameter.

The acknowledged rule is that clearance should never be less than 3/32 of an inch, i.e. 2.38 mm. Given manufacturing tolerance, pipe may have an actual inside diameter slightly smaller than the figure indicated in API standards. This is why these standards define the diameter of the gage (drift).

B. Borehole wall/outside of casing clearance

Two types of clearance should be considered: between the pipe per se and the hole, and between the coupling outside diameter and the hole.

The two are very important since they affect:
- how readily the string can be lowered into the borehole,
- how thick the ring of sealing cement is between the formation and the string.

Actually, a borehole is never a perfect cylinder. As the string is run into the hole, it must automatically undergo deformation to follow the curve of the hole. This is why it is necessary to keep enough clearance to allow ready deformation. The less cylindrical the hole and the deeper the well, the more clearance becomes a prime concern.

Additionally, the ring of cement must be thick enough to provide a strong, leak-proof bond between the pipe body and the walls of the hole. The quality of the cement job depends primarily on how well the cement flows through the annulus.

Experience has yielded a rule of the thumb giving the minimum hole diameter to enable a string of pipe to be run in and cemented (**Table 2.5**).

TABLE 2.5

String (in)	Coupling diameter (in)	Min. hole diameter (in)	Coupling/hole clearance (in)
18 5/8	19 3/4	22	2 1/4
16	17	18 1/2	1 1/2
13 3/8	14 3/8	16	1.650
11 3/4	12.866	14	1.134
10 3/4	11 3/4	12 3/4	1
9 5/8	10 5/8	11 5/8	1
7	7 21/32	8 5/8	31/32
5 1/2	6.050	6 3/4	0.700
5	5.491	6 1/8	0.634
4 1/2	5	5 5/8	5/8

This table should be considered only as a general recommendation. A number of constraints may mean a slight divergence from the figures cited. For instance, the 8 1/2" diameter is very often used to run in a 7" string because this bit size is the only one that can pass through the preceding 9 5/8" casing when the casing seat depth requires the use of thick pipe.

The minimum admissible clearance can be seen to decrease as the pipe and hole diameter become smaller.

2.1.2.5 Determining the nominal production string diameter

The last casing run into the borehole is the production string. The following parameters must be taken into account when its nominal diameter is chosen:

- What completion equipment will be used?
- What will the reservoir's flow rate be? The tubing diameter will depend on the flow rate.
- Should the option of drilling deeper be kept open?
- What is the degree of uncertainty about the depth?

A. *Exploratory wells*

Here the aim of drilling is to discover one or more objectives that are likely to be productive. As a secondary consideration, it is a good idea to include a tubing string to deal with any production.

In addition, the final diameter slated for the borehole must be large enough to enable further drilling through the string, if the last string in the program has to be run in before the objective has been reached due to technical reasons. It must even allow an extra, smaller-diameter string to be run in if need be.

In this case, considerations of safety and deepening prevail.

The diameter of the last string in an exploratory well is usually 7". This pipe diameter will allow further drilling with a 6" or 57/8" bit and addition of an extra 41/2" string or liner.

B. *Development wells*

The purpose of these wells is to reach a producing formation with known characteristics at a known depth as economically as possibly.

Here considerations of well completion therefore determine the diameter of the last string (**Fig. 2.7**).

In particular, the choice of tubing diameter(s) will determine the diameter of the production casing.

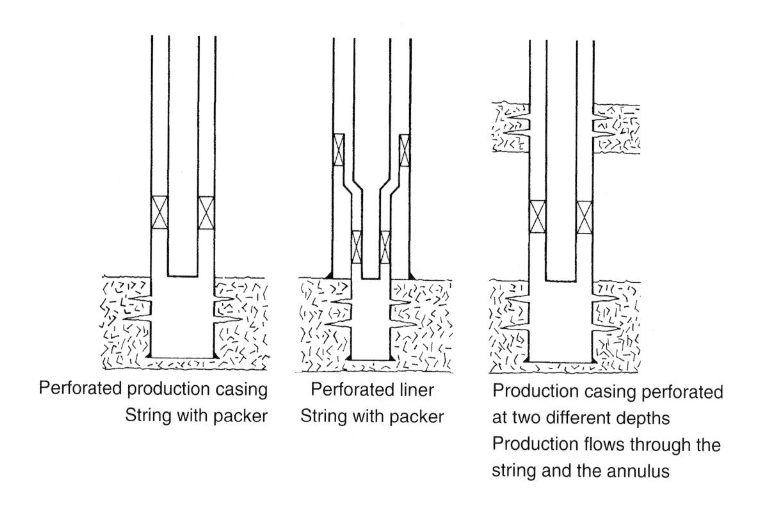

Fig. 2.7

Examples of completion (*Source: Complétion et reconditionnement des puits, Chambre Syndicale de la Recherche et de la Production du Pétrole et du Gaz Naturel, Editions Technip, Paris, 1985).*

API standards specify five production tubing diameters. The average flow rate can be estimated for each, given acceptable head losses:

Tubing (in)	2 3/8	2 7/8	3 1/2	4	4 1/2
Average oil flow rate (cu. m./day)	100	300	500	1000	1500

Accordingly, production strings will have a diameter of 7" for high-output wells or wells with multiple completion (two or three tubings). More moderate producers will be equipped with 5 1/2" or 5" strings.

Cased-hole or open-hole completion will depend on pay zone characteristics:

- open hole (**Fig. 2.8c**):
  - consolidated formation,
  - one-phase fluid,
- cased hole (**Figs. 2.8a and 2.8b**):
  - unconsolidated formation,
  - several fluids.

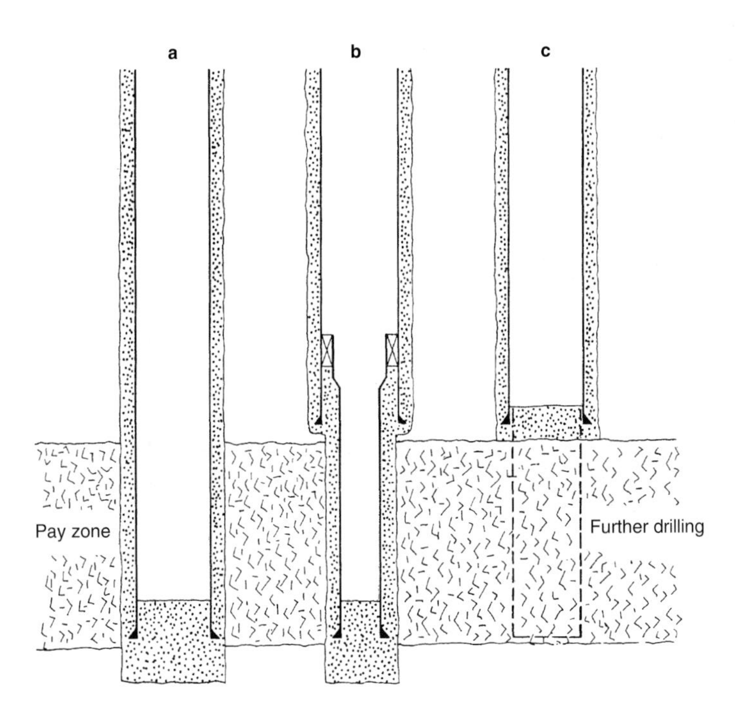

Fig. 2.8

Examples of completion (*Source: Complétion et reconditionnement des puits. Chambre Syndicale de la Recherche et de la Production du Pétrole et du Gaz Naturel, Editions Technip, Paris, 1985).*

Once the diameter of the last string has been defined, the diameter of the last drilling phase is deduced taking advisable clearance into account. Then going up the hole from the objective, a casing is assigned to each technical difficulty encountered. Step by step, the surface is reached and the diameter of the first drilling phase is deduced.

2.1.2.6 Determining casing shoe depths

Whenever a casing has to be set, it involves high costs. In addition, drilling must be resumed with a smaller diameter bit. It is therefore a penalty to be paid so that the objective can be reached. A good driller establishes a program that allows the objective to be reached with a minimum number of casing strings and tries to set the casing shoe as deep as possible. This is why determining the depth of each casing shoe, i.e. the depth where each drilling phase is stopped, is one of the main parameters in well architecture.

Geological and mechanical data are used as the basis to do this. The determining factor is to guarantee that operations can progress unhampered.

The choice involves calculated risks and is based on expert examination and interpretation of geological data, rock mechanics and the mechanical properties of the tubular goods that will be used.

A. Geological criteria

Lithologic criteria are the most important.

The shoe's sealing function

As mentioned earlier, the purpose of casing is to cover the formations that are penetrated. This keeps the borehole walls from caving in and prevents communication between different formations: the ones that have been cased and the ones that are going to be drilled.

To achieve this, casing shoes must be located in formations with well-nigh zero permeability, usually clayey beds or massive limestone or anhydrite layers.

Lithologic variations

A casing may also be set as the result of a change in lithology that may require a different and/or better-suited type of drilling mud for technical or economic reasons. This is true, for instance, of massive salt formations (only a few meters thick). Salt-saturated drilling mud is the most appropriate drilling fluid for them.

A casing will therefore be run in, usually to the top of the salt-bearing bed. The mud will then be changed to continue the drilling phase in the salt.

B. Criteria related to the mechanical properties of rock or casing

In order to set up a casing program, the following must be taken into account:
- the fluid pressures in the formations penetrated (pore pressure),
- the formations' capacity to withstand the fluid pressures,
- the mechanical properties of the pipe.

Pressures

Under this heading, a distinction is made between:

- **Normal pressures** when the pressure gradient is hydrostatic. This means that the pressures encountered correspond to the weight of a column of water from the surface to the depth under consideration. The pressure gradient is approximately 1 bar/10 m.
- **Abnormal pressures** when the pressure gradient is usually higher. This is due to trapping of fluids lighter than water, such as hydrocarbons, or to undercompaction of sediments. The gradient can go up to a value corresponding to the weight of overlying formations (a gradient of 2.3 bar/10 m), i.e. the geostatic gradient.

Drillers often use the plot on a pressure-depth diagram. It gives an explicit picture of the pressure system expected as the well is drilled.

Stability of formations

Evaluating how well formations hold up to the mud and the filtrate is done by pressure tests called "leak-off tests" (**Fig. 2.9**). The purpose is to determine the pressure at which a formation starts to allow drilling fluid to be injected into it, or yields and breaks down with continuous pumping. With formation breakdown, there are no drilling fluid returns, a phenomenon known as lost circulation.

The leak-off test should especially be performed when a particularly porous and permeable bed is encountered. This is the type of bed most likely to be invaded and weakened, or even fractured by the filtrate.

Leak-off test procedure

- Pressure is built up with the high-pressure pumping unit at a slow, constant flow rate, volumes are measured and the corresponding pressures are plotted on a graph.
- During pressure buildup, the intercasing annulus is checked and if pressure increases, it is bled off. Pumped volumes are compared to theoretical volumes.
- Pumping is stopped either:
  - when the test pressure is reached (P_t),
  - when three or four consecutive points stray from the normal straight pressure buildup line on the graph (the diverging point is then P_i), or
  - if the formation fracture pressure is accidentally reached (P_f).
- After pumping has ceased, the pressure is recorded for three to five minutes when the formation or cement job at the shoe are tested. Pressure decline is plotted on the graph versus time.
- After pressure has been recorded, it is slowly bled off and the volume of mud returns is measured.
- The result of a pressure test, an injectivity test or accidental fracturing of the uncased part of the hole or the cement is expressed in terms of equivalent test density, injectivity density or fracturing density at the shoe or weak point.

Well:	Date:
Shoe depth: Z_s = 2722 m	Depth: Z = 2751 m
Mud weight: d = 1.26	Weak point depth: Z_i = 2722 m
Start-up: P_i = 127.6 kg/cm^2	kg/cm^2
Injectivity density: $d_{inj} = d + \frac{P_i}{Z_i} \cdot 10 = 1.73$	Signature: Date:

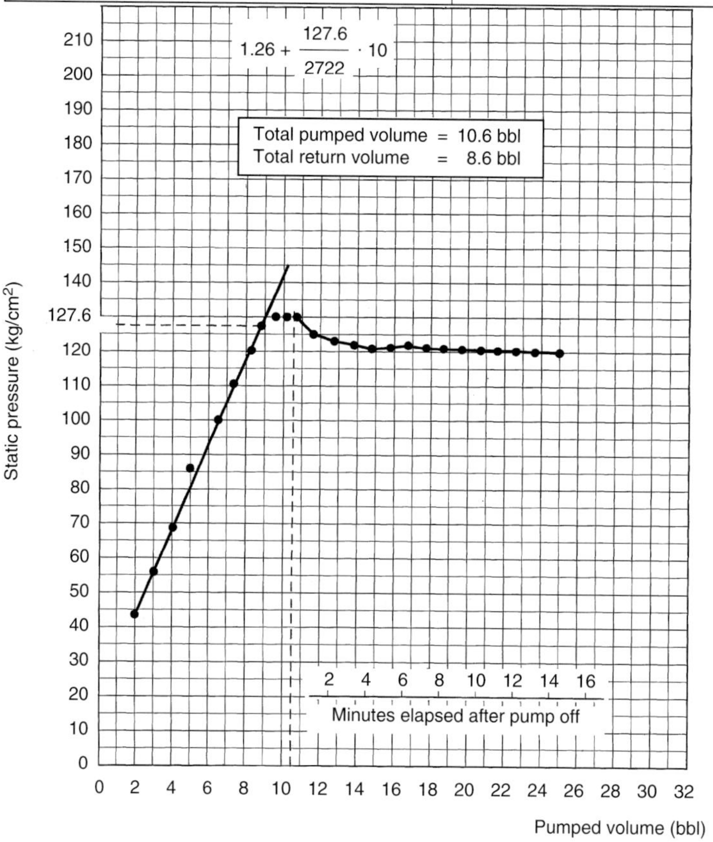

Fig. 2.9

Typical recording of a leak-off test (*Source: SNEAP*).

The leak-off test gradients will be plotted on the same pressure-depth diagram mentioned earlier.

The rule to obey is:

Formation pressure < Leak-off test pressure or LOTP

This rule is the most important criterion in determining the depth of the casing shoe. Mud losses during drilling may cause an uncontrolled kick in the worst-case scenario and this is a risk that should never be run.

With the pressure profile and the leak-off test pressures, the casing shoe depth is determined. Assumptions are made as to the hazards encountered.

This evaluation principle is illustrated by a (depth-pressure) graph (**Fig. 2.10**).

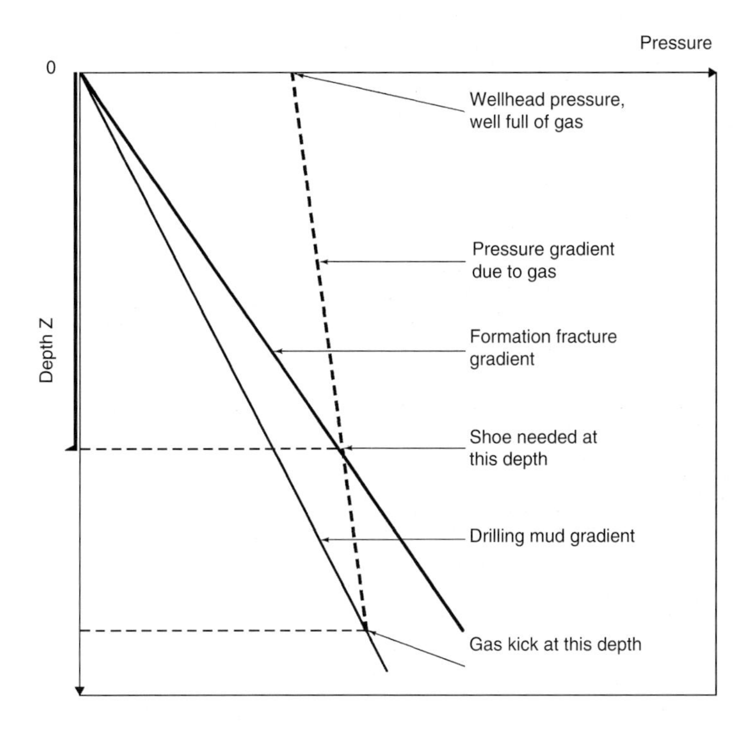

Fig. 2.10

Diagram of pressure versus depth : choosing the casing shoe depth.

For this example, the following assumptions were chosen:

- the gas kick is at depth Z,
- the density of the kick is known,
- the gas is considered at the balance pressure of the mud.

Different assumptions can of course be put forward depending on existing information on geology, the drilling objectives and the project's economics. It is impossible to cover all the risks — too many protection casings would be required. As a result, any inadequacy in architectural strength will be offset by monitoring during drilling (mud logging). Mud logging is an important exploration tool, but also serves to control kicks or lost circulation — and thereby limits the adverse effects considerably (see the chapter on kick control).

2.1.2.7 Determining cementing height

The height of cement required depends on the function and type of casing string.

A. *Surface casing*

This string is cemented over its total height to the extent possible. This enables it to support the safety equipment and remain firmly anchored on the surface.

B. *Intermediate casing string*

The shoe must be properly cemented and this requires a cement height of at least 150 m. The annulus must be cemented when specific reasons dictate:

- if the formation contains corrosive fluids,
- if the formation contains fluids that need to be protected.

The annulus of intermediate strings will be cemented in one stage from the shoe or in two stages depending on the context.

C. *Production casing*

Here the cement operation is very important. This is because the cement must not only anchor the string but also provide as good and long-lasting a seal possible between the borehole and the casing.

Additionally, the cement height will depend on:

- the position of the string in relation to the pay zone: whether the hole is open or cased,
- the thickness of the pay zone,
- the thickness of the permeability barrier.

2.1.2.8 Calculating the mechanical properties of pipe

Calculation involves the maximum stresses that the pipe will be subjected to when it is run into the borehole and cemented, and during later drilling phases and possible production. These external stresses may never be experienced. Differences in pipe calculations can come from differences in safety coefficients and assumptions. Assumptions about external stresses will be different according to the type of string.

A. *Conductor pipe*

This string is not really casing and undergoes low stresses. In particular, its bursting strength is negligible since a well can not be closed before a surface casing is set.

B. Intermediate or surface casing

a. Collapse

Maximum stress conditions are the result of:

- The hydrostatic pressure of the drilling mud in the annulus. Some flowing formations can also be considered in calculating the collapse stresses.
- Total mud loss during a later drilling phase, i.e. the well is empty. But the empty-well assumption is often highly constraining for very deep wells. When it is applied to the letter it means heavy nominal pipe weights and limited choice in the next drilling bit diameter. When this is the case, the assumption must be reconsidered.

Pipe is chosen with or without taking into account a lowered collapse strength due to tensile stresses (ellipse of biaxial yield stress).

The common safety coefficient is 1.125, but coefficients of 1 or 1.1 are also widespread. A coefficient of 0.85 is sometimes used for cemented zones.

b. Tension

A safety coefficient of 1.60 to 1.75 will be adopted for when the pipe is being lowered into the well in order to have enough extra tensile capacity. The coefficient is in relation to the weakest section of the pipe (in the pipe itself or in the coupling). Local tension overloads due to the pipe bending in well bore curves must also be included.

c. Bursting

The bursting pressure used is the difference between the annular and inside pressures. Annular pressure will be chosen equal to the pressure of a hydrostatic column of water.

For inside pressure, the situation inside the pipe must be defined:

1. If there is mud inside, a maximum wellhead pressure is usually taken and the pressure profile is calculated.
2. If the assumption of a well full of gas is used, the depth and pressure of the reservoir need to be known.

The pressure may be limited by the casing shoe fracture pressure. Another possibility is to choose the assumption of circulating a gas kick out of the well.

The bursting pressure must be compatible with the BOP operating pressure on the surface and with the formation breakdown pressure at the shoe depth. The safety coefficient is 1.1.

C. Production casing

a. Collapse

Maximum stress conditions are the result of:

- the same annulus factors as were mentioned for the protection casing,
- total mud loss for a production well.

However, for great depths the situation will be confined to a partially empty well. The same approach holds true for a gas development well.

For an development oil well, the minimum pressure expected in the well at pay zone depth will be taken. The safety coefficient is 1.125, and a figure of as little as 0.85 can also be used for cemented zones.

Because of the importance of tension, the influence of tensile forces on collapse strength must be taken into consideration.

b. Bursting

In the same way as for the intermediate casing, a column of water will be considered as annular pressure.

For the inside pressure, an attempt will be made to know the pressure profile on the basis of the formation pressure and the characteristics of the fluid inside:

- mud,
- gaseous effluent,
- completion fluid.

The safety coefficient is 1.1.

c. Tension

Intermediate casing conditions can be applied to a production casing. Variations in tension on a cemented string (buckling), due to an increase in temperature or a variation in pressure, will be calculated and may impose production casing anchoring tensions.

Conclusions on establishing a drilling and casing program

The final choice of a drilling and casing program will be based on the considerations discussed in the preceding paragraphs. There are also a number of commonsense rules that should be kept in mind:

- An exploratory drilling program would incur exaggerated costs if it was designed to allow rational development in any and every eventuality. It should be remembered that few of these wells discover commercially attractive reservoirs.
- Small drilling diameters (under 5 3/4") limit fishing possibilities and, more important, allow for little investigation into penetrated formations. Some well logging instruments can not be run in.
- The program must always specify the height of cement behind the strings. One of the basic objectives of the casing operation is to isolate the formations behind the string from the ones that the bit will be drilling through in the next phase. Another is to prevent communication between the reservoirs sealed off by the string.
- For a development well, the most important decision is setting the production casing with the pay zone either cased or open hole. The choice is based on considerations involving the reservoir, safety and the production mode. It is recommended to choose a diameter for this string that will facilitate all the testing, stimulation and final completion operations.

SOME EXAMPLES OF DRILLING AND CASING PROGRAMS

Development:

1. Parentis (oil)

Drilling (in)	Casing (in)	Tubing (in)	Depth (m)
20	16		20/50
15	11 3/4		200/300
9 7/8	7		2250
6		2 3/8	2500

2. Meillon–Pont d'As (gas)

Drilling (in)	Casing (in)	Tubing (in)	Depth (m)
24	18 5/8		150
17 1/2	13 3/8		1000
12 1/4	9 5/8		2700/3000
8 1/2	7		4500
5 3/4	5 (liner)	5	5000

3. Handil (oil)

Drilling (in)	Casing (in)	Tubing (in)	Depth (m)
26	20		200
17 1/2	13 3/8		700 (KOP)
12 1/4			1050
then widened			
17 1/2			
12 1/4	9 5/8		2650

4. Villeperdue (oil)

Drilling (in)	Casing (in)	Tubing (in)	Depth (m)
12 1/4	9 5/8		220
6	4 1/2	2 3/8	1850/1900

Exploration:

5. North France

Drilling (in)	Casing (in)	Tubing (in)	Depth (m)
36	30		36
26	18 5/8		300
17 1/2	13 3/8		800
12 1/4	9 5/8		2250
8 1/2	7 (liner)		4500

6. Southwest France

Drilling (in)	Casing (in)	Tubing (in)	Depth (m)
26	20		200
17 1/2	13 3/8		1250/1300
12 1/4	9 5/8		3200/3400
8 1/2	7		3700/4000
6	5 (liner)		4375/4575

7. W. Texas, USA

Drilling (in)	Casing (in)	Tubing (in)	Depth (m)
36	26		26
24	20		475
17 1/2	13 3/8		(possibility not used)
12 1/4	9 5/8		3220
8 1/2	7 5/8		4850
6 5/8	5 1/2		5797
	(poor cement job)		

8. Gulf Coast, USA

Drilling (in)	Casing (in)	Tubing (in)	Depth (m)
15	10 3/4		
9 7/8	7		
6	5		

2.2 FURTHER STUDIES

The drilling and casing program along with the geological cross-section are the fundamental documents in planning how and with what equipment the well will be drilled. Eventually, an estimate of the cost price of the drilling operation can be made.

All these estimates and forecasts are usually included in extra programs added on to the well program.

2.2.1 Drilling bit program

As diameters have been defined in the drilling program, the drillability of formations must be examined to determine the bits and their optimum parameters.

In exploration, an attempt can be made to relate the formations to be drilled with other similar ones after discussing their possible drillability with the geologists. In any case, the choice of the type of bit will be defined on the well site. Accordingly, a more than adequate supply of bits must be on hand.

In development, the rate of penetration, the type, number and footage of bits can be deduced from analyzing bit performance in previous wells.

2.2.2 Drilling fluid program (see the chapter on drilling fluids)

The program defines the type and characteristics of the drilling fluids to be used phase by phase. Also included are possible conversions and any weighting up that might be necessary. The program therefore depends on the geology, architecture and objective of the well and on the logs and other operations required for the pay zone. The necessary volumes and flow rates will be deduced from it. This will serve as a guide in the choice of the rig and in calculating consumption of drilling fluid chemicals, water and oil. Ecological constraints increasingly determine the choice of the type of drilling fluid according to the amount and type of discharges and local regulations.

2.3 CHOOSING THE DRILLING RIG

The hoisting and pumping capacity must be studied along with drilling fluid capability and safety equipment such as BOP, choke and manifold. Then all these points must be checked against the drilling and casing program, along with the drilling contractor's safety rules and the type of drilling contract sought.

2.4 BUDGET OF THE OPERATION

All these studies must be based on the bottom line: economic feasibility. Technical choices must always remain a compromise between the best technical solution and the least cost.

Chapter 3

DOWNHOLE EQUIPMENT

The architecture of an oil well has been discussed, and now the means of building it will be described. There are two types of equipment or apparatus:

- downhole equipment that is run into the well, i.e. drilling bits and drill string, are covered in this chapter, and
- the complete drilling rig that remains on the surface and is mainly used to operate the downhole equipment (Chapter 4).

3.1 DRILLING BITS AND THE PARAMETERS FOR THEIR USE

3.1.1 Roller cone bits (Based on *Oilfield Catalog*, Hughes Tool Division, Hughes Tool Company) (Fig. 3.1)

A roller cone bit is made up of three main parts: the cones, the bearings and the body of the bit. Each cone has concentric rows of teeth that interfit with the rows of teeth in the adjacent cones. The teeth can be made of steel machined in the cone or tungsten carbide inserts cold-pressed into holes drilled in the cone. The cones are mounted on bearing shafts that are an integral part of the body of the bit.

The size and thickness of the component parts of the bit depend on the type of formation to be drilled. Bits for soft formations require little weight, have smaller bearings, less thick cone shells and thinner legs. This design leaves more room for long thin cutters.

Bits for hard formations work with more weight, have stubbier cutters, bigger bearings and sturdier bodies.

3.1.1.1 Cone geometry

To understand how cone geometry can affect the way the teeth cut the rock formation, consider the cone for soft formations as schematically represented in **Fig. 3.2**. It has two basic conical angles and neither of them has its apex in the center of the bit. Since the cones have to rotate around the axis of the bit, they skid at the same time as they rotate and provide an effective gouging and chiseling action for drilling soft formations.

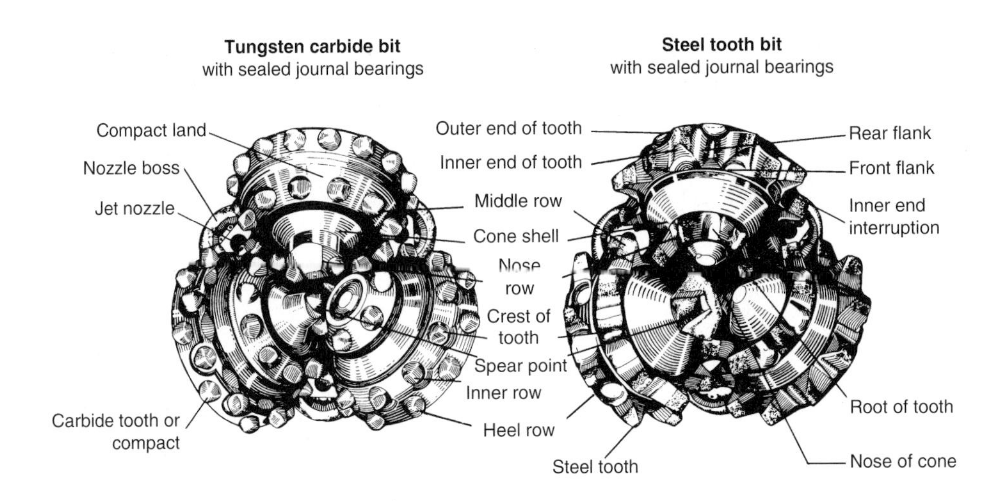

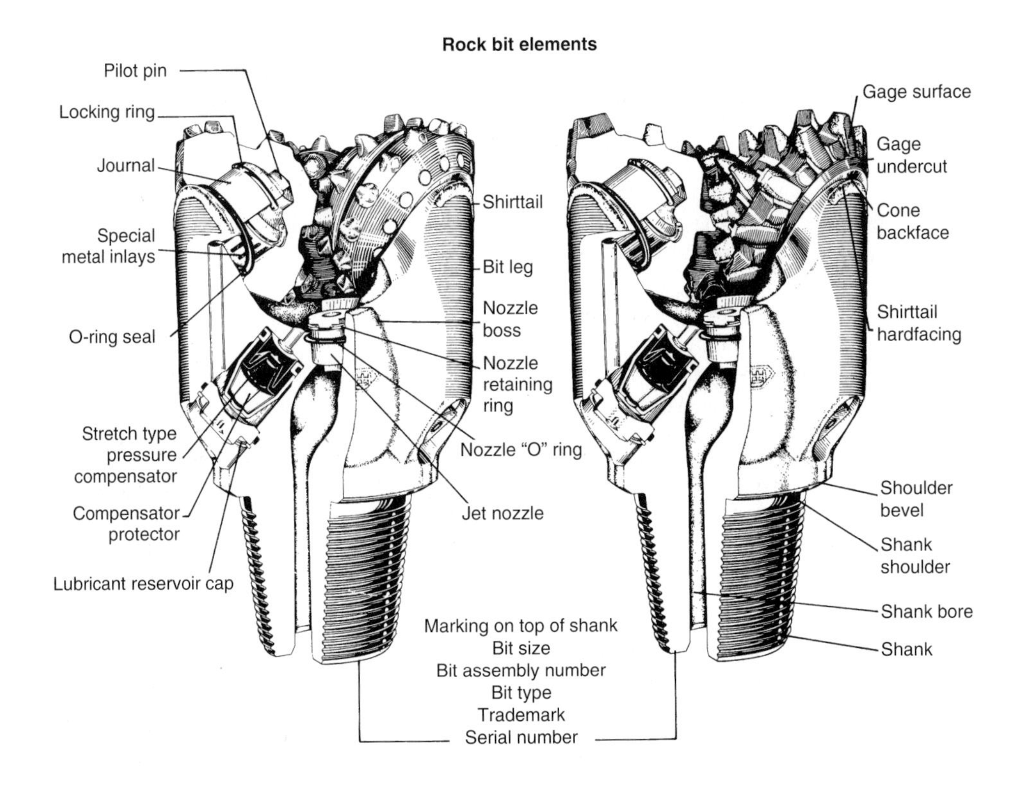

Fig. 3.1
Three-cone bit terminology *(Source: Hughes Tools)*.

This effect is enhanced to improve the rate of penetration in soft formations by offsetting the axes of the cones in relation to the axis of rotation of the bit (**Fig. 3.2**). Cones on bits for hard formations move in what is closer to real rotation, with little offset. As a result, they break down the rock formation chiefly by crushing it.

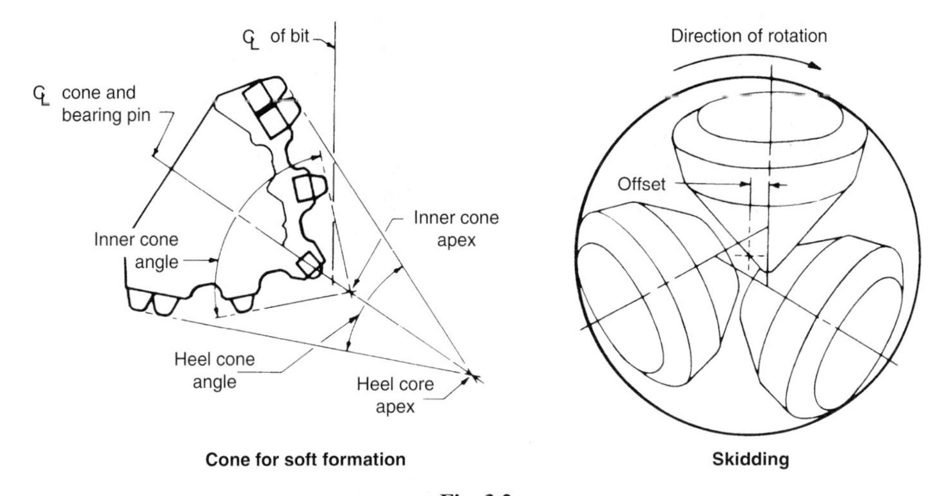

Fig. 3.2

Design of rock bit cones *(Source: Hughes Tools).*

3.1.1.2 Bearing design

There are currently three types of bearings:

- nonsealed ball or roller bearings,
- sealed ball or roller bearings,
- friction bearings.

Nonsealed, nonlubricated ball or roller bearings are mainly used today in bits for spudding in a well, when the lifetime of the bit is less of a constraint since tripping times are short. They are also advisable in instances where high rotational speeds are required (**Fig. 3.3**). When the drilling fluid is either air, gas or foam, nonsealed ball and roller bearings are also used. However, these bits are designed so that part of the drilling fluid is channeled to cool the bearings.

The sealed ball or roller bearing was designed for insert bits. The system is currently used chiefly on toothed bits and its lifetime is at least the same as that of the cutters.

The friction, or journal bearing was developed so that its lifetime would match that of the carbide cutters. It contains no moving parts, there is just a bearing shaft fitted to the inside bore of the cone. The contact areas between the shaft of the bearing and the bore of the cone are carbide treated. They are coated with special metals and specially treated to better withstand wear and tear and keep from seizing up.

3.1.1.3 Lubricating system (Fig. 3.4)

Sealed-bearing bits have their lubricating system in each arm. The system comprises a grease reservoir, a rubber membrane compensator and a leak-proof channel. The compensator equalizes the pressure in the bearing between the pressure of the drilling fluid and the pressure of the grease encapsulated by the manufacturer.

3.1.1.4 Cutters

Bits with steel teeth are used when spudding in a well, in soft formations, at high rotational speeds and in zones where the bed thickness makes insert bits antieconomical. Depending on anticipated use, cones with milled teeth may have tungsten carbide protection on the teeth.

Bits for soft formations are designed with long, widely spaced teeth to help penetrate in the formation and tear off larger cuttings.

Bits for medium and medium-hard formations have more closely spaced teeth. Each tooth also has slightly larger angles to withstand the load required to overcome the resistance of the formation.

Hard formations have high compressive strength and are usually very abrasive. Bits designed to drill these formations are equipped with sturdy, closely spaced teeth and thick cone shells so as to withstand heavy weights.

These bits are designed without a skidding effect. The three cones rotate practically on the bottom, thereby reducing abrasive wear on the teeth. The tungsten carbide insert bit was originally designed to drill very hard, abrasive silica or quartzite formations.

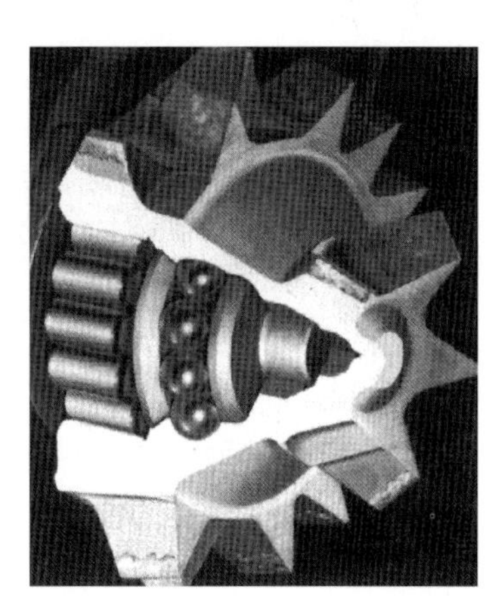

Fig. 3.3
Ball and roller bearings
(Source: Hughes Tools).

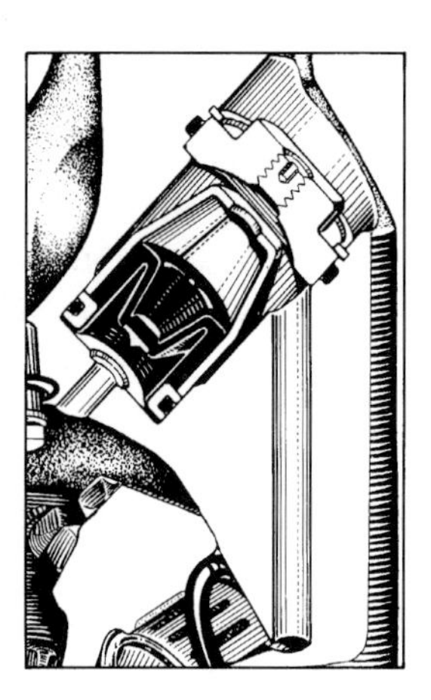

Fig. 3.4
Bearing lubrication system
(Source: Hughes Tools).

Due to the relatively short lifetime of toothed bits, this type of formation proved to be very expensive to drill. Today, because of the progress in metallurgy and in the shape of the inserts, bits have been developed that are suitable for drilling economically in a much wider drillability range.

Cylindrical sintered tungsten carbide inserts are machined into a number of very different shapes (**Fig. 3.5**). Then they are set in holes precision-machined in the cones. This assembly gives a cutting structure that withstands abrasion wear and compression stresses well.

The ovoid shape is the sturdiest and is designed for the crushing and chipping action required in drilling very hard formations. Ogive inserts are a little more pointed to drill slightly softer formations. The cone also has a sturdy profile that is well suited to the crushing and chipping type of drilling action. It is used with a scraping action to drill medium-hard formations.

The chisel shape is used in medium and softer formations for maximum penetration owing to a gouging and scraping action. Specific chisel shapes are chosen according to the formation and the geometrical features of the bit.

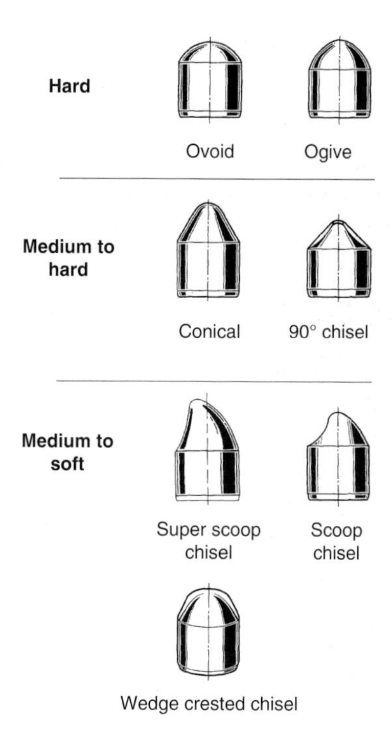

Fig. 3.5
Examples of carbide insert shapes *(Source: Hughes Tools)*.

3.1.1.5 Bit hydraulics

What is termed a "conventional" rock bit has a single drilling fluid passage down the axis of the bit. The fluid washes the bit from the inside of the cones, but is less effective than in jet bit nozzles. The conventional bit is usually seen now only in large-diameter wells. The jet bit has three nozzles, each located between the arms of two adjacent cones (**Figs. 3.1 and 3.6**). The orifice sizes on the nozzles can be changed for maximum effectiveness. Turbulent flow is effective for velocities ranging from 80 to 150 m/s. The purpose of fluid streams is to keep the cones clean, cool down the bearings and especially to sweep formation cuttings toward the annulus. On some bits, elongated nozzles (**Fig. 3.7**) improve cleaning efficiency.

Fig. 3.6
Slanted nozzle *(Source: Reed Tool Co.).*

Fig. 3.7
Elongated nozzle *(Source: Reed Tool Co.).*

3.1.1.6 Classification (Fig. 3.8a)

The International Association of Drilling Contractors (IADC) classifies rock bits by means of three numbers and one letter. A Hughes ATJ22 bit is designated by the code 5-1-7 (G):

5 = a bit with inserts for soft to medium-soft formations,
1 = the lowest hardness,
7 = gage inserts around the perimeter and friction-sealed bearings,
G = reinforced gage inserts.

	Series	Formations	Types	Standard roller bearing	Roller bearing Air-cooled	Roller bearing Gage protected	Sealed roller bearing	Sealed roller bearing Gage protected	Sealed friction bearing	Sealed friction bearing Gage protected
				1	2	3	4	5	6	7
Milled tooth bits	1	Soft formations with low compressive strength and high drillability	1							
			2							
			3							
			4							
	2	Medium to medium hard formations with high compressive strength	1							
			2							
			3							
			4							
	3	Hard semi-abrasive and abrasive formations	1							
			2							
			3							
			4							
Insert bits	4	Soft formations with low compressive strength and high drillability	1							
			2							
			3							
			4							
	5	Soft to medium formations with low compressive strength	1							
			2							
			3							
			4							
	6	Medium hard formations with high compressive strength	1							
			2							
			3							
			4							
	7	Hard semi-abrasive and abrasive formations	1							
			2							
			3							
			4							
	8	Extremely hard and abrasive formations	1							
			2							
			3							
			4							

Additional letter

A = bits with friction-sealed bearings suited for air drilling
C = jet bits with a central nozzle
D = special bits for deviated wells
E = jet bits with elongated nozzles
G = bits with reinforced protected gage
J = bits with slanted nozzles
R = bits with reinforced welds for cable-tool drilling
S = bits with standard steel teeth
X = bits with wedge-shaped inserts
Y = bits with conical inserts
Z = bits with inserts other than wedge- or cone-shaped

Fig. 3.8a IADC table for classification of rock (roller) bits *(Source: Drilling Data Handbook, Editions Technip, Paris 1989).*

3.1.2 Diamond bits

There are three types of diamond bits:
- with natural diamonds (**Fig. 3.8b**),
- with PDC or polycrystalline diamond cutters,
- with TSP or thermally stable polycrystalline diamond cutters.

These three types differ in the nature of the cutting elements which crush the rock in a mechanically different way. Natural diamonds abrade the rock and work like a large grindstone. Polycrystalline diamonds shear the rock and cut shavings like a metal lathe tool. All these drilling bits are one-piece and this makes them quite sturdy.

Natural diamonds

PDC

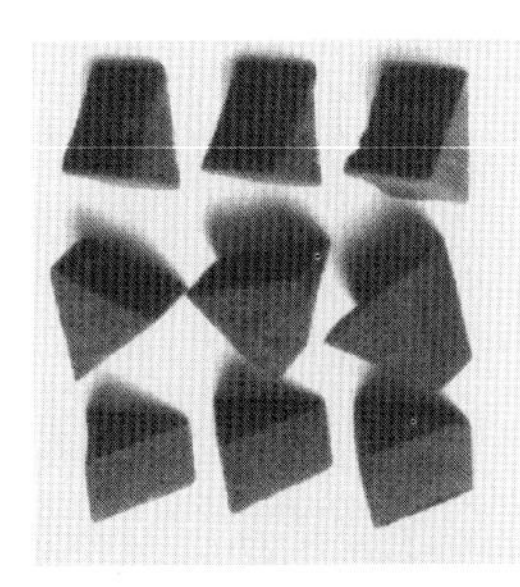

TSP

Fig. 3.8b
Cutters for one-piece bits
(Source: Diamant Boart Stratabit).

3.1.2.1 Natural diamond bits (Fig. 3.9)

Diamond (crystallized carbon) is the hardest known mineral. Its compressive strength is approximately 80,000 bar (tungsten carbide: 50,000; C.45 steel: 15,000). Its melting point is very high (3650°C), but it turns into graphite on the surface around 1450°C. Diamonds are measured by weight in units called carats (1 carat = 0.2 gram). For drilling bits, the size of the stones is measured in the number of stones per carat and usually varies from 2 to 12.

The size, type and number of diamonds on a given bit depend on the planned rate of penetration, the size of cuttings and the homogeneity of the formation. When hard formations are drilled at a slow penetration rate, the cuttings are very small and more easily swept out of the hole than at fast penetration rates. Small diamonds can be used to get maximum contact on the working face without hindering cuttings removal. On the other hand, softer formations need less load on the working face. Larger diamonds are used to tear out a larger chunk of rock and leave more room for the cuttings to pass. Though there are general rules for selecting the size of diamonds, the experience gained with previous bits is the best guide.

The bits are still manufactured by hand. The process begins with machining a graphite mold (**Fig. 3.9**). Inside the mold the rows and locations for each diamond are drawn along with the design of the drilling fluid passages. Each diamond location is drilled out with a small mill. The water courses or waterways for the drilling fluid are built in relief in the mold with a special putty. The diamonds are then set one by one in each small hole where they are fixed in place. The mold is filled with tungsten carbide powder with an added binder whose formulation is each manufacturer's secret. The melting point of the binder is situated between 400 and 1400°C depending on its formulation and varies according to the hardness required in the final matrix.

After a steel core has been placed in the mold, the whole thing is put into a furnace. The temperature must not exceed the beginning of diamond graphitization. Then the thread is machined.

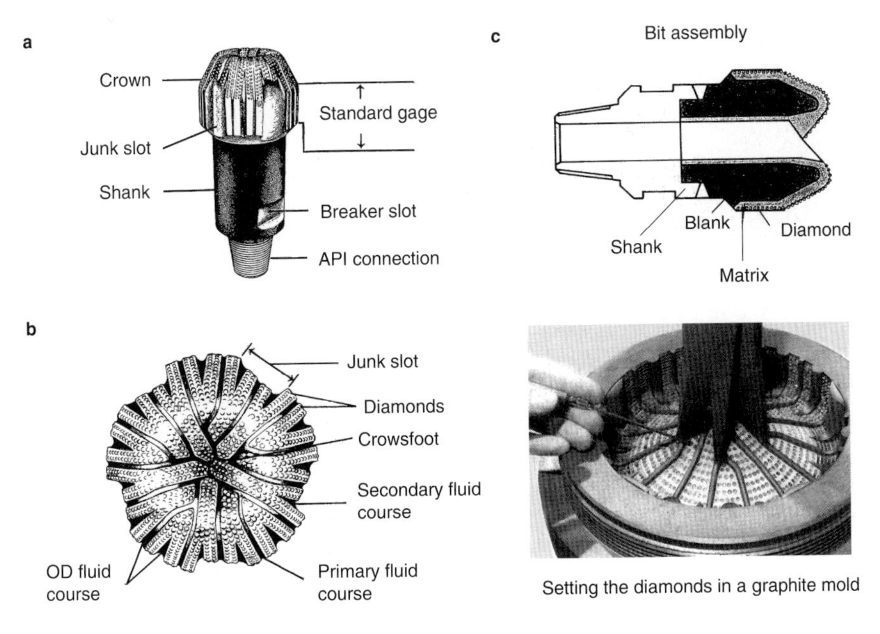

Fig. 3.9

Natural diamond bit terminology *(Source: Hughes Tools)*.

The three basic bit profiles-round, tapered, short or long-fit most applications. There are an almost unlimited number of variations within these categories. The shape that is best suited to a specific case is a practical trade-off between the maximum rotating life of the bit and the maximum rate of penetration. The round nose gives more force in hard formations, while the elongated nose shape promotes faster cutting.

Diamond compound bits are commonly used in mining and are fairly specific in that the diamonds are not set in the matrix. The diamond compound is made up of tiny diamonds and a metallic binder. The diamonds are distributed throughout the mass in a layer 5 to 10 mm thick. When the bit is in operation, worn diamonds come unstuck and further layers of diamonds are exposed.

Use of diamond bits

As for any bit, the decision to drill with a diamond bit must be based on a cost analysis. Certain drilling situations suggest that a diamond bit can be used economically:

1. When the lifetime of roller bits is very short because of wear and tear on bearings or teeth, or because teeth have been broken.
2. When the rate of penetration is very slow (1.5 m/h or less), because mud density is high, or because the rig has inadequate hydraulic power.
3. With a six-inch diameter or less, when the lifetime of roller bits is short.
4. When the hole angle is increased in a directional well.
5. When the weight on the bit is limited.
6. In turbodrilling, where high rotational speed promotes penetration of a diamond bit.

The use of a diamond bit is limited in certain very hard fractured formations where the diamonds can be exposed to sharp impacts. Formations containing flint or pyrite shorten the lifetime of the diamond bit when pieces come loose and roll under the bit, damaging the diamonds.

3.1.2.2 PDC bits (Fig. 3.10)

Bits with polycrystalline diamond blanks or compacts have either a steel body or a matrix. Steel bodies are machined and then covered with tungsten carbide to slow down erosion. Matrix bodies are manufactured from the same tungsten carbide material as natural diamond bits.

It was *General Electric* that developed the synthesis process for the diamond compacts (Stratapax) made of a synthetic diamond deposit a few tenths of a millimeter thick on a tungsten carbide stud. The compacts are then set in the surface of the bit so as to provide maximum shear to each cutter. The major drawback of the PDC is that it can not withstand temperatures above 800°C. This means synthetic diamonds can not be set in a carbide matrix in the same way as natural diamonds. They must be set in the matrix by brazing in molded holes. In bits with steel bodies, PDC compacts are also brazed on cylindrical studs that will then be set in drilled holes.

The development of these products has constantly widened the applications for PDC bits and the suitable formation hardness range is located between medium-hard and soft.

Limitations that can be cited are strength in abrasive formations and hydraulic effectiveness in keeping the PDC compacts clean.

3.1.2.3 TSP bit (Fig. 3.11)

The latest development in diamond bits is the use of thermally stable polycrystalline diamonds. In contrast with PDC, TSP diamonds have undergone treatment to eliminate incompatible elements from the standpoint of thermal dilation and can therefore withstand the temperatures required to manufacture a carbide matrix. Their triangular shape has sharp edges for shearing the rock. TSP diamonds are often used in conjunction with natural diamonds.

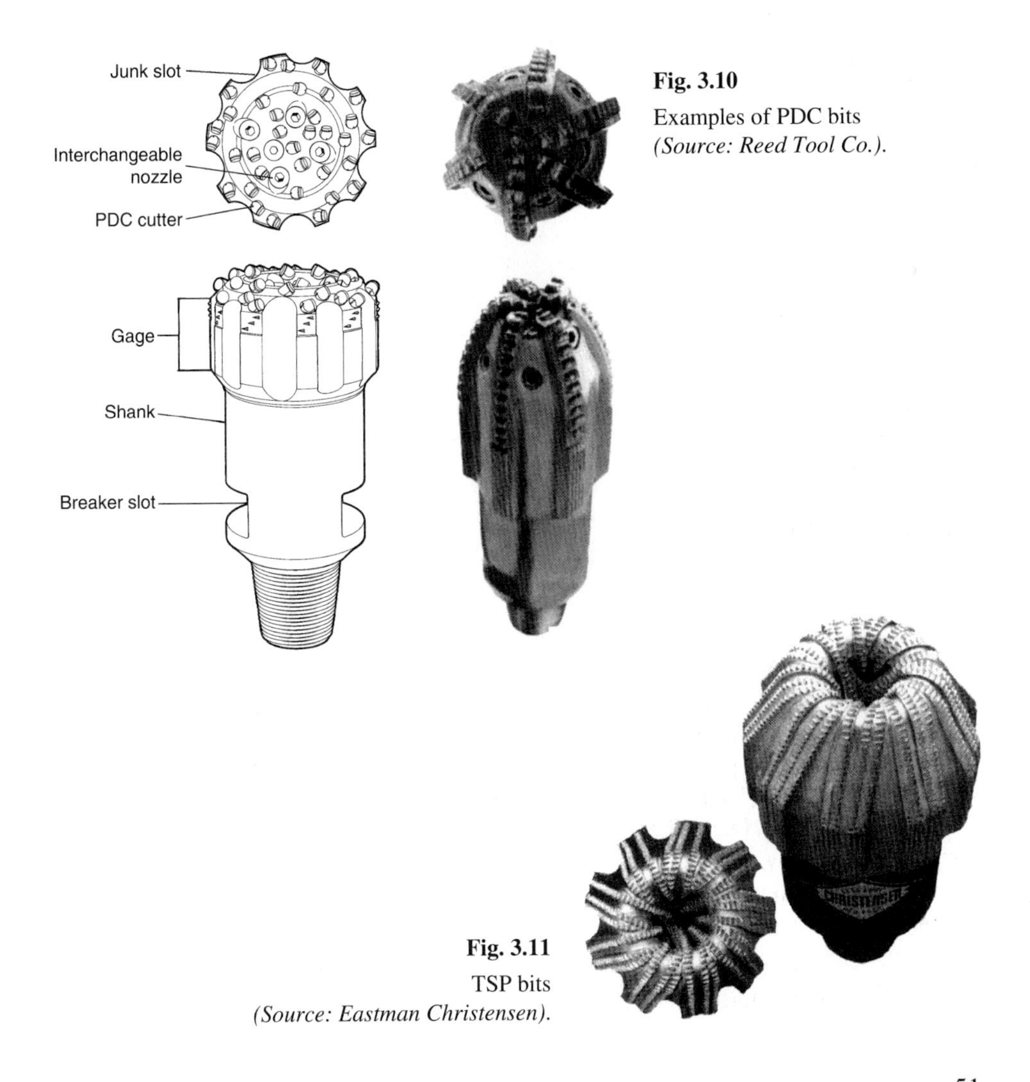

Fig. 3.10
Examples of PDC bits
(Source: Reed Tool Co.).

Fig. 3.11
TSP bits
(Source: Eastman Christensen).

3.1.2.4 IADC classification

Classification is based on four characters and is therefore consistent with the classification of rock bits. Since the code of bit condition uses the same bit wear designation for three-cone rock bits and bits with nonmoving cutter elements, daily reports, bit wear reports and data bases are filled out in the same way.

The first character (D, M, S, T, O) defines the type of cutting element: diamond, PDC and sintered body, PDC and steel body, TSP, others.

The second character (a number from 1 to 9) defines the type and general shape of the profile.

The third character involves hydraulics in general.

The fourth character (a number from 0 to 9) defines the size of cutting elements and their density on the bit.

3.1.3 Parameters for using drilling bits

What is termed drilling parameters are a number of factors that condition how fast the well is deepened.

They are classified into two categories:

- mechanical parameters which involve the type and shape of the bit, the weight on the bit and the rotation speed,
- hydraulic parameters, i.e. flow rate, pressure, type of drilling fluid and its characteristics (density, viscosity, filtrate).

3.1.3.1 Mechanical parameters

The formations that are drilled through have widely different degrees of hardness. Penetration may vary from 100 meters/hour to several dozen centimeters per hour. To break down rock formations with such drillability differences, the bits used will obviously have to work according to different principles.

A. Selecting the bit

Because formation hardness is so varied and since there are so many different types of bits, it is not easy to choose the best one for the formation that is being drilled.

In exploratory drilling, the formations are an unknown factor and they often change suddenly. This hardly ever allows a perfect match between the bit and the rock (only under exceptional conditions). Suppositions can only be made and they are often invalidated by the facts.

Here cooperation between the rig geologist and the company man is a must. The geologist is qualified to interpret any change in formation and forecast the nature and thickness of the new formation. The supply of bits depends totally on this information, but some extra bits must nevertheless be on hand to handle any unexpected needs.

In contrast, the formations are known in development drilling. The nature and thickness of the beds to be drilled can therefore be anticipated with a fair degree of certainty and the bit can be chosen to match the type of rock. There are several types of bit capable of drilling a given formation properly. A comparative study carried out while the first wells are being drilled on a field will allow further development wells to be drilled faster and therefore more cost-effectively. The best choice will be the bit that gives the lowest price per foot of hole drilled under good technical conditions.

Accordingly, in development drilling a bit program is established that takes all the accessible information into account:

- logging results,
- bit records,
- lithologic description.

The program should be strictly complied with (except in the event of something unexpected such as a bit pulled out of the hole with an abnormal degree of wear).

Drillers always tend to choose bits for harder formations than they actually need. Since the bit is sturdier, there is less risk of stray rollers and cones.

B. Cost price per meter of hole drilled

The following factors are involved in the calculations:

- P_b = net cost of the bit,
- P_h = rig rental price per hour,
- T_t = tripping time,
- T_r = rotating time,
- m = number of meters drilled.

The cost price is given by the formula:

$$P_m = \frac{P_b + P_h (T_r + T_t)}{m}$$

Example:

A bit A for hard formations has drilled 41 m in 17 h. In the same formation, a bit B for softer formations has drilled 35 m in 12 h. If the rig rental price is \$700 per hour, bits A and B cost \$1500 and a complete round trip lasts 4 h:

$$P_m \text{ for bit A} = \frac{1500 + 700\,(17+4)}{41}$$
$$= \text{approximately \$395 / m}$$

$$P_m \text{ for bit B} = \frac{1500 + 700\,(12+4)}{35}$$
$$= \text{approximately \$360 / m}$$

It can be seen that for less footage and rotating time, bit B was more economical than bit A.

Example of comparison between a roller bit and a diamond bit:

Well number 1

Three roller bits have drilled 310 m in 126 h, i.e. an average 103 m in 42 h per bit.
Average cost of a bit = \$3000
Rig rental per hour = \$1000
Average tripping time = 12 h.

This gives:

$$P_m = \frac{P_b + P_h\,(T_r + T_t)}{m}$$

$$= \frac{3000 + 1000\,(42 + 12)}{103}$$

$$= \text{approximately } \$550\,/\,\text{m}$$

Well number 2 (in the same field)

A diamond bit has drilled 330 m in 117 h.

The net cost of the bit is \$41,500 (the net cost is the purchase price minus the salvage value of the serviceable diamonds left).

The cost price per meter will be:

$$P_m = \frac{41{,}500 + 1000\,(117 + 12)}{330}$$

$$= \text{approximately } \$515\,/\,\text{m}$$

C. *Abnormal wear*

When there is no change in the type of formation, bits that have been pulled out of the hole provide valuable information for selecting a new one.

A bit can be in poor condition due to the following reasons:

- teeth worn to such an extent that they can no longer break the formation rock down,
- bit diameter (or gage) lost through wear and subsequently a smaller diameter hole drilled,
- bearings deteriorated due to an excessively high load (locked roller cone).

In an initial approach the evidence suggests that:

- a bit whose bearings alone are worn should be replaced by a bit for softer formations,
- a bit with a dramatic drop in rate of penetration, that is pulled with its teeth completely worn down but without too much looseness in the cones, should be replaced by a bit for harder formations,
- a bit with considerable gage loss, but normal wear on teeth and bearings, should be replaced by one that is well protected around the outside and has minimum offset.

D. Weight on the bit

Laboratory studies and well-site tests have shown that if the weight applied to the bit is increased at constant rotation speed, there is an improvement in the rate of penetration. The increase is almost directly proportional to the weight on the bit if the drilling fluid can manage to keep the bit clean enough. The basic rule is a weight of one ton per inch of bit diameter in soft formations and three tons in hard ones.

However, penetration rate can reach a maximum if the weight on the bit reaches the load limit which embeds a tooth entirely in the rock. Above this threshold, any extra weight is supported by the body of the cone which is pressed against the formation. As a result, bearing life is shortened with no corresponding increase in penetration rate.

The harder the formation, the higher the load limit is, and it may not be possible to reach it. In addition to the load limit, the weight on a bit is limited by its design as such, i.e. the type of bit:

- a bit with long teeth for soft formations will automatically have relatively small bearings, i.e. capable of withstanding lower loads,
- in contrast, a small-toothed bit for hard formations will have considerably larger bearings that can withstand high loads.

The weight on the bit can also be limited by the weight of the drill collars used in the drill string. If more weight is exerted on the bit than can be supplied by the drill collars, there is a danger of:

- pipe buckling and failure,
- destabilizing soft formations,
- doglegs in the borehole where formations have high-angle dips.

E. Rotational speed

The weight on the bit is closely related to the rotational speed. The two parameters can not be increased indefinitely at the same time without causing extremely severe stress conditions in the drill string and the drilling bit.

Studies show that the rate of penetration grows in direct proportion to the rotational speed in soft formations, but not in hard formations. In hard formations, the penetration rate no longer increases above an optimum speed.

The harder the formation, the lower the maximum rotational speed is. The more weight there is on the bit, the lower the optimum rotational speed.

The rotational speed may also be limited by:

- Vibrations in the drill string (resonance phenomenon which may cause fatigue and failure).
- Certain types of bits (with inserts) that act by crushing the rock. Here the weight factor is more important. Excessive rotational speed would cause penetration rate to decline and tungsten carbide inserts to be damaged.
- Drag in deviated wells.

Example:
A three-cone bit with steel teeth for soft formations: 150 to 250 rpm
Insert bit: 50 to 100 rpm

F. Insert bits

The way they are used is a little different than for bits with steel teeth. Even in hard formations, the weight on the bit does not reach three tons per inch of bit diameter.

Hughes recommends the following rule:

$$WN = \text{constant}$$

with W in 10^3 lbs and N in rpm, the constant depends on the type of bit and its diameter.

A 121/4" bit, type J33 has a WN constant of 5900. At 100 rpm, the optimum weight will be:

$$W = 59{,}000 \text{ lb or } W = 27 \text{ t.}$$

G. New bits and changing bits

When a new bit is run in, it is advisable not to exert the maximum weight immediately. If the bit that was pulled is of the same type but worn, the pattern left on the working face is different. If the maximum weight is suddenly applied, the uneven load distribution on the nose of the number 1 cone and the heels of the three cones in particular could cause these parts to deteriorate quickly.

This is even more important when the pulled bit has lost gage. In addition to the risks mentioned above, the new bit may get stuck in the smaller-than-gage diameter hole.

This is why the first centimeters with a new bit should be drilled with little weight and slow rotation. When the hole is under gage, the last few meters drilled by the preceding bit must be carefully redrilled. If the bit is of a different type, a new bottomhole profile must be cut out.

This is why reporting bit wear on the bit record is so important.

The decision to pull a bit must not be arbitrary, but the only indications the driller has are:

- a declining rate of penetration,
- regularly increasing rotational torque.

However, torque surges do not always indicate a worn bit. They may come from formation dip, well profile or stabilizers placed in the drill collar string.

3.1.3.2 Hydraulic parameters

A. Flow rate

The statement was made earlier that rate of penetration increases in direct proportion with the weight on the bit. For a given flow rate, if the weight on the bit is gradually increased, there may come a time when rate of penetration is no longer proportional to the weight.

It may even decrease when more weight is applied to the bit. When the direct proportion between penetration and weight no longer holds true, the phenomenon is termed the balling up point.

Beyond this point, there is no longer a high enough flow rate to clear all of the cuttings out as they are produced by the teeth of the bit. Part of the weight on the bit is then supported by the cuttings.

The balling up point can be delayed by increasing the fluid velocity at bit nozzles, thereby keeping the bit gage area cleaner.

The flow rate therefore has an influence on rate of penetration, but only up to a certain limit. Above this value it may, on the contrary, cause detrimental hole enlargements and wall erosion in the section at drill-collar level.

The basis for calculating the flow rate is the cuttings settling velocity in the annulus. The real problem of determining the settling velocity can be complex because of the large number of parameters involved:

- shape and equivalent diameter of cuttings,
- state of fluid flow and settling conditions,
- drilling fluid rheological parameters.

In practice, annular mud velocities of 20 to 25 m/min have often proved to give good results.

As a general rule, the following can be recommended:

- for soft formations: annular velocity of 30 to 40 m/min,
- for hard formations: annular velocity of 25 to 30 m/min.

For commonly used flow rates, the corresponding annular velocities can be calculated:

Drilling diameter (in)	Flow rate (l/min)	Annular velocity (m/min)
17 1/2	3000 to 4000	21 to 28(1)
12 1/4	2000 to 2600	32 to 41(1)
9 7/8	1500 to 1900	42 to 53(1)
8 1/2	1000 to 1600	43 to 69(1)
6	600 to 800	51 to 68(2)

(1) with 5" pipe
(2) with 3 1/2" pipe

Even though the driller wants to delay the onset of bit balling, he must not use a flow rate such that:

- There is a caving risk in soft formations.
- Hole walls in unconsolidated formations are eroded because of high mud flow rates. Erosion can be troublesome when a gage hole is needed to achieve maximum effectiveness from a stabilized drill string.

- Pressure losses increase in the annulus producing the same effect at the bit depth as an increase in mud density, with a resulting drop in rate of penetration.
- Formations are in danger of fracturing due to the increased pressure losses added to hydrostatic pressure, with the resulting lost circulation.

B. *Jet nozzles*

Jet bit nozzles help sweep out the bottom of the hole and also induce sufficient turbulence to keep the bit teeth clean. Minimum fluid velocity is considered to be approximately 80 m/s, with the range generally from 80 to 150 m/s.

As an illustration, the nozzle size can be calculated which corresponds to this range for each drilling diameter and an average flow rate:

Drilling diameter (in)	Average flow rate (l/min)	Nozzle size for velocity at orifice (in 1/32")	
		80 m/s	150 m/s
17 1/2	3500	22 – 22 – 22	16 – 16 – 17
12 1/4	2300	18 – 18 – 18	13 – 13 – 13
8 1/2	1300	13 – 14 – 14	9 – 10 – 10
6	700	10 – 10 – 10	7 – 7 – 7

Of course this can not be applied independently. It must be considered in the context of both the required hydraulic power and the risks of nozzle diameters that may cause plugging by cuttings or lost circulation materials.

C. *Density*

Laboratory studies have confirmed what every driller has experienced personally: an increase in mud density means slower rate of penetration.

It is easy to understand that a rock formation is more difficult to break up when the hydrostatic pressure acting against it increases.

The influence of density is illustrated by the curve in **Graph 3.1** which shows the variation in rate of penetration V versus the differential pressure, ΔP, for shales (after W.C. Maurer). V_o is the penetration that would exist with zero ΔP, and ΔP is the difference between the bottomhole pressure and the formation pressure. Bottomhole pressure is the sum of hydrostatic pressure and annular pressure losses.

D. *Viscosity*

Testing has shown the detrimental effect of viscosity on rate of penetration. Logic suggests that the more viscous the liquid is, the less readily it can enter the pores and the small fractures produced by the bit and the less it can contribute to breaking up the formation.

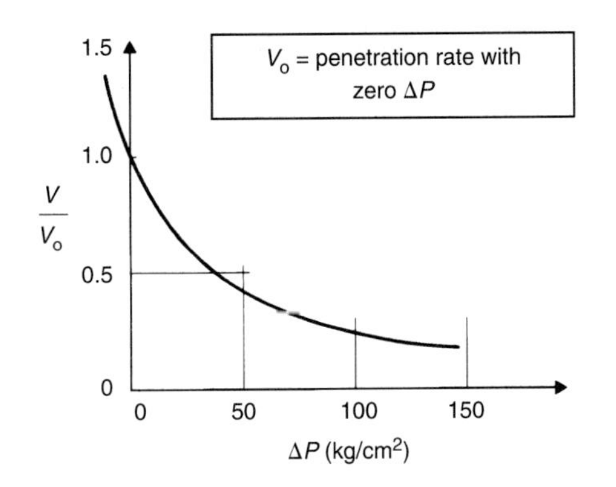

Graph 3.1

Influence of the differential pressure at the working face on the rate of penetration.

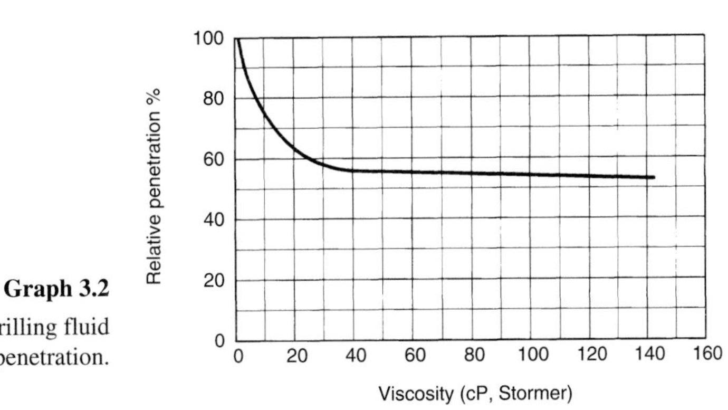

Graph 3.2

Influence of drilling fluid viscosity on rate of penetration.

The curve in **Graph 3.2** shows the influence of viscosity.

E. Filtrate

The effect of this parameter is not very well known. The low densities and viscosities that usually go along with substantial filtrate would seem to bring about the improved performance that has been recorded.

F. Proportion of oil in the drilling mud

Experience has shown that oil added to drilling mud boosts the rate of penetration. Furthermore, the lubricating effect of the oil extends the lifetime of bit bearings.

3.2 THE DRILL STRING

3.2.1 Introduction

The drill string is the mechanical assemblage connecting the rotary drive system on the surface to the drilling bit (**Fig. 3.12**).

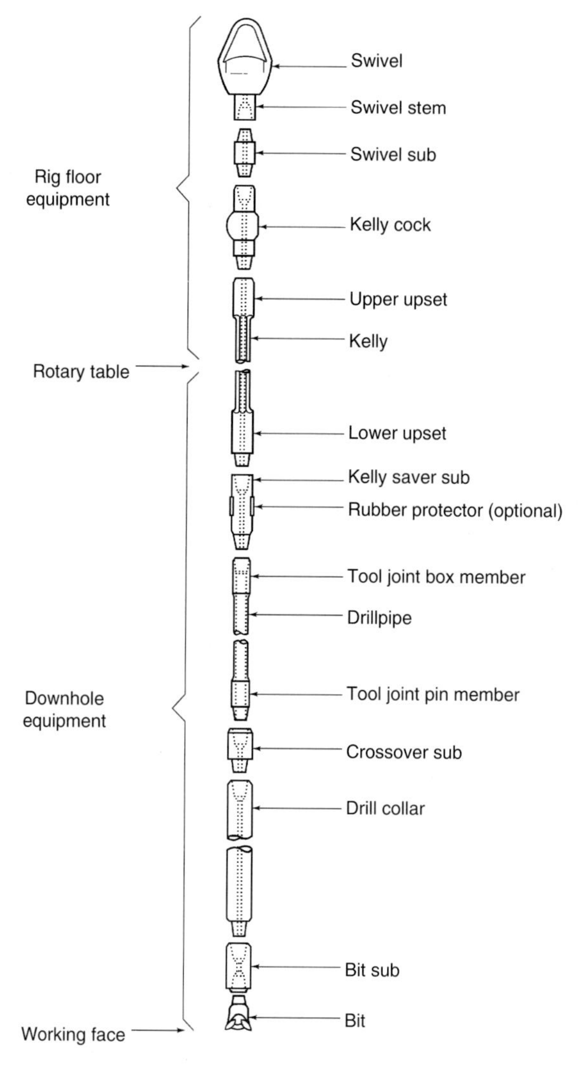

Fig. 3.12

Terminology of the drill string *(Source: The drill string–Unit 1, Lesson 3, Petex, IADC).*

It fulfills the following functions:
- transmits the energy required to break up the rock formation by whatever drilling bit may be used,
- guides and controls the trajectory of the bit,
- exerts a compressive force on the drilling bit, or weight on the bit,
- allows drilling fluid to circulate with minimum pressure losses.

The main downhole equipment is:
- drill collars, and
- drillpipe.

Ancillary equipment includes drill collar stabilizers, shock absorbers and crossover subs.

On the surface, the drill string hangs from the hook by the swivel and the rotary drive comes from the kelly. Two safety valves can close off the inside of the string.

3.2.2 Drill collars

Their function is an essential one in the drill string, since they condition proper use of the drilling bit. They are first and foremost a steel weight whose mass provides the force to press the drilling bit onto the formation. A number of operating constraints are involved in using them:
- the drilling diameter,
- minimum pressure losses (as little as possible),
- easy handling and transportation,
- buckling strength,
- rigidity.

The above factors have led to the following sizing and manufacturing technologies:

3.2.2.1 Dimensions (Fig. 3.13)

The smallest inside diameter must be sought as it will give the heaviest drill collar. However, it must allow:
- measurement instruments to be run down inside the pipe, and
- acceptable pressure losses.

Usual inside diameters are between 2 and 3 inches depending on the outside diameter of the drill collar.

The outside diameter will range between a minimum and a maximum that can be defined as follows:
- The maximum diameter depends on: the drilling diameter, whether an overshot can grapple it, whether a washover pipe can fit over it, the allowable fluid velocity in the annulus and the risks of sticking by differential pressure.
- The minimum diameter is directly related to the rigidity of the drill collar assembly.

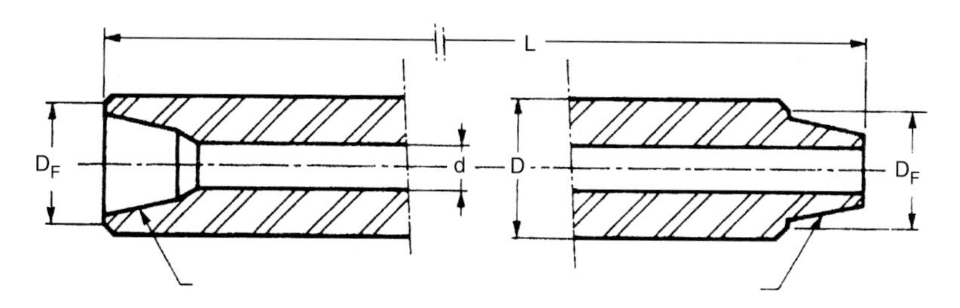

Drill collar No. ([1])	Outsider diameter D		Bore d		Length ±0.15 L (m)	Diameter at bevel ±0.4 D_F (mm)	BSR
	(in)	(mm)	+1/16 to 0 (in)	+1.6 to 0 (mm)			
NC23-31	3 1/8	79.4	1 1/4	31.8	9.1	76.2	2.57
NC26-35 (2 3/8 IF)	3 1/2	88.9	1 1/2	38.1	9.1	82.9	2.42
NC31-41 (2 7/8 IF)	4 1/8	104.8	2	50.8	9.1	100.4	2.43
NC35-47	4 3/4	120.7	2	50.8	9.1	114.7	2.58
NC38-50 (3 1/2 IF)	5	127.0	2 1/4	57.2	9.1	121.0	2.38
NC44-60	6	152.4	2 1/4	57.2	9.1 or 9.4	144.5	2.49
NC44-60	6	152.4	2 13/16	71.4	9.1 or 9.4	144.5	2.84
NC44-60	6 1/4	158.8	2 1/4	57.2	9.1 or 9.4	149.2	2.91
NC46-62 (4 IF)	6 1/4	158.8	2 13/16	71.4	9.1 or 9.4	150.0	2.63
NC46-65 (4 IF)	6 1/2	165.1	2 1/4	57.2	9.1 or 9.4	154.8	2.76
NC46-65 (4 IF)	6 1/2	165.1	2 13/16	71.4	9.1 or 9.4	154.8	2.05
NC46-67 (4 IF)	6 3/4	171.5	2 1/4	57.2	9.1 or 9.4	159.5	2.18
NC50-70 (2 1/2 IF)	7	177.8	2 1/4	57.2	9.1 or 9.4	164.7	2.54
NC50-70 (2 1/2 IF)	7	177.8	2 13/16	71.4	9.1 or 9.4	164.7	2.73
NC50-72 (2 1/2 IF)	7 1/4	184.2	2 13/16	71.4	9.1 or 9.4	169.5	2.12
NC56-77	7 3/4	196.9	2 13/16	71.4	9.1 or 9.4	185.3	2.70
NC56-80	8	203.2	2 13/16	71.4	9.1 or 9.4	190.1	2.02
6 5/8 REG	8 1/4	209.6	2 13/16	71.4	9.1 or 9.4	195.7	2.93
NC61-90	9	228.6	2 13/16	71.4	9.1 or 9.4	212.7	2.17
7 5/8 REG	9 1/2	241.3	3	76.2	9.1 or 9.4	223.8	2.81
NC70-97	9 3/4	247.7	3	76.2	9.1 or 9.4	232.6	2.57
NC70-100	10	254.0	3	76.2	9.1 or 9.4	237.3	2.81
NC77-110	11	279.4	3	76.2	9.1 or 9.4	260.7	2.78

([1]) The drill number consists of two parts separated by a hyphen. The first part is the connection number in the NC style. The second part, consisting of 2 (or 3) digits, indicates the drill collar outside diameter in units and tenths of inches. The connections shown in parentheses in Col. 1 are not a part of the drill collar number; they indicate interchangeability of drill collars made with the standard (NC) connections as shown. If the connections shown in parentheses in column 1 are made with the V-0,038R thread form, the connections and drill collars, are identical with those in the NC style. Drill collars with 8 1/4 and 9 1/2 inches outside diameters are shown with 6 5/8 and 7 5/8 REG connections, since there are no NC connections in the recommended bending strength ratio range.

Fig. 3.13

Cylindrical drill collars: dimensions and threads (API Spec 2)
(Source: Drilling Data Handbook, Editions Technip, Paris, 1989).

These considerations are illustrated in the table (**Fig. 3.14**) that shows drill collar selection according to drilling diameter. The unit length is standardized at 30 feet or 9.144 m. The unit weight of common drill collars can be found in **Fig. 3.15**.

Drilling diameter	Drill collar outside diameter	Drill collar inside diameter
24 to 12 1/4	9 1/2	3
9 7/8	7 3/4 to 8	2 13/16
8 3/4 to 8 1/2	6 3/4	2 13/16
6 3/4 to 6	4 3/4	2 1/4

Fig. 3.14

Selecting drill collar diameters according to drilling diameters (in inches).

Drill collars	Weight/m (kg/m)	Unit weight (30 ft) (kg)
11 1/4 × 3	467.6	4276
9 1/2 × 3	323.2	2955
8 × 2 13/16	223.1	2040
7 3/4 × 2 13/16	207.4	1896
6 3/4 × 2 13/16	149.8	1370
4 3/4 × 2 1/4	69.6	636

Fig. 3.15

Weight of drill collars.

3.2.2.2 Metal used in drill collar manufacture

Conventional drill collars are machined out of bars of 4165H (U.S.) or 42 CD4 (NF) steel which has good properties after tempering and quenching.

API Standard (SPEC 7)

DC inside diameter (in)	Minimum yield strength	Minimum tensile strength
3 1/8 to 6 7/8	110,000 psi (785 MPa)	140,000 psi (967 MPa)
7 to 10	100,000 psi (689 MPa)	135,000 psi (931 MPa)

The steels are generally chromium-molybdenum alloys that comply with mechanical specifications after tempering and quenching while retaining a hardness compatible with machining.

Also to be mentioned are nonmagnetic drill collars that are required for drilling directional wells. Azimuth measurement instruments can only be run into a drill string with drill collars that do not influence the measurement of the magnetic field.

Originally, these drill collars were made of K. Monel alloy (over 60% nickel), but the price was too high and the threads seized up too often. The K. Monel alloy has been replaced by a chromium-manganese austenitic iron alloy or similar.

The bore is drilled out by two cutting tools working from each end toward the middle.

3.2.2.3 Drill collar profiles

A. *Slick drill collars*

This is the simplest shape for drill collars, they have the nominal outside diameter over the total length.

These drill collars need to be screwed to a lifting sub (**Fig. 3.16a**) to be handled and held by safety clamps when they are being supported by the slips on the rotary table. To eliminate the need for lifting subs and safety clamps and waste less time assembling drill collars, manufacturers have introduced the double-recess ZIP profile with:
- a slip recess for safety, and
- an elevator recess for lifting (**Fig. 3.16b**).

The recess shoulders need to be protected from wear and tear by the formation during drilling. Accordingly, the ZIP profile has had further surface treatments, or hardfacing. Hardfacing consists in depositing tungsten carbide grains on the outside diameter.

In the drilling business the current trend is not to use elevator recesses which are not effective enough compared to their useful lifetime and the extra cost involved.

Care should be taken not to use overly rough hardfacing surfaces as they can cut casing like a very effective milling tool.

B. *Spiral drill collars*

This type of drill collar, illustrated in **Fig. 3.16b**, is used to lessen the risk of differential pressure sticking in permeable formations.

Three spiral grooves are machined into the outside DC surface to:
- reduce the area in contact with the walls of the borehole,
- keep the drilling fluid circulating well around the body of the drill collar.

The reduction in spiral drill collar weight is evaluated at 4% of slick drill collar weight.

C. *Square drill collars*

These drill collars are very rigid and help keep the hole straight, since they are used with a clearance of 1/32" between the walls of the hole and the ribs of the drill collar. However, they are high priced and hard to manage and so drillers have replaced them with oversize cylindrical drill collars or multiple stabilizers.

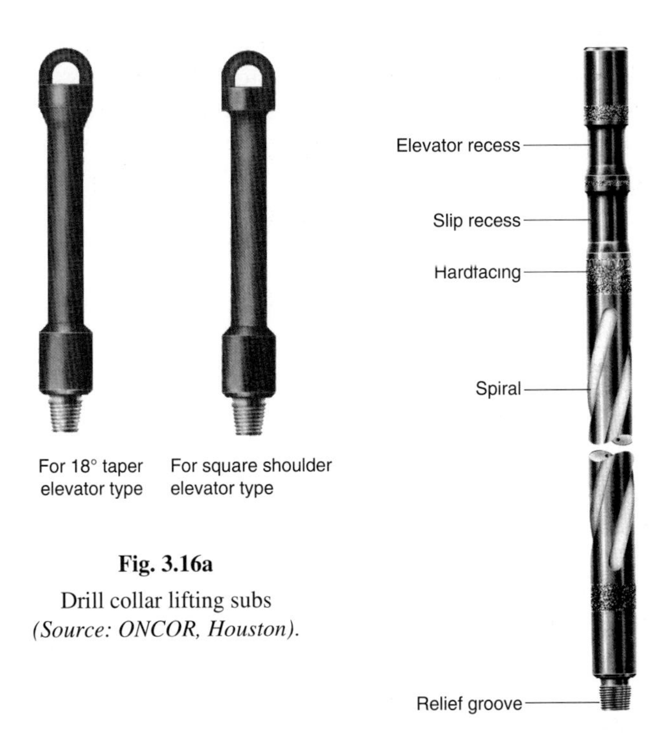

Fig. 3.16a
Drill collar lifting subs
(Source: ONCOR, Houston).

Fig. 3.16b
Example of spiral drill collar
(Source: SMF International).

3.2.2.4 Threads (rotary shouldered connections) (Fig. 3.17)

The threads connecting drill collars together are made to API standards. They are tapered for two reasons:

- this shape provides the strongest connection between two lengths of pipe of the same wall thickness,
- it allows great ease and speed in screwing and unscrewing (self-alignment, connection is made up without rotating as many times as the number of thread turns).

The threads used at the very beginnings of rotary drilling were the normal ones commonly used in mechanics at the time. This is why they were called "regular".

Threads such as: FH (full hole), IF (internal flush) and NC (numbered connection) can also be mentioned. NC threads are the current standard in common nominal dimensions. The shoulder of the thread transmits most of the stresses that drill collars are subjected to. It is the only metal/metal seal for the drill string's internal pressure. The make up torque is what provides the required compression prestressing. The connections are bent back and forth and fatigue occurs after a certain number of weeks or months of duty. Drill collar threads usually show incipient failure in the form of microcracks. The only remedy is to cut off the damaged thread and make a new one on the drill collar. These microcracks can be detected by ultraviolet magnetic inspection. After the thread has been thoroughly cleaned to remove any trace of lubricant and oxidation, the end of the drill collar is magnetized. A liquid (kerosene) that contains very fine-grain fluorescent magnetic particles is vaporized onto the thread. The particles are attracted by the lips of the crack(s). When examined under ultraviolet rays, the fluorescent particles show up very clearly and thereby outline the crack(s). This type of inspection is of course not valid for nonmagnetic drill collars.

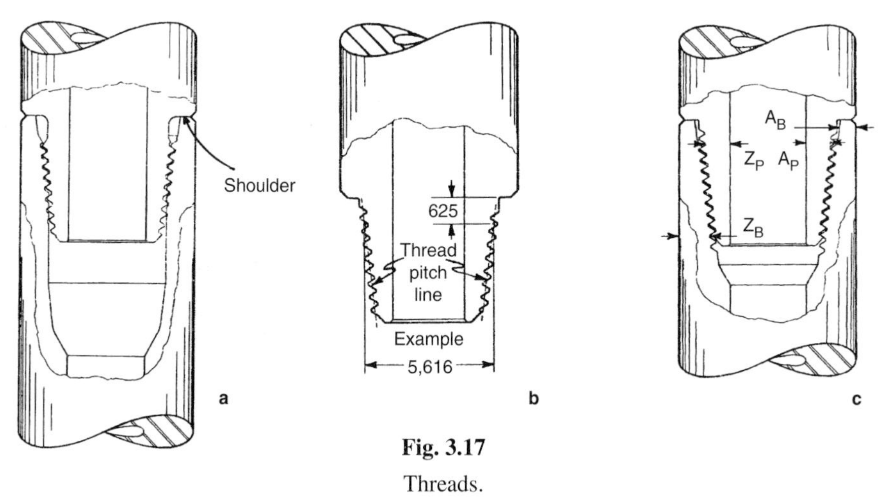

Fig. 3.17

Threads.

a. Drill collar connection. **b.** Drill collar thread. **c.** Critical stress concentration zone.
Z_B and Z_P: danger zones for bending stress. A_B and A_P: danger zones for axial stress.
(Source: Proper field practices for drill collar strings, Garret & Wihn, SPE 5124).

3.2.2.5 Example of predetermining a drill collar assembly

The drilling program chosen to serve as an example includes a 121/4" phase with a rock bit that requires WOB of 2 tons per inch of bit diameter. The drill collars available on the well site are 91/2", 8" and 63/4". The drilling mud has a density of 1.18. **Fig. 3.18** shows the external stresses on the drill collar assembly connected to the bit.

The equation for the stability of this assembly can be written as:

$$P_{DC} + P_2 S = P_1 S + WOB$$

$$WOB = P_{DC} - (P_1 - P_2)\, S$$

$$P_{DC} = L \times S \times d_s$$

$$P_1 - P_2 = L \times S \times d_m$$

P_{DC} = drill collar weight
WOB = weight on the bit (formation reactive force)
S = drill collar cross-section
P_1 = hydrostatic pressure at depth Z_1
P_2 = hydrostatic pressure at depth Z_2

$$Z_1 - Z_2 = L$$

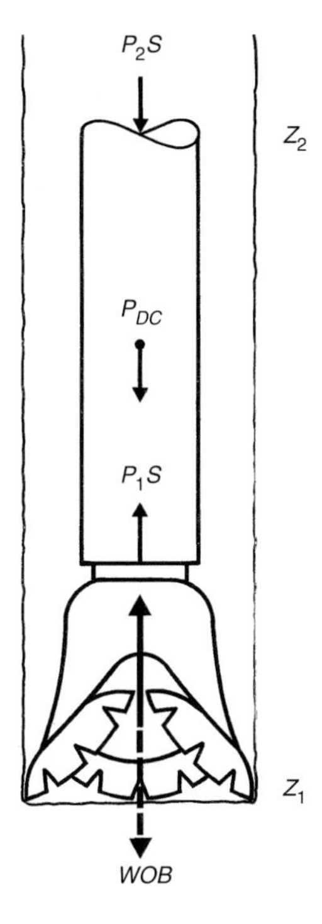

Fig. 3.18
Principle of using drill collars.

with

d_s = density of the steel
d_m = density of the mud

$$WOB = L \times S\,(d_s + d_m)$$

$$= LP_{DC}\left(1 - \frac{d_m}{d_s}\right)$$

$Sd_s = P_{DC}$ weight per unit of length

$$k = 1 - \frac{d_m}{d_s}$$

k is called the buoyancy coefficient (see *Drilling Data Handbook*, 6th edition, A37, p.41, Editions Technip, Paris, 1989).

In the example, $k = 0.849$.
Required WOB: 2 t $\times$ 12.25 = 24.5 t.
Drill collar 91/2": P_{DC} = 323.2 kg/m

$$L = \frac{24{,}500}{323.2 \times 0.849} = 89.29 \text{ m}$$

As mentioned earlier in the introduction, the drillpipe connects the drill collars to the earth's surface. However, drillpipe is not designed to withstand compression. To avoid compressing the bit with more than the available drill collar weight, a number of extra drill collars are added for safety (**Fig. 3.19**). A safety margin between 10 and 20% is commonly taken.

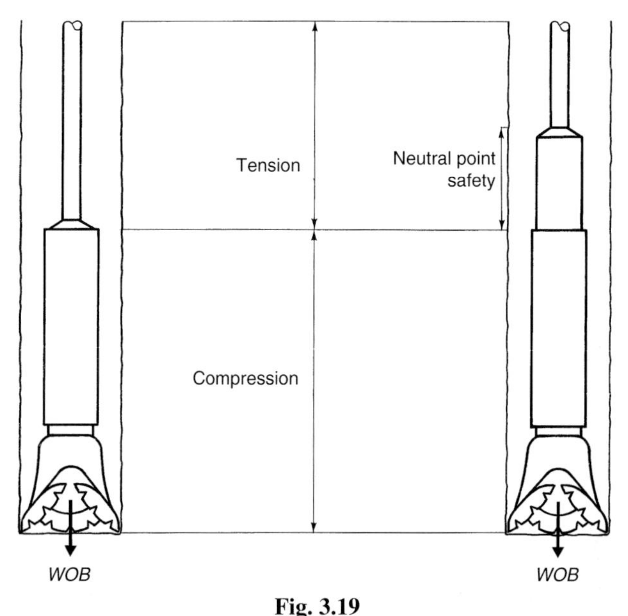

Fig. 3.19

Principle of using drill collars: neutral point.

So that there will not be too great a difference in rigidity between drill collars and drillpipe, the recommendation —especially for large drilling diameters— is to used several drill collar dimensions. The most rigid will be placed above the bit, moving up to more flexible before the drillpipe. The critical zone of the first part of the drillpipe string can also be mechanically protected by placing heavyweight drillpipe above the drill collars (**Fig. 3.20**).

Heavyweight 5" drillpipe has roughly the same outside dimensions as 5" drillpipe, except for extra central thickness and extra long tool joints. But what makes it different is the 3" inside diameter rather than around 4" as for conventional drillpipe. As a result, they are heavier and withstand buckling better.

In deviated boreholes, the theoretically available weight on the bit is the weight in the drilling mud of the drill collars multiplied by the cosine of the angle of inclination. Horizontal boreholes require drill collars to be used only to guide and keep to the trajectory. Weight is exerted on the bit by all the heavyweight drillpipe and this part of the drill string must go up to a lower-angle inclination where gravitational forces can act.

3.2.3 Drillpipe (Fig. 3.21)

This is seamless steel pipe with a die-cast extra thickness called an upset on each end. There are internal upsets (IU), external upsets (EU) and internal and external upsets (IEU).

The IU has extra thickness extending inward, the EU extending outward and the IEU extends both ways.

The upsets serve to weld the tool joints, or connections, that enable the lengths of drillpipe to be screwed together.

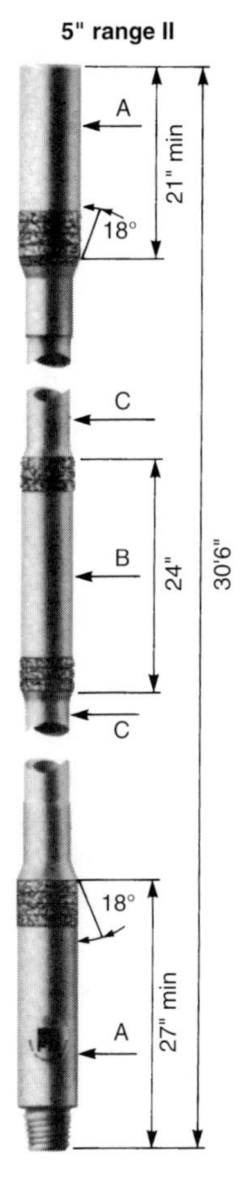

Fig. 3.20
Heavyweight drillpipe
(Source: SMF International).
A: 61/2". B: 51/2". C: 5". Id: 3".

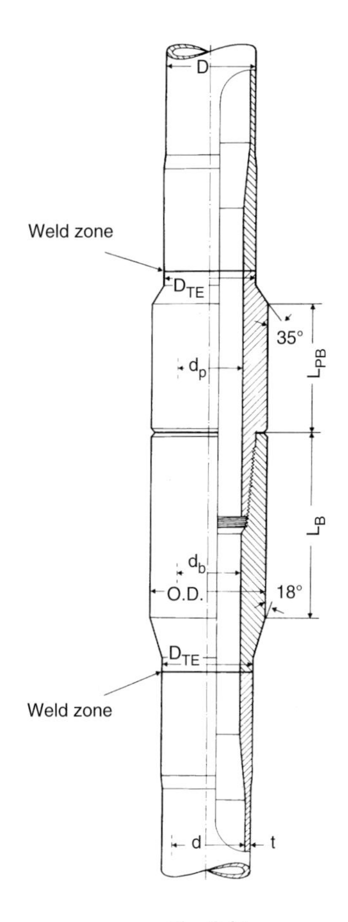

Fig. 3.21
Tool joint *(Source: Vallourec).*

3.2.3.1 Drillpipe characteristics

API has set standards for four grades of steel for all types of drillpipe:

Grade	Yield strength		Tensile strength
	Min	Max	Min
	psi (MPa)	psi (MPa)	psi (MPa)
E	75,000 (517)	105,000 (724)	100,000 (690)
X-95	95,000 (655)	125,000 (862)	105,000 (724)
G-105	105,000 (724)	135,000 (931)	115,000 (793)
S-135	135,000 (931)	165,000 (1138)	145,000 (1000)

The standard grade is grade E. When the standard pipe's mechanical strength is no longer sufficient, drilling must then be continued by adding higher-grade pipe (X, G or S).

Pipe lengths are classified as follows by *API:*
Range 1 = 18 to 22 ft.
Range 2 = 27 to 30 ft.
Range 3 = 38 to 45 ft.

Oilwell drillers use range 2 which is stacked in threes on heavyweight rigs and in twos on lightweight rigs.

The nominal diameter of drillpipe is the outside diameter of the pipe body in inches: 23/8, 27/8, 31/2, 4, 41/2, 5, 51/2 are the standard sizes.

The most commonly used sizes in oilwell drilling are 31/2" and 5" (very often 41/2" in the U.S.). These two diameters can be used to drill most programs. The inside diameter is not a direct dimension, it is a value calculated on the basis of the nominal weight of the pipe body expressed in lb/ft by *API.*

Standard 5" pipe has a nominal weight of 19.50 lb/ft. It is also available in 26.60 lb/ft.

Standard 31/2" pipe weighs 13.30 lb/ft and can also be found in 15.50 lb/ft.

The higher the nominal weight is, the larger the cross-sectional area and the greater the mechanical strength. The pipe is of course heavier too. So that the lengths of pipe can be screwed together, they have connections called tool joints. Modern-day tool joints are separate parts that are friction welded onto the drillpipe body upsets (**Fig. 3.22**).

Tool-joint steel has a yield strength of 120,000 psi whatever the grade of pipe (E, X, G, S) it has been welded to. Friction welding technology on parts of this size has improved the

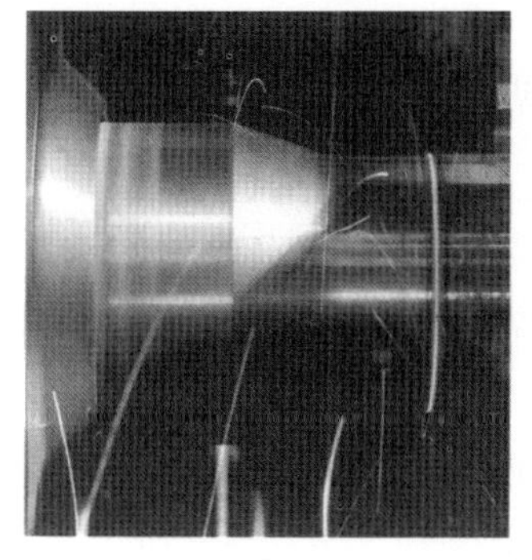

Friction welding *(Source: Mannesmann W.A.G.).*

Box and pin tool-joint *(Hughes Tools).*

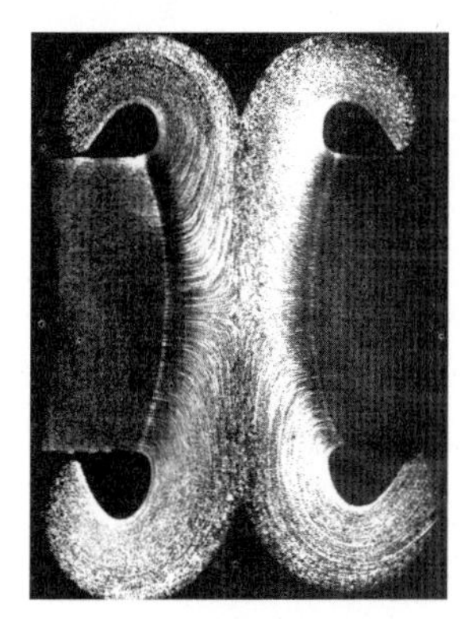

Weld cross-section *(Drilcon Industries Ltd.).*

Fig. 3.22

Friction welding of tool joints.

strength of this assembly of two different alloys. The threads are similar to drill collar threads: NC 50 for tool joints with 61/2" OD for 5" pipe and NC 38 for 43/4" tool joints and 31/2" pipe. Drilling a 6" phase after having run in a 7" casing or liner requires a 31/2" drill string whose tool joints have enough clearance to pass through.

The 18-degree shoulder of the box tool joint is used to hoist the drill string with the elevator clamp and the hook. There are also tool joints with 90-degree shoulders (square shoulders), but they have been almost totally phased out.

As the drillpipe is used it gets worn by dragging against the formation. The box tool joints lose thickness and so do the pipe bodies. Decreased wall cross-sections automatically entail a loss of mechanical resistance. This is why *API* has classified used pipe into:

Class I .... new pipe, all dimensions match nominal values,
Premium .... 80% of the thickness remaining,
Class II .... 70% of the thickness remaining.

The drilling contractor must regularly have his drill strings inspected. He must get a certificate complying with API standards on the class of wear for the pipe he uses. He will only be allowed to use it as specified for the mechanical properties corresponding to the classes.

Example (*Drilling Data Handbook*, 6th edition, pp. 61-62, Editions Technip, Paris, 1989)

Mechanical properties	Drillpipes	Class I	Premium	Class II
Tensile yield strength (10^3 daN)	31/2 – 13.30 Grade E	120.8	94.4	81.6
	5 – 19.50 Grade E	176	138.6	120.3
Torsional strength (daN.m)	31/2 – 13.30 Grade E	2520	1950	1680
	5 – 19.50 Grade E	5580	4380	3790
Bursting pressure (MPa)	31/2 – 13.30 Grade E	95.1	87	76.1
	5 – 19.50 Grade E	65.5	59.9	52.4

3.2.3.2 Selecting pipe

Earlier, 5" (or 41/2") pipe was seen to be used up to the 81/2" drilling phase. Beyond this phase, after cementing the 7" casing, 31/2" drillpipe must be used.

Given the tensile yield strength (see table above), calculations can give the maximum depth reached under certain conditions and for a given type of pipe:

$$T_e \times 0.9 = (L_{dp} \times p_{dp} + L_{dc}\, p_{dc}) \times k$$

T_e = tensile yield strength reduced by 10%,
L_{dp} = length of drillpipe,
L_{dc} = length of drill collars,
p_{dp} = nominal weight of drillpipe,
p_{dc} = nominal weight of drill collars,
k = buoyancy coefficient.

The safety margin or extra tensile capacity is the difference:

$$T_e \times 0.9 - (L_{dp}\, p_{dp} + L_{dc}\, p_{dc})\, k = M$$

Drillers have to check that the value of M is and remains sufficient given the drilling conditions anticipated in the program.

Example:

L_{dc} = 250 m
p_{dc} = 149.4 kg/m
k = 0.847
p_{dp} = 31.24 kg/m, 5" pipe – 19.50 Grade E
T_e = 139,000 daN, Premium class

$$L_{dp} = \frac{(T_e \times 0.9 - M - kL_{dc}\,P_{dc})}{kP_{dp}}$$

$$= \frac{(139{,}000 \times 0.9 \times 1.02 - 50{,}000 - 0.847 \times 250 \times 149.4)}{0.847 \times 31.24}$$

$$= 1737 \text{ m}$$

If the driller wants to keep the same tensile safety margin to continue drilling, he needs to use pipe with a yield strength T_e greater than 139 10^3 daN. Higher grade pipe — X, G, or S — is then added. If the well is to be drilled ultradeep and choice is limited, higher grade and heavy nominal weight pipe can be chosen. But it is always preferable to use pipe with the largest inside diameter available to minimize pressure losses. The table below gives orders of magnitude for a circulating fluid with a flow of 1000 l/min:

Pipe ID \ Type of fluid	A) Vp = 10 cP, d = 1.08	B) Vp = 30 cP, d = 1.20
5" – 19.50 – E (ID.TJ: 3 3/4)	4.7 bar/1000 m	6.4 bar/1000 m
3 1/2" – 13.30 – E (ID.TJ: 2 11/16)	36.2 bar/1000 m	48.8 bar/1000 m
Drill collar ID 2 13/16"	3.30 bar/100 m	4.5 bar/100 m

Since pressure losses are approximately proportional to the square of the flow rate, inside pressure losses can be evaluated for the following examples of drill strings:

- Drilling 12 1/4" with DP 5" to 2500 m
  DC 9 1/2" length = 200 m
  Mud type A: flow 22,000 l/min
  Inside P = (6.60 + 10.8) $(2.2)^2$ = 84.2 bar
- Drilling 6" with DP 3 1/2" to 3000 m
  DC length: DC 4 3/4" × 2 1/4" = 200 m
  Mud type B: flow 800 l/min
  Inside P = (25.9 + 136.6) × $(0.8)^2$ = 104 bar.

3.2.4 Ancillary equipment (Fig. 3.12)

The kelly and its accessories are mainly dealt with here.

3.2.4.1 The kelly

This is the rotating link between the rotary table and the drill string. It is also a pipe that supports the total weight of the drill string. Its chief functions are to:

- connect the swivel to the uppermost length of drillpipe,
- transmit rotary torque to the drill string,
- allow the drill string to move longitudinally while keeping rotation under way,
- convey the drilling fluid from the swivel into the drill string.

The kelly cross section may be square, hexagonal or even triangular. A safety valve called the upper kelly cock is screwed onto the uppermost connection to close off the inside of the drill string in the event of a kick. The lower kelly valve is manual and fulfills the same function in addition to preventing the drilling mud from running onto the rig floor when a length of pipe is being added. The kelly is the longest of the tubulars in the drill string: 40 to 54 ft, since the driller needs to be able to drill a little farther than the length of a drillpipe component in order to add on an extra one.

3.2.4.2 Crossover subs

There are two solutions for connecting all the components of the drill string. Either the same thread must be used — this is impossible due to the wide variety of nominal dimensions — or adapters must be sandwiched in: crossover sub, X sub or just sub. The drilling contractor needs the widest range possible of subs to cope with all changes in drill collars, motors, stabilizers, etc.

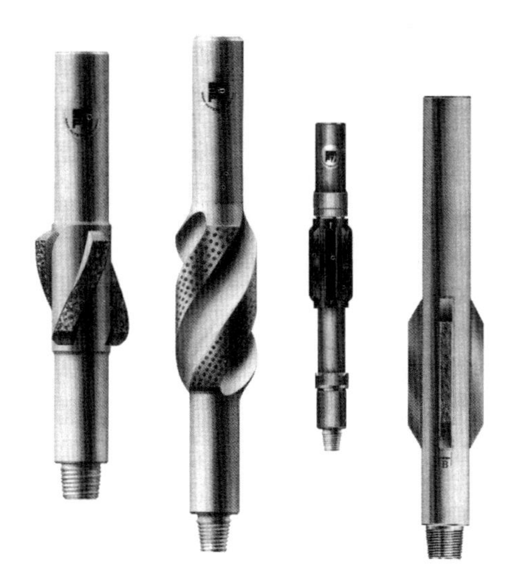

Fig. 3.23
Examples of stabilizers *(Source: SMF International).*

3.2.4.3 Stabilizers (Fig. 3.23)

As the name suggests, they are included in the drill string, more precisely at drill collar level, to control the bit and keep it on the right trajectory, whether vertical or deviated. The shapes and makes vary depending on the formation, the abrasiveness and the service required. Three stabilizers on a BHA (bottomhole assembly, see Chapter 10: Directional Drilling) seem to be the maximum since they generate considerable friction in the borehole.

3.2.4.4 The shock absorber

This looks like a drill collar and is screwed on just above the bit. The purpose is to damp (or filter) the vibrations generated by the rotation of the drilling bit.

Chapter 4

THE DRILLING RIG

4.1 INTRODUCTION

Now that the principles of modern drilling methods have been discussed, the material resources, operating techniques and personnel required to drill the actual oilwell will be dealt with.

The drilling rig, or more comprehensively the well site, includes the following:

- production of primary energy,
- expendable product storage and warehousing,
- facilities for handling waste discharges,
- shelters,
- the derrick,
- the pumping facilities and tanks.

Figure 4.1 serves as a reminder and gives a schematic picture of rig functions: hoisting, pumping and rotation.

A drilling rig can be classified initially by its maximum drilling depth rating:

Lightweight rigs .... 1,500 to 2,000 m
Intermediate rigs .... 3,500 m
Heavyweight rigs .... 6,000 m
Ultraheavy rigs .... 8,000 to 10,000 m

Drilling depth capacity means weight on the hoisting hook, based on the weight of drill strings and casings.

On the basis of commonly acknowledged tripping times, the maximum power that the drawworks will need to develop can be evaluated.

This is why when a drilling rig is due to be selected, the only thing of interest is the power rating of the drawworks. This characteristic fits in with the depth rating and English-speaking drillers even have a practical rule of the thumb for it: every 100 feet of borehole requires 10 horsepower at the drawworks. For the rig categories listed above, this gives:

Lightweight rigs........................... 650 hp
Intermediate rigs .... 1300 hp
Heavyweight rigs .... 2000 hp
Ultraheavy rigs.............................. 3000 hp

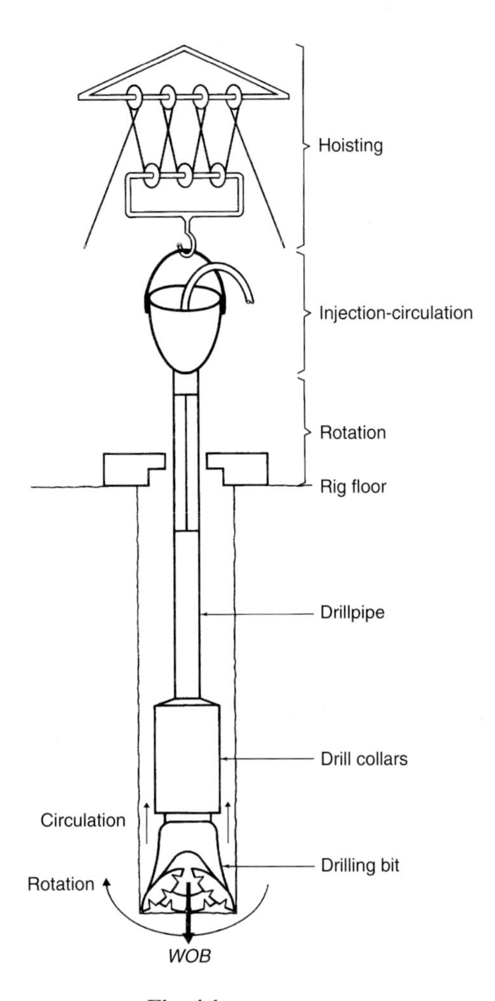

Fig. 4.1

Principle of the hoisting function.

The other functions (pumping, rotation) are sized according to the conventional drilling and casing program for a well at the appointed depth.

Present-day organization in companies involved in oil prospecting is based on the specific nature of tasks. The owner of the well will benefit from producing it, but will require an exploration permit delivered by the relevant local authorities to do so. When several oil companies are on the same permit, they are said to be associated in a joint venture. Generally one of the partners will serve as operator. Its function will be to carry out the exploration program as set out in the permit specifications. The program includes geological analysis, a geophysics survey and exploratory wells.

The operator's technical role is therefore very important in drilling exploratory wells. Once the engineering has been completed (drilling and casing program, determination of fluids and location, etc.), the operator calls for tenders from drilling service companies.

The drilling contractor's function is to rent a complete drilling rig along with its operating personnel. Commercial and technical relationships between operators and drilling contractors can vary widely depending on the basis of payment stipulated in the contractual document. The system may be a day work contract, a turnkey contract, a footage contract or an incentive contract.

4.2 HOISTING EQUIPMENT

The equipment includes:
- the hoisting tower structure,
- the drawworks and its accessories,
- the drilling line,
- the control panel.

4.2.1 The hoisting tower structure

There are three main types of structures: the derrick, the mast and the trailerized guyed mast.

4.2.1.1 The derrick

This type is the oldest and derives directly from the tower that used to be built of wood. It is in the shape of a sharply pointed pyramid with one "foot" in each corner resting on the angles of a square. The square is the rig floor (**Fig. 4.2**).

An upper platform, the water table, is what holds the drilling line sheaves, or crown block. Another platform between the two (about 85 ft high) is where the derrickman stands to rack up the lengths of drillpipe or drill collars.

The metal structure of the derrick can be welded or bolted together. There are practically no more derricks for onshore drilling because dismantling and reassembling operations are long, dangerous and therefore no longer economically viable at all.

In contrast, mobile rigs offshore use this construction technique because it is economical and well suited to offshore conditions. Here no dismantling is needed to move from one site to another, since it is the whole rig that is moved.

A distinction is made between dynamic derricks and standard ones. The dynamic type is installed on floating supports such as ships and semisubmersibles. These derricks are subjected to extra dynamic stresses due to rolling, pitching and heaving of the support and to winds. The space available between the rig floor and the crown block must be higher to handle the wave-induced vertical movements of the floating support.

The specifications of two types of Dreco derrick are given below as an illustration.

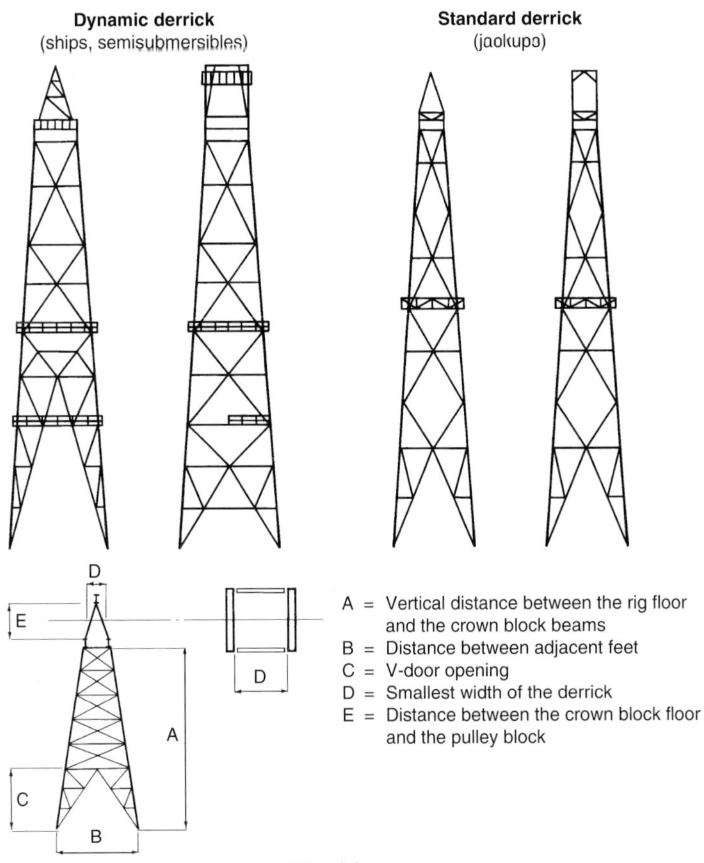

Fig. 4.2

Examples of derricks.

A. *Dynamic derricks*

Height	160 ft (useful hook stroke length)
Base..	40 ft
Crown block platform..................	18 ft
V-door..	60 ft
Maximum hook load.....................	1,000,000 lb

a. With 60% of the drill string racked up, conditions of use are:

800,000 lb hook load,
50 knots wind speed,
10 degrees of roll, with a period of 10 s,
3 degrees of pitch, with a period of 7 s,
5 ft of heave, with a period of 8 s,
4 × 24 t tensioners,
Traveling block at maximum height.

b. Waiting on weather with maximum drill string racked up:

250,000 lb hook load (traveling block tied down to the rig floor),
70 knots wind speed,
15 degrees of roll, with a period of 10 s,
4 degrees of pitch, with a period of 7 s,
6 ft of heave, with a period of 8 s,
4 × 24 t tensioners.

c. Under extremely severe conditions:

250,000 lb hook load,
100 knots wind speed,
30 degrees of roll, with a period of 10 s,
6 degrees of pitch, with a period of 7 s,
7 ft of heave, with a period of 8 s.

B. *Standard derricks (jackups)*

Height .... 147 ft
Base .... 30 ft
Top .... 8 ft
V-door .... 34 ft

a. Drilling

1,000,000 lb static hook load,
No racking,
85 mph wind speed.

b. Extremely severe conditions

115 mph wind speed,
No racking.

c. Running in casing

700,000 lb hook load,
85 mph wind speed,
Maximum racking.

d. During towage

250,000 lb hook load,

20 degrees of roll, with a period of 10 s.

The main specification differences involve the free height available (an extra 13 ft to allow for the heave compensator) and the different hook loads according to specific conditions.

4.2.1.2 The mast (Figs. 4.3a and b)

The mast is a structure shaped like a very pointed A. It has the particular feature of being rotary jointed at the base so that it can be assembled or dismantled horizontally and then pulled to an upright position using the drawworks and a special hoisting cable. This type of drilling tower is well suited to onshore drilling rigs requiring a good deal of mobility. The racking board is in a cantilever position and lengths of pipe are racked on a floor, called the setback, that is separate from the mast structure.

Technical specifications are identical to those for derricks:

- maximum hook load given the reeving system,
- free height available in the mast,
- width at the base,
- resistance to wind with and without racked drill string.

Capacities are comparable to those for derricks.

There are other less common types of masts that meet installation requirements on an offshore development platform where a conventional mast can not be placed in a horizontal position due to lack of room. The solution is to use a folding mast (**Fig. 4.4**) or a telescoping mast. The telescoping mast has two sections that fit together and are dismantled and laid down horizontally, taking up only half as much room.

4.2.1.3 The guyed trailerized mast (Figs. 4.5 and 4.7a)

This type of mast is used with lightweight rigs and workover rigs (for production wells). It telescopes in two or three sections and the working position is at a forward slant. This requires a specific guying system for each rig. The guying layout that is generally used is shown in **Fig. 4.6**.

Mast specifications (see Fig. 4.7a)

The drilling contractor must check to see that all cables and fittings are in a good state of repair and that anchor points are up to standards. The nominal capacity of this type of hoisting mast is based on precise tension values for guy wires. Most of these truck-mounted rigs have racking capacity only for doubles. The *Cardwell* specifications in **Fig. 4.7b** show the areas where this type of mast can be used. It is exceptional to exceed a depth of 3000 m because there is not enough free height available under the substructure to accommodate the BOP stack required for this depth.

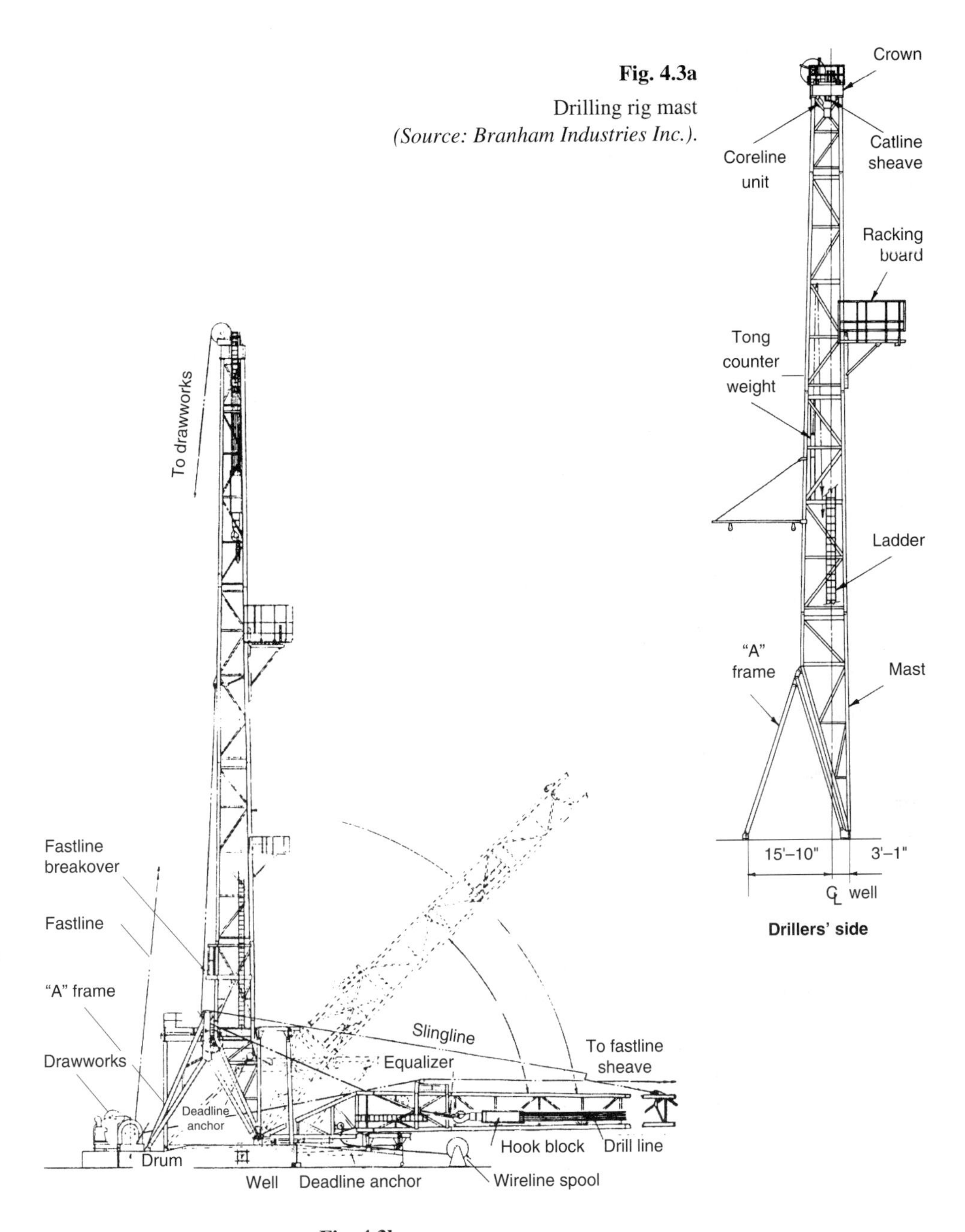

Fig. 4.3a

Drilling rig mast
(Source: Branham Industries Inc.).

Fig. 4.3b

Schematic diagram showing how a mast is erected
(Source: Branham Industries Inc.).

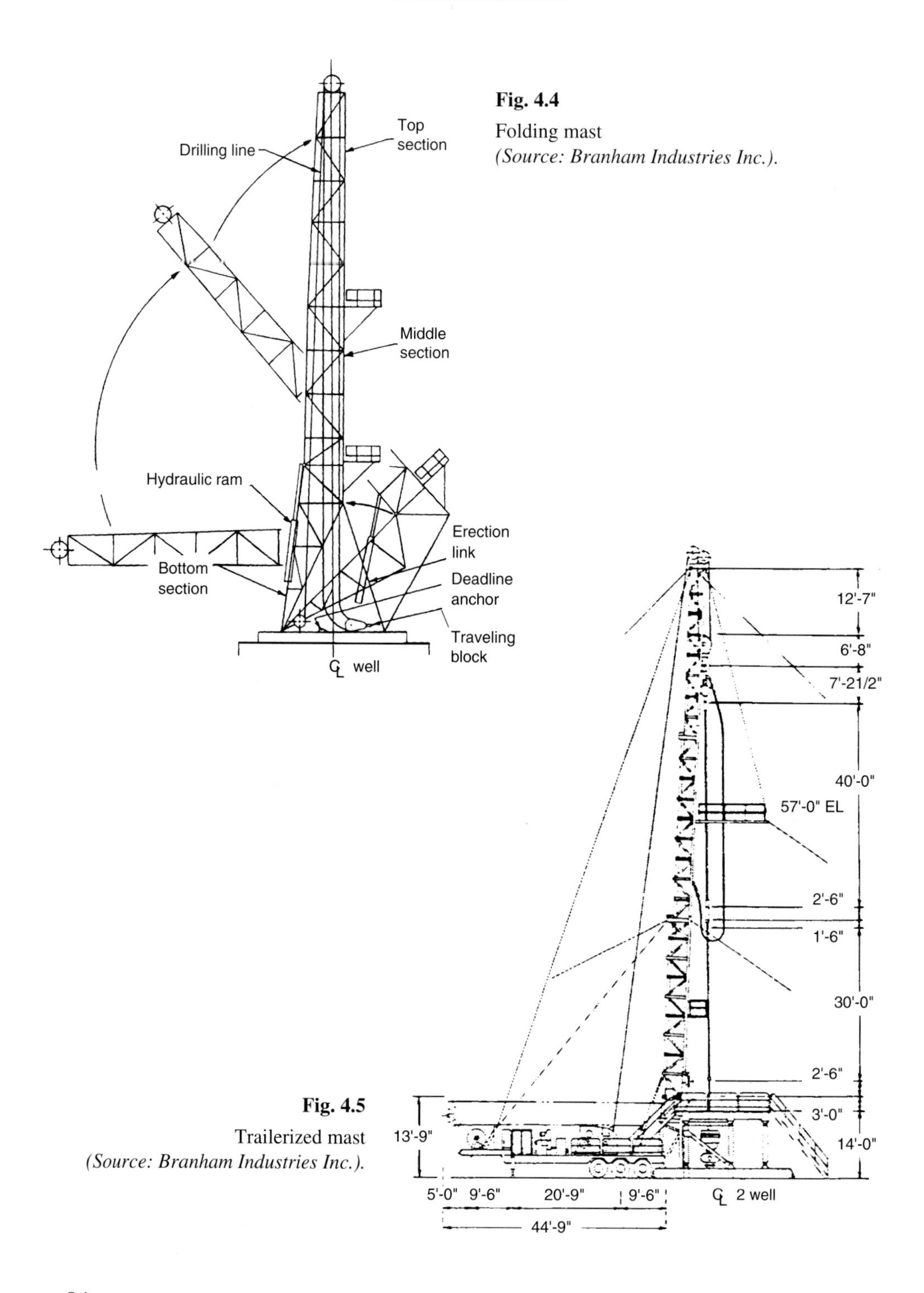

Fig. 4.4
Folding mast
(Source: Branham Industries Inc.).

Fig. 4.5
Trailerized mast
(Source: Branham Industries Inc.).

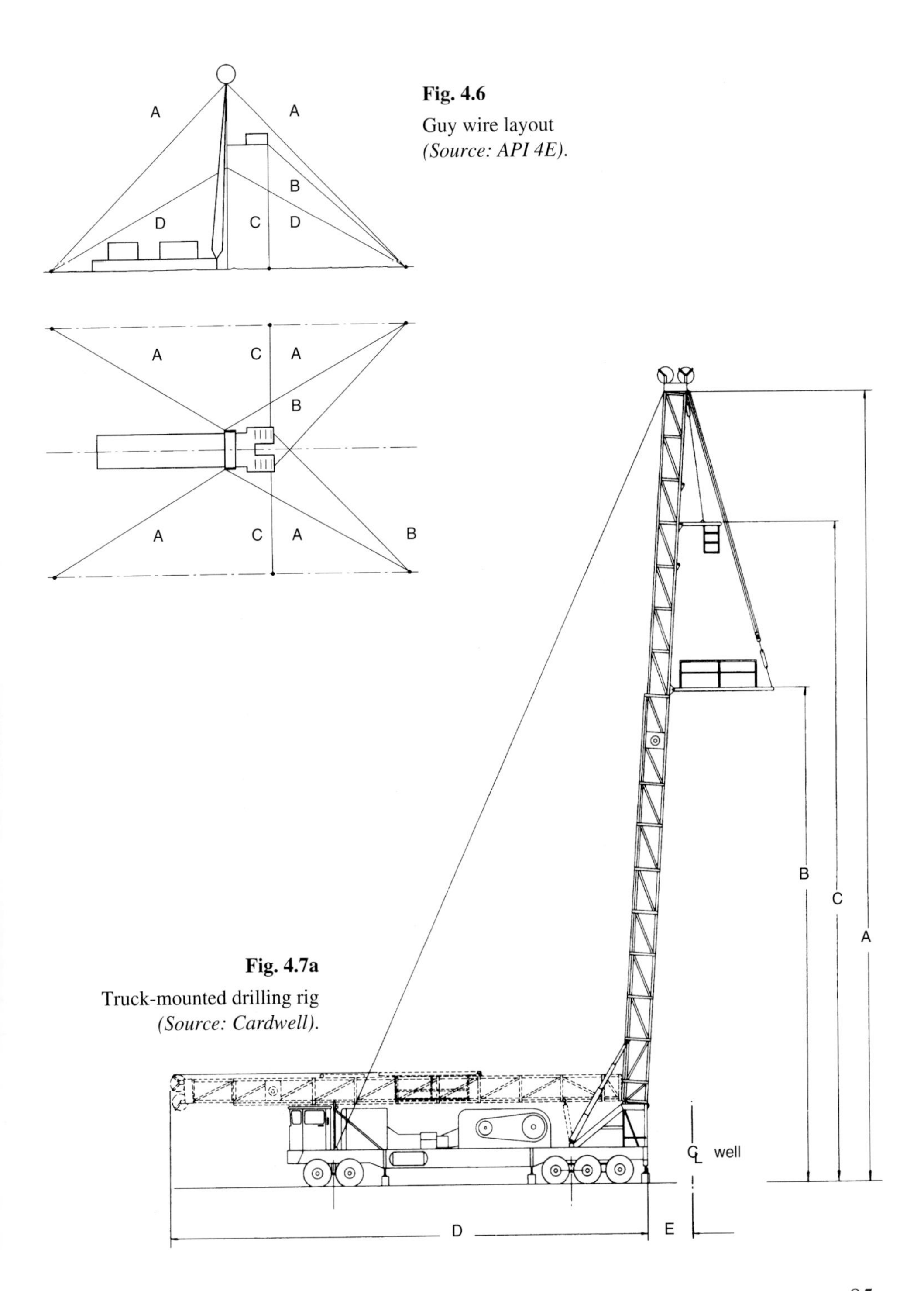

Fig. 4.6
Guy wire layout
(Source: API 4E).

Fig. 4.7a
Truck-mounted drilling rig
(Source: Cardwell).

Mast specifications (see **Fig. 4.7a**)

Model	Hook load (lbs)	Board capacity				A	B	C	D	E
		41/2" DP	31/2" DP	27/8" OD tubing	7/8" rods	(feet-inches)				
65' single pole	105,500 4 lines					67-0			38-8	
120-69	100,000 4 lines	2,500'		7,200' singles	7,700' doubles	69-0	25-4 29-4	53-10 55-9 57-8	41-0	4-4
140-69	140,000 6 lines	2,500' 3,600'		7,200' 10,200' singles	7,700' 9,900' doubles	69-0	25-4 29-4	53-10 55-9 57-8 59-7	41-4	4-4
140-91	140,000 6 lines	3,850'		8,400' doubles	9,450' triples	91-0	54-3 55-0 55-9	77-0 78-0 79-0	55-5	4-4
180-96	180,000 6 lines	4,800'		12,480' doubles	11,550' triples	96-0	54-5 59-2 63-11	78-2 82-11 87-8	57-0	4-4
215-96	215,000 6 lines	7,200'		16,200' doubles	11,550' triples	96-0	54-8 59-11 65-2	78-0 82-9 87-6	58-0	4-4
250-103	250,000 8 lines	7,200'		16,200' doubles	11,550' triples	103-0	54-0 58-0 63-6 68-6	79-4 84-1 88-10	60-6	4-4
250-108	250,000 8 lines	7,200'		16,200' doubles	11,550' triples	108-0	58-0 63-6 68-6	79-4 84-1 88-10	65-6	4-4
255-108	270,000 8 lines	8,000' doubles	10,000' doubles	20,000' doubles		108-0	61-9 66-0 71-0		66-9	5-0
300-112	300,000 8 lines	10,000' doubles	12,000' doubles	24,000' doubles		112-0	64-0 69-0 74-0		68-8	5-0
375-118	375,000 10 lines	12,000' doubles	14,000' doubles	25,000' doubles		118-0	64-0 69-0 74-0		71-8	5-0
450-118	450,000 10 lines	14,000' doubles	17,000' doubles	30,000' doubles		118-0	64-0 69-0 74-0		74-0	5-0

Fig. 4.7b

Example of mobile rig mast characteristics *(Source: Cardwell)*.

4.2.1.4 Substructures

These structures serve to raise the rig floor to leave room for wellhead assemblies and BOP stacks. They can be separate from the hoisting mast. Here they consist of box-like structures piled up on either side of the wellhead. The rig floor is assembled on top of the boxes and the hoisting mast sits directly on the box substructure (**Fig. 4.8**). Most intermediate-capacity masts are an integral part of a hoisting assembly with an elevating substructure where the drawworks and racking floors are folded at ground level by girders articulated in the shape of a parallelogram. Once the mast has been erected by the drawworks, the floor is pulled into an unfolded position using the drilling line (**Fig. 4.9**).

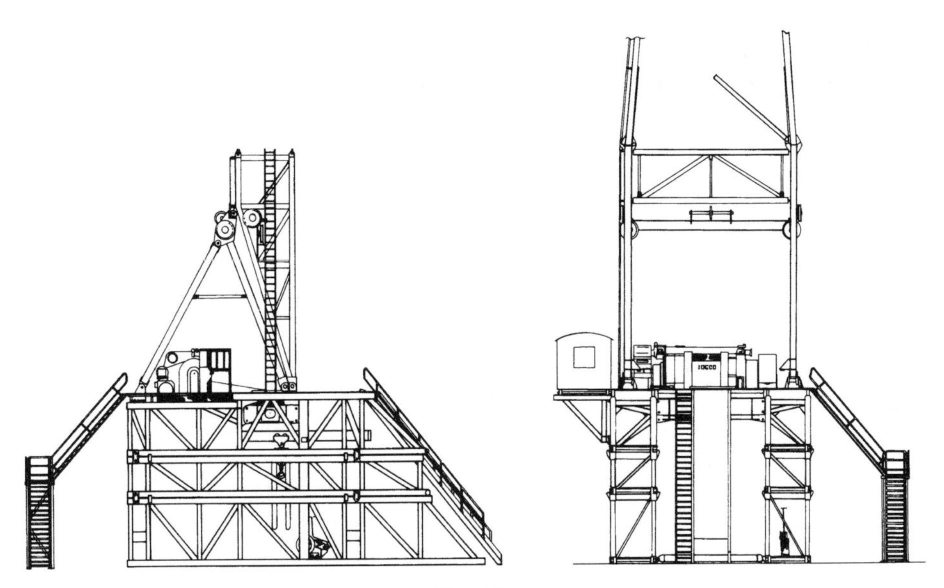

Fig. 4.8
Box substructure *(Source: IRI)*.

Fig. 4.9
Folding substructure *(Source: Branham Industries Inc.)*.

The Dreco slingshot substructure can also be mentioned. The mast is raised in its vertical position on the one-piece rig floor (**Fig. 4.10**).

Fig. 4.10
Slingshot substructure *(Source: Dreco Inc.)*.

4.2.2 Hoisting mechanics

4.2.2.1 The drilling line reeving system

Figure 4.11 gives a schematic idea of how the drilling line is strung up and of the component parts of the system.

A. Deadline

The drilling line is secured to a specific deadline anchor (**Fig. 4.12**) which measures the tension on that end of the line. It also allows new lengths of line to be run into the system in order to relieve the worn parts of the line by moving them from critical wear points on the pulleys of the crown block or the traveling block. Slipping the line, then cutting it off helps lengthen the lifetime of the drilling line.

B. Crown block

The crown block is the set of pulleys that the drilling line passes through. It is supported by the top platform of the drilling mast or derrick (**Fig. 4.13**). The load on the crown block, and in turn on the mast or derrick, is greater than the load on the hook. The reeving system is such that there are two more lines on the crown block, the deadline and the fastline which is connected to the drawworks drum. If the reeving system has ten lines and the load on the hook is 150 t for example, the line is under 15 t of tension in static conditions. But the crown block supports 150 t + 2 × 15 t, i.e. 180 t.

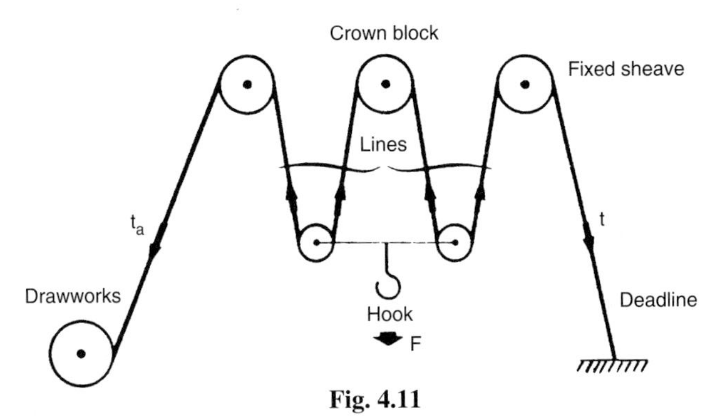

Fig. 4.11

Reeving the drilling line *(Source: Drilling Data Handbook, Editions Technip, Paris, 1989).*

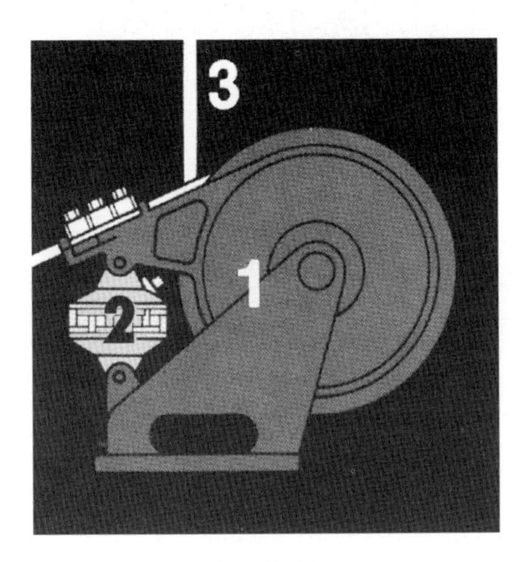

Fig. 4.12

Deadline tie down *(Source: Totco).*
1. Deadline anchor. **2.** Hydraulic tension measurement cell. **3.** Drilling line.

Fig. 4.13

Crown block *(Source: National Supply Co.).*

C. *The traveling block and hook*

The two are usually manufactured in a solid block (**Fig. 4.14**), i.e. the pulleys, or sheaves, and the hook are assembled in a compact package. The hook has a shock absorber to lessen stresses when the load is picked up and make screwing connections easier. The elevator bails are connected to two side hooks.

D. *The drilling line*

Drilling line has a metal core with six steel wire strands braided, or cabled, around it. The lay of the wires made into strands is the opposite of the lay of the strands on the core of the wire rope (normal or regular lay). This makes the drilling line stiffer but somewhat less prone to rotate. **Fig. 4.15** shows two common types of construction along with the breaking strength references. The steel may be of three grades: PS (plow steel), IPS (improved plow steel) and EIPS (extra improved plow steel). Diameters vary widely depending on the type of rig, but generally do not exceed 1.5 inches. The drilling line requires attention and care. To evaluate the wear it can withstand, the toolpusher computes the daily drilling line service which is the product of the load times the distance traveled. The total line service is expressed in tons per kilometer or per mile, and will be used as reference points to initiate maintenance operations such as slipping and cutting off drilling line.

E. *The fastline*

This is the end of the drilling line that is reeled up on the drawworks drum.

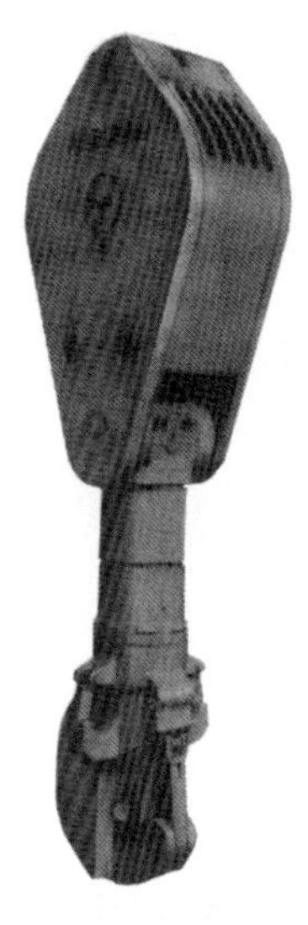

Fig. 4.14

Traveling block
(Source: National Supply Co.).

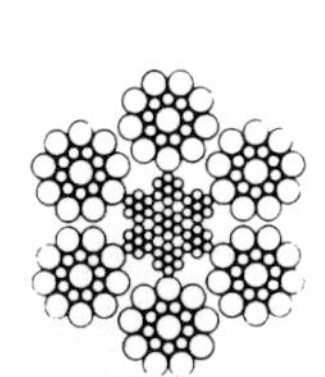

6 × 19 Seale
with IWRC

Wire diameter	Nominal breaking strength (t) EIPS
1"	51.7
1 1/8"	65.0
1 1/4"	79.9
1 3/8"	96.0
1 1/2"	114.0
1 5/8"	132.0
1 3/4"	153.0
2"	198.0

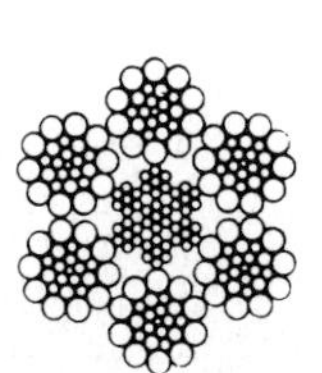

6 × 26 Warrington
Seale with IWRC

	Nominal breaking strength (t)	
Wire diameter	IPS	EIPS
3/4"	25.6	29.4
7/8"	34.6	39.8
1"	44.9	51.7
1 1/8"	56.5	65.0

Fig. 4.15

Drilling line properties
(Source: National Supply Co.).

4.2.2.2 The drawworks

The drawworks is the heart of the drilling rig. As mentioned earlier, it is the capacity of the drawworks that characterizes a rig and indicates the depth rating for the boreholes that can be drilled.

The different mechanical parts are:

- A grooved drum where the drilling line will be reeled up (**Fig. 4.16**). There are brake rims on the edges of the drum where the brake bands are mounted. The brake controls the lowering speed of the load hanging from the hook (**Fig. 4.17**). The system is highly reliable but does not have enough capacity to absorb all the energy produced by a string of casing lowered to great depths. All drawworks are equipped with an auxiliary device, which is mounted on the drum shaft, to slow down the load. This will be described later on.
- A gearbox behind the drawworks enables the driller to select from two or three gear ratios. Two ratios are sufficient when the drawworks is electrically driven. Here regulating the variation in rotation speed is well controlled. **Fig. 4.18** shows an inside view of the gearbox. Two parallel shafts are connected by pairs of sprockets and chains. There are the same number of pairs as gear ratios. One of the sprockets in each pair can rotate freely around the shaft when the dog clutch system is disengaged. Engaging a gear means mechanically moving the dog clutch system so that it blocks the rotation of the sprocket in relation to its shaft. The secondary shaft then rotates at

Fig. 4.16

Drawworks drum *(Source: National Supply Co.).*

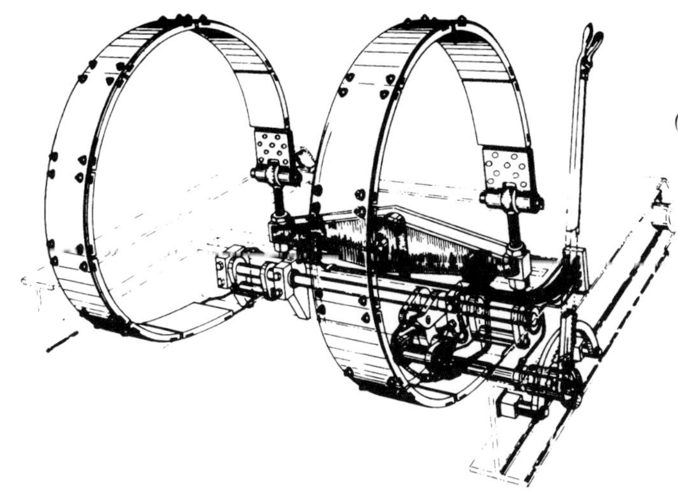

Fig. 4.17
Brake bands and brake lever *(Source: National Supply Co.).*

Fig. 4.18
Drawworks transmission gearbox.

the speed corresponding to the selected reduction ratio. The secondary shaft also causes the drawworks drum to rotate by means of two pairs of sprockets and chains located on either side of the gearbox housing. These two extra ratios (low and high drum drives) are engaged by Airflex-type air clutches (**Fig. 4.19**). A schematic diagram of the complete power transmission layout for a medium power range drawworks is given in **Fig. 4.20**.

Fig. 4.19

Air clutch (Airflex type) *(Source: Mid Continent).*

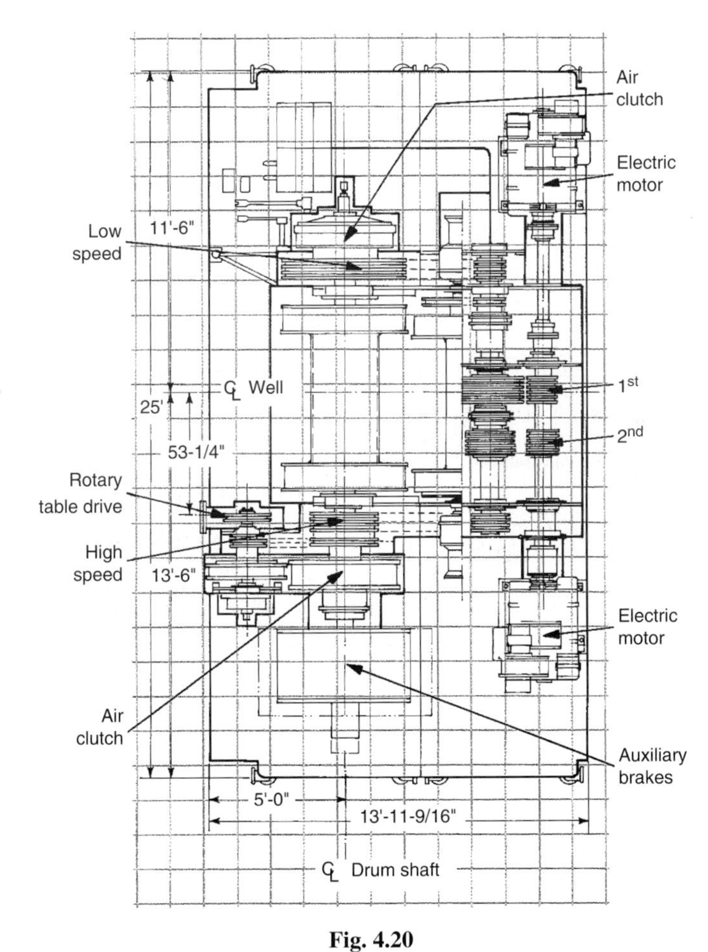

Fig. 4.20

Drawworks power transmission.

4.2.2.3 Auxiliary brakes

The braking capacity of the band system is not dynamically adequate when heavier loads are lowered into the well. This is why there is an added slowdown brake incorporated in the drawworks drum axis on all rigs. Two types of mechanisms are used:

- The hydrodynamic brake. The operating principle is to convert the mechanical energy produced by lowering a load into heat by means of a rotor that is made to rotate by the drawworks drum. The amount of mechanical energy that can be absorbed depends on the rotation speed and on the volume of water circulating in the working chamber. In order to adapt the deceleration to the load, the driller regulates the level of water in a small tall surge tank located in the water cooling circuit. The tank adjusts the amount of fluid in the brake and varies the braking torque (**Fig. 4.21**).

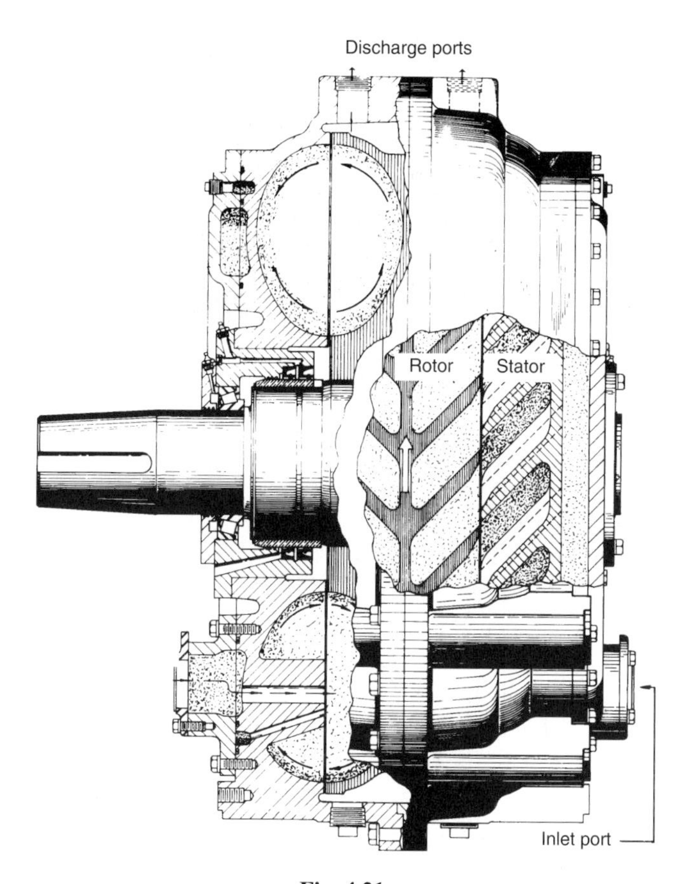

Fig. 4.21

Hydrodynamic brake (Parkersbourg type) *(Source: PARMAC).*

The system is reliable and requires very little maintenance, but it has major drawbacks: it provides little braking at slow speeds and regulation is too inflexible. As a result, its use is confined to lightweight drilling rigs.

- The eddy-current brake (**Fig. 4.22**) which includes a driven element (rotor) and a stationary member which provides a controllable and adjustable magnetic field. The rotor cuts the lines of the magnetic field. The electromagnetic forces induced in the rotor tend to oppose the rotary movement. The eddy currents produced in the rotor generate heat by Joule effect. The heat is dissipated by a water circulation system. The amount of braking torque is related to the intensity of the magnetic field produced by coils and as a result this type of brake very flexible to operate.

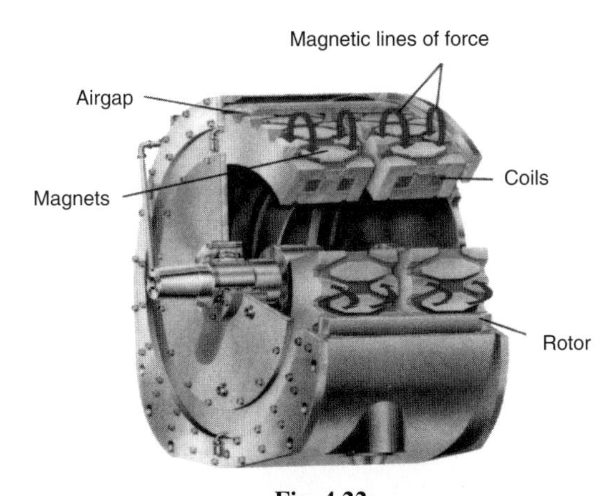

Fig. 4.22

Eddy-current brake (Elmagco type) *(Source: Baylor)*.

4.2.2.4 Rig floor tools and equipment

There are two types of tools and equipment: the ones used for hoisting and the ones used for screwing, making up and breaking out the drill string.

- Hoisting tools: The drilling hook has an ear-shaped device on either side for the bails that support an elevator (**Fig. 4.23**). For each nominal dimension of pipe there is a type of elevator. It is common to use lifting subs that are screwed to the drill collar thread in order to hoist drill collars. The upper part of the sub has the same dimensions as drillpipe so that elevators do not have to be switched.
  To hang the drill string on the rotary table, slips are placed in the master bushing (**Fig. 4.24**). For safer handling of slick drill collars, i.e. drill collars without any recess, a clamp is used on top of the slips. To make the crew's work easier on the rig floor, some pipe slips are pneumatically powered and can be operated directly by the driller (**Fig. 4.25**).

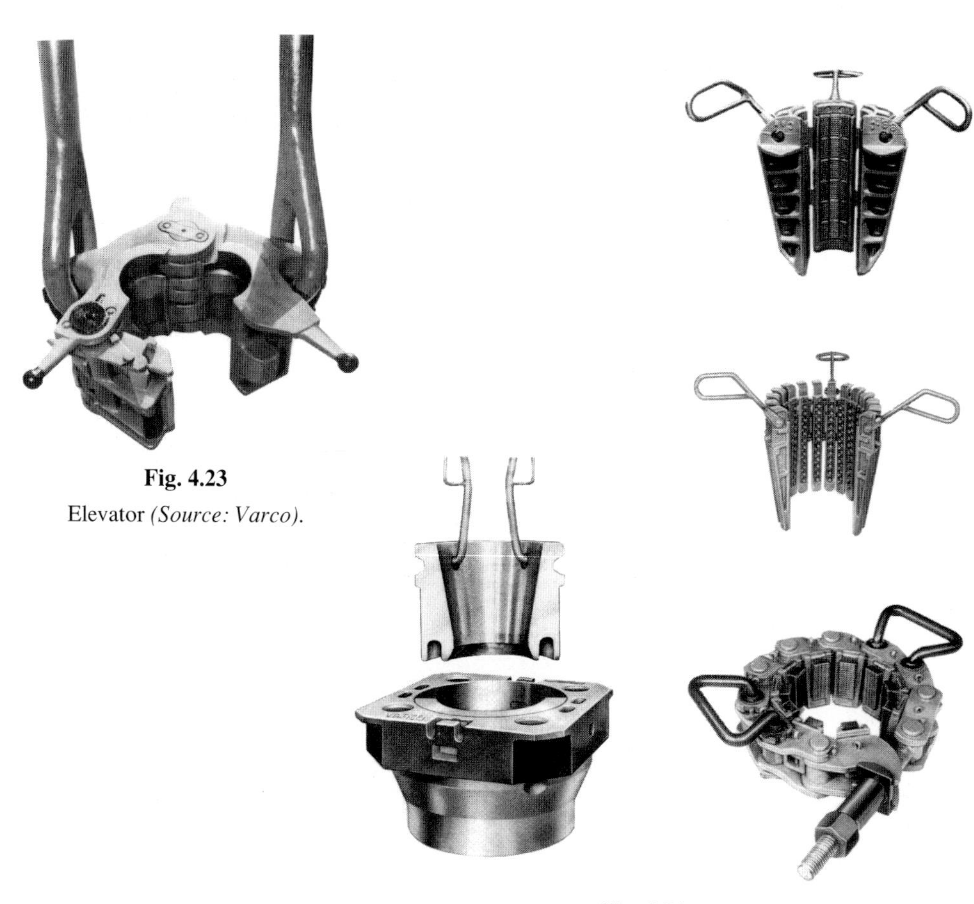

Fig. 4.23
Elevator *(Source: Varco).*

Fig. 4.24
Examples of slips, safety clamp and master bushing *(Source: Varco).*

Fig. 4.25a
Pneumatic power slips *(Source: BLM).*

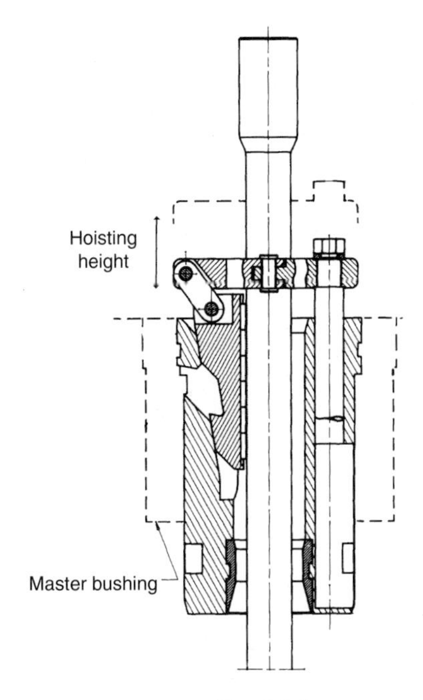

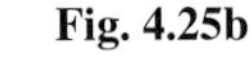

Fig. 4.25b

Pneumatic power slips in cross-section *(Source: BLM).*

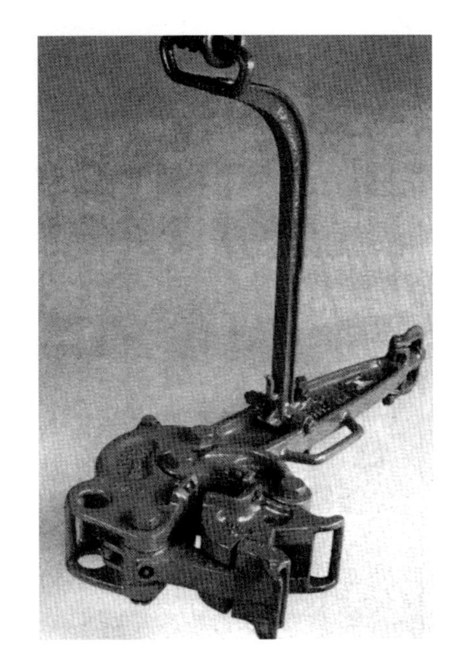

Fig. 4.26

Manual make up tongs *(Source: Varco).*

- Screwing tools: Makeup and breakout torque is still very commonly applied with multiple-jaw tongs (**Fig. 4.26**). The backup tong is secured to a stationary point by line or chain. The other is connected to a head that is rotated by the cathead transmission. The pulling force exerted by the cathead provides torque on the tubular by means of the lever arm corresponding to the length of the tong. Two crew men are required to place and remove these tongs (**Fig. 4.27a**).

The first phase in screwing pipe together, i.e. putting the two ends together until the box and pin shoulders meet, should be performed as quickly as possible. For quick connections, a chain wrapped around the pipe is still very commonly used. One end of the chain is held tight by a crew man (**Fig. 4.27b**). The other end rolls up on the spinning cathead controlled by the driller. The winch effect rotates the pipe as the chain spins.

The spinning chain method is quick but requires a lot of skill and good coordination. It is hazardous for the crew man who is in charge of pulling and holding the laps of chain around the pipe. Safety would dictate using only pneumatic power tongs (**Fig. 4.28**).

Any and all offshore rigs, whatever the type, along with heavyweight onshore rigs are equipped with hydraulic robot tongs (**Fig. 4.29**) that can also operate on the mousehole. They are rail mounted so they can be rolled off to leave the rotary table clear when necessary.

Fig. 4.27a

Tongs are placed on a connection *(Source: National Supply Co.)*.

Fig. 4.27b

Pipe is screwed together with the chain before it is made up *(Source: Sumitomo France SA)*.

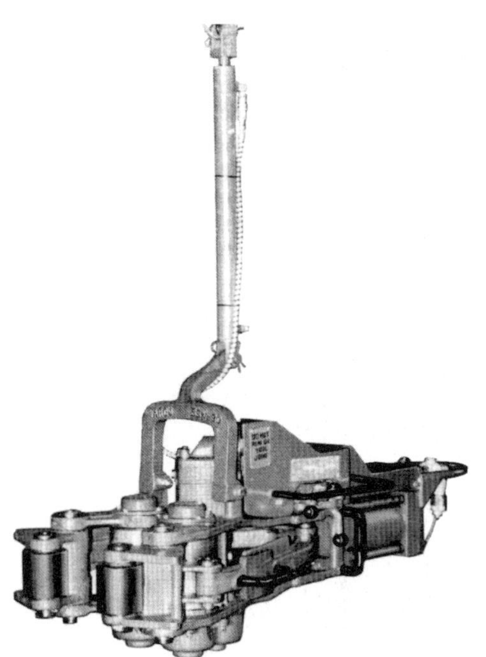

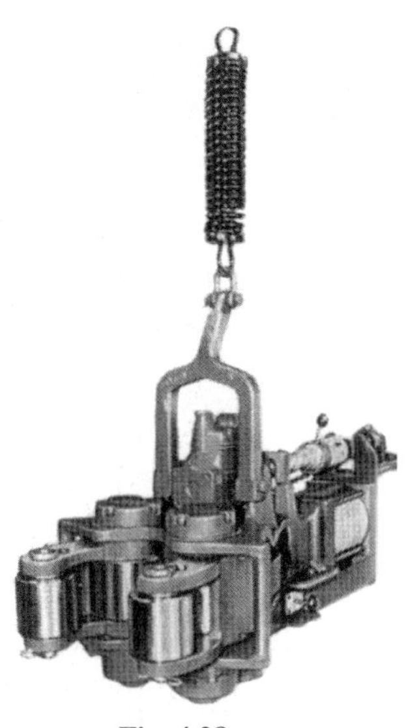

Fig. 4.28

Makeup tongs *(Source: Varco)*.

Fig. 4.29

Robot tongs to assemble and make up pipe *(Source: BLM)*.

4.3 ROTATING EQUIPMENT

4.3.1 The rotary table (Fig. 4.30)

This mechanical apparatus is very simple and requires only little maintenance, and this is what makes it so attractive for the working conditions on a rig. The main bearing supports the maximum load in static conditions or at slow rotation speed. During drilling (above 50 rpm) in fact, the weight of the drill string is hanging from the hook. Rotary table maintenance consists in checking the level and quality of the oil in the lubrication system. The nominal size is characterized by the through diameter where the master bushing is installed. The master bushing serves to hold the drill string by means of the slips and to drive the kelly drive bushing during drilling. Nominal sizes in inches may be as follows: 17 1/2, 20 1/2, 27 1/2, 37 1/2 and 49 1/2.

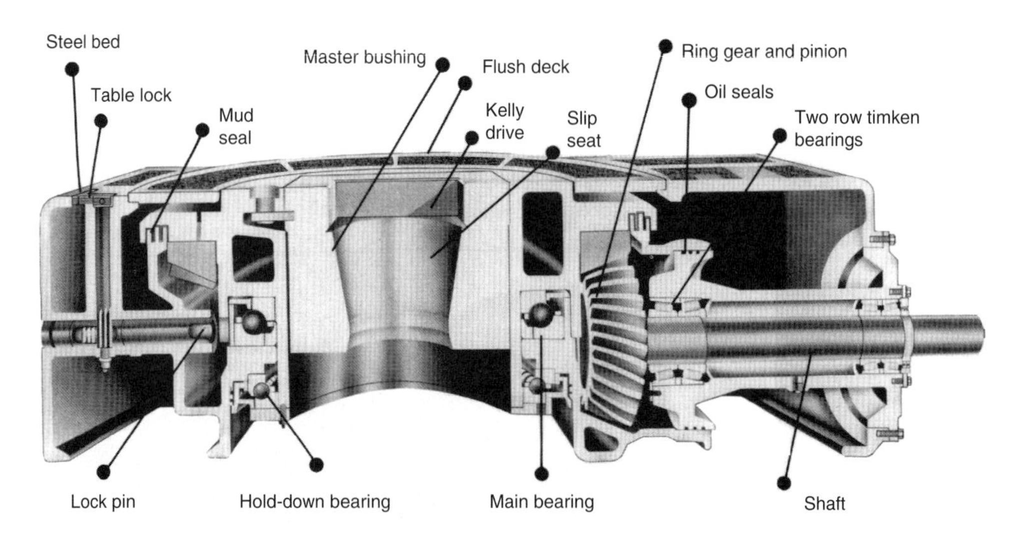

Fig. 4.30

Rotary table terminology *(Source: Skytop Brewster)*.

The rotary table is driven by means of a sprocket and chain by the drawworks (the gear ratios of the gearbox can then be used). It can also be driven by means of an electric motor independent of the drawworks transmission on heavyweight rigs.

4.3.2 The kelly

The components connected to the kelly will be discussed from a more general standpoint (**Fig. 4.31**).

4.3.2.1 The kelly

The kelly can have a square, hexagonal or triangular cross-section. It is rotated by the table and by means of the kelly drive bushing that it fits into. The kelly bushing is equipped with four horizontal axis rollers so designed as to transmit torque to the kelly and in turn to the drill string screwed onto the lower kelly coupling (**Fig. 4.32**). The whole assembly can slide freely up and down along the kelly's stroke length. With a total length of 40 ft or 54 ft, the kelly has a useful stroke length of 37 ft or 51 ft respectively.

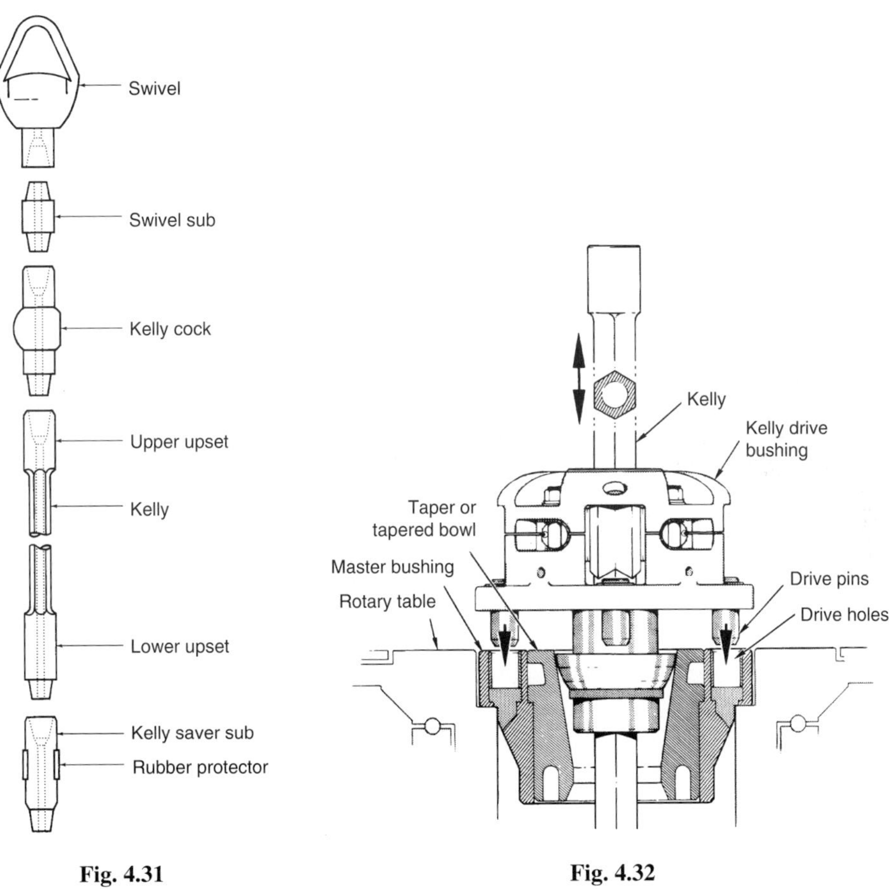

Fig. 4.31
Kelly terminology
(Source: API SPEC 7).

Fig. 4.32
Kelly drive bushing *(Source: Petex).*

To control any incipient blowouts that might occur through the inside of the drill string, safety valves are mounted on each end of the kelly (lower kelly valve and upper kelly cock) (**Fig. 4.33**). The two valves are operated by rotating ninety degrees with a wrench that is kept on the rig floor. The lower valve must have a small enough diameter so that it can be run into the borehole that is being drilled.

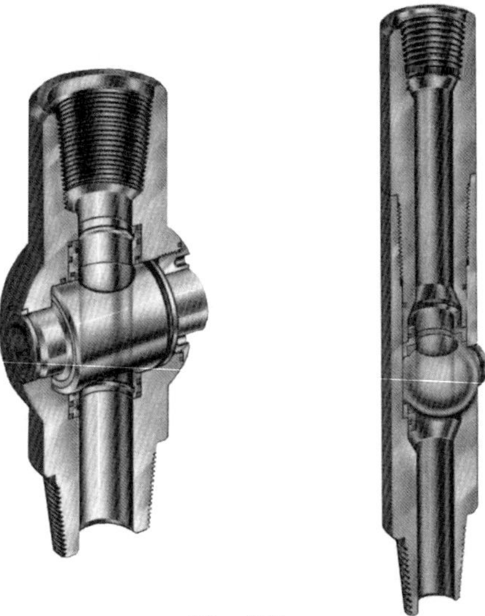

Fig. 4.33
Kelly valves *(Source Control Flow Inc.)*.

4.3.2.2 The kelly saver sub

Each time a length of drillpipe is added, i.e. after the useful length of the kelly has been drilled, the kelly must be unscrewed from the drill string, then screwed back onto it. Since this is done often, the drill string is screwed and unscrewed from a low-cost coupling rather than from the threads on the bottom of the kelly itself. Because the sub rotates inside the BOPs, a rubber protector is placed on the outside sub diameter to protect them from wear.

4.3.2.3 The swivel (Figs. 4.34a and b)

This component hangs from the lifting hook by its bail. It is designed for both the maximum drill string load and for the maximum rotational speed. Additionally, a rotating seal allows drilling fluid to be injected under pressure by the mud hose connected to the swivel gooseneck.

It should be noted that all connections above the useful section of the kelly have a left-handed thread so that they are not broken out by the rotary table turning to the right.

Fig. 4.34a

Swivel *(Source: National Supply Co.).*

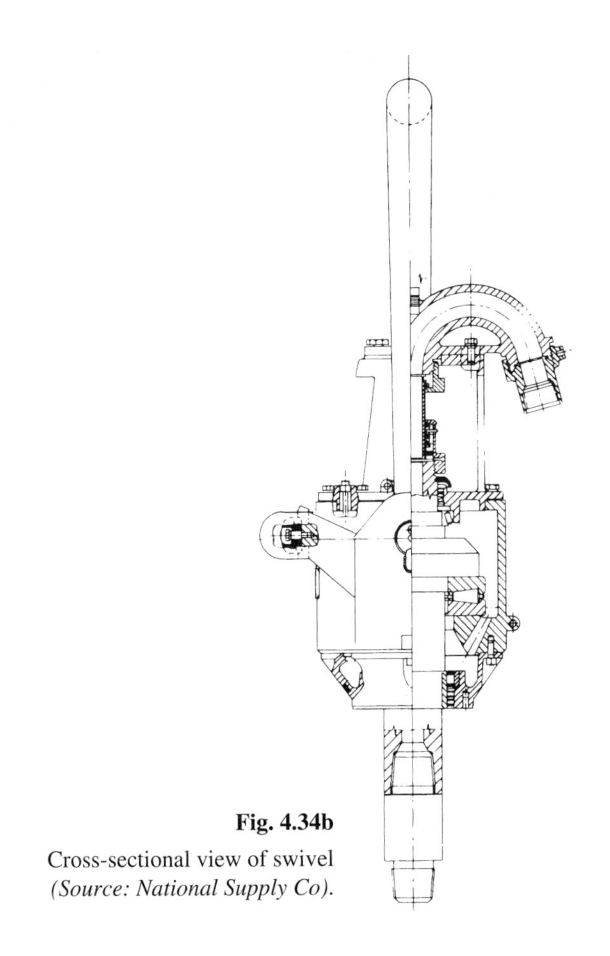

Fig. 4.34b

Cross-sectional view of swivel *(Source: National Supply Co).*

4.3.3 The power swivel (Fig. 4.35)

As the name indicates, this is a swivel that fulfills the same functions as conventional ones but also provides mechanical transmission to the rotary string. It may be driven in the same way as independent rotary tables, i.e. by direct-current electric motor or by hydraulic motor. The hydraulic option is less conventional in design and requires installation of a unit with specific hydraulic power.

Though the advantages of the power swivel as described later on are very attractive, installing one entails a number of constraints:

- a dolly, or trolley beam guide system, must be installed in the derrick to absorb reactive torque,
- the structure must be reinforced because of the extra torsional stress,
- the derrick must be built up since the power swivel is longer than a conventional one,

- extra hoses and electric cables need to be on hand on the rig,
- the weight above ground increases considerably,
- extra investment is required and maintenance in particular is a much heavier job than with a rotary table and kelly system.

However, the advantages of the system dictate using a power swivel when costly development operations are involved, such as in the North Sea:

- no need to handle a kelly,
- can be reconnected to the drill string at any mast height during tripping,
- drilling with thribbles is possible,
- the drill string can be pulled out while rotating and circulating: back reaming,
- extra-long core samples can be taken,
- no need to unrack the drill string between two development wells when the rig can be moved with the mast erect and drill string racked up on the rig,
- static torque can be applied for a much longer time (only when the swivel is hydraulically driven).

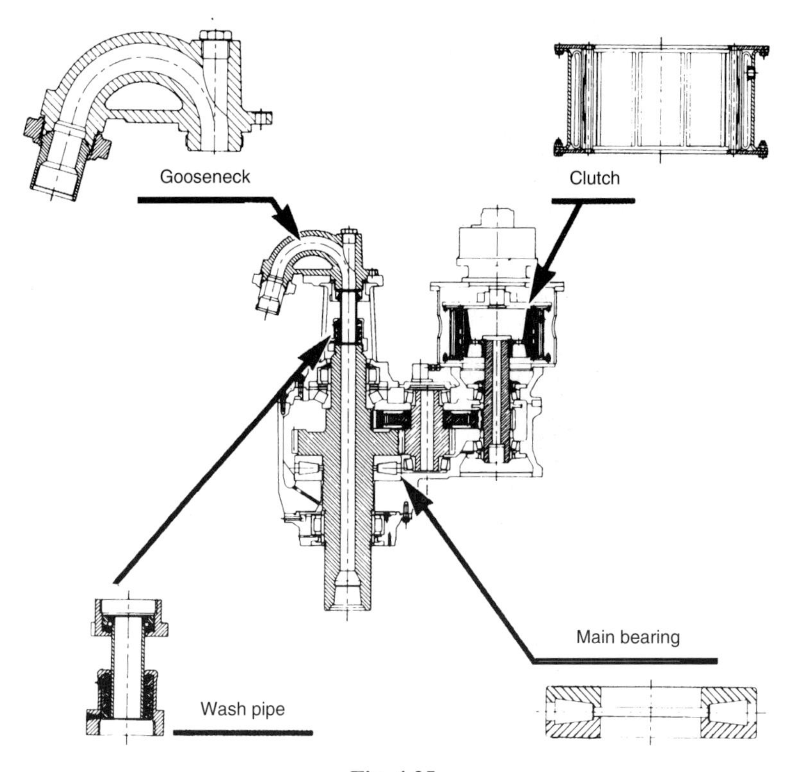

Fig. 4.35

Power swivel: main components *(Source: ACB).*

Figure 4.36 illustrates all the ancillary components of the power swivel. The degree of complexity is a good indication of the maintenance requirements.

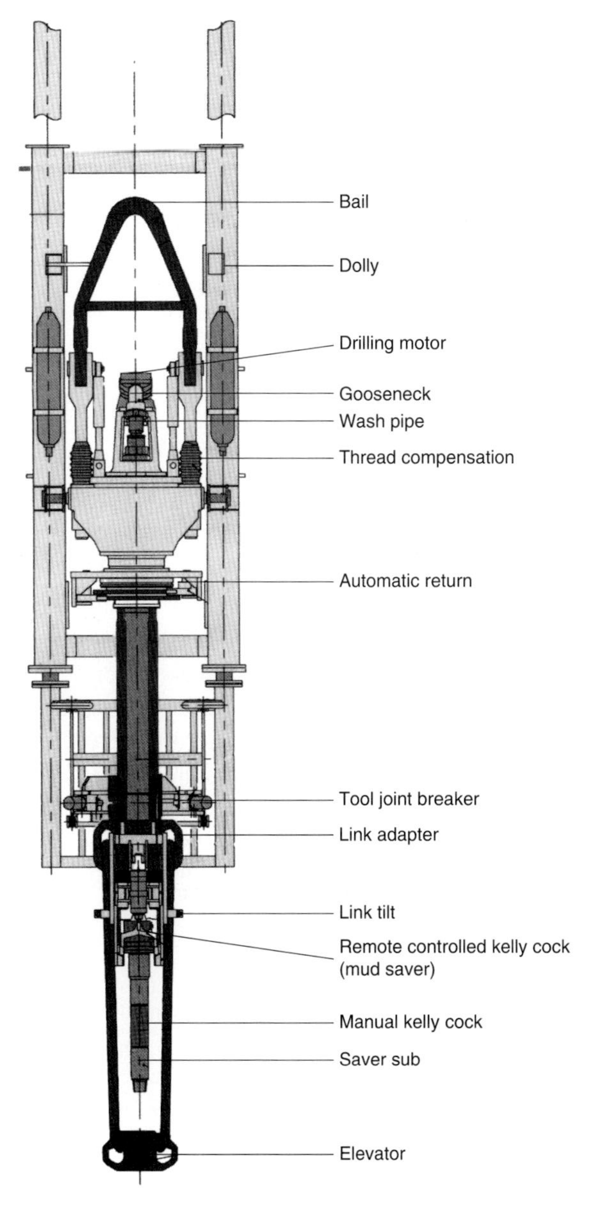

Fig. 4.36
Power swivel assembly *(Source: ACB)*.

4.3.4 Pumping equipment

4.3.4.1 Conventional requirements

Drilling fluid pumps must provide the flow rates required during the different drilling phases. The flow rates are selected by the drilling engineer according to the following criteria:

- annular velocity of the mud (in the annulus between the borehole walls and the drillpipe),
- cleaning the face of the bit,
- maximum lag time for cuttings to reach the surface,
- type of flow in the annulus,
- borehole wall stability,
- drilling with a downhole motor.

The resulting maximum flow rates are:

3500 l/min with a 17 1/2" bit diameter,
2500 l/min with a 12 1/4" bit diameter,
1500 l/min with a 8 1/2" bit diameter,
600 l/min with a 6" bit diameter.

Pump discharge pressure is directly related to the pressure losses in the circulating system (surface casing, drillpipe, drill collars, annulus), to the pressure drop in the drilling bit nozzles or in a downhole motor, and to the flow rate and physical characteristics of the fluid (density and viscosity).

Modern circulating pumps have an operating pressure of 5000 psi, i.e. 35 MPa, but users restrict the range to approximately 25 MPa for reasons of operational safety and maintenance. This means that heavyweight rigs need to be equipped with two 1600 hp pumps (1200 kW). Offshore, where there is no skimping on installed power, there are often three 1600 hp pumps. A lightweight rig could be equipped with two 800 hp pumps. Two independent pumping systems are a must, though they must be able to work at the same time, so that one of the two can be kept on standby and guarantee that the mud can always be circulated in the well.

4.3.4.2 Circulating pumps (Fig. 4.37)

Mud pumps are of the reciprocating piston-type, with the reciprocating movement of pistons and rods produced by a conventional crankshaft and connecting rod system. They are positive-displacement type pumps and provide a flow rate that depends directly on the cylinder capacity and the crankshaft rotation rate. In order to adjust the flow rate, the crew counts the cycles per minute, defined by pump strokes per minute.

Modern mud pumps are triplex and single-acting (**Fig. 4.38**). The three pistons move in dismountable cylinder liners, suck up the fluid by a suction pipe, then discharge it into a discharge pipe through a discharge valve.

Fig. 4.37
Triplex drilling mud pump *(Source: National Supply Co.)*.

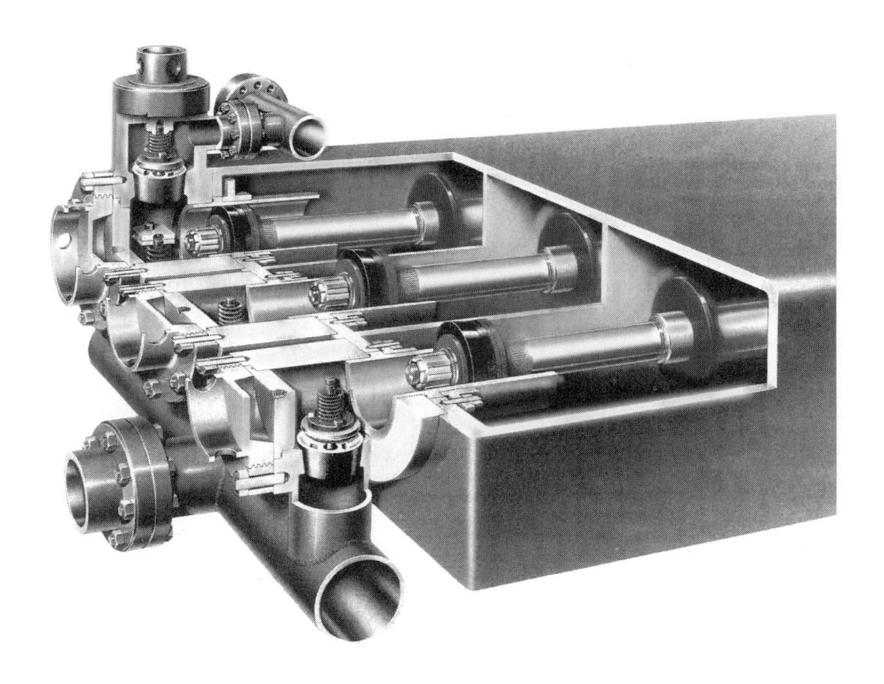

Fig. 4.38
Cutaway view of a triplex pump *(Source: Mission Drilling Products)*.

An important feature of this type of pump is that the cylinder capacity can be altered by changing the diameter of the cylinder liner. When a maximum flow rate is required, the pump is equipped with the largest size liner. When maximum pressure is required (5000 psi – 35 MPa), a liner with a smaller cross-section is usually necessary.

This type of pump provides an irregular instantaneous flow rate which means that a pulsation dampener must be installed on the discharge end. A pressure relief valve is also needed.

Triplex pumps can not usually suck up fluid directly from the mud pits because of volumetric efficiency and cavitation risks due to high linear piston velocities. As a result, they are boosted, or supercharged, by a centrifugal pump.

4.3.5 Power and power transmission

The installed capacity on a drilling rig has constantly increased to meet the needs of modern drilling techniques. The table below gives a breakdown of power ranges.

Depth reached (m)	Hook load (t)	Drawworks rating (kW)	Pump rating (kW)	Total installed capacity (kW)
6000-9000	400-600	1500	2000-2600	3000-3750
4000-6000	300-400	1100	1800-2000	2250-3000
3000-4000	200-300	750	1100-1800	1850-2250
900-3000	100-170	300-525	750-1100	1100-1850

4.3.5.1 Prime movers

The steam engine was replaced by the diesel engine long ago as a prime mover, but production platforms often use gas turbines. Some drilling sites are even connected to the electric power distribution network. Although electricity offers major advantages, e.g. low-cost, noiseless energy, the self-reliance of the drilling site is jeopardized and in many instances this is unacceptable. Additionally, the way the drilling site operates requires peaks of power which are not admissible on the distribution network. This technical snag means a specific infrastructure must be incorporated in the site. It has quite a substantial impact of the cost of the rig and furthermore, safety dictates that there must also be an independent emergency system to function in the event of a network power outage.

Diesel engines provide maximum distribution flexibility for both heavy and lightweight rigs.

4.3.5.2 Power transmission systems

The three basic systems are mechanical, hydraulic and electric.

A. Mechanical transmission (Fig. 4.39)

Several diesel engines are parallel-operated through interconnection by a chain and clutch system called a compound. They are fitted with torque converters. The driller assigns the engines to rig components depending on his needs: during drilling, one or two engines for pumping, one for the rotary table transmission; during tripping all the engines can be assigned to the drawworks.

This type of transmission allows easy maintenance and use, but it does lack flexibility in utilization and location. It is now used only for lightweight, truck-mounted rigs. Here the size of all components (engine, transmission) is such that everything is kept on board and requires no complex mechanical dismantling and reassembly at the end of each drilling operation. In this type of setup, each mud pump has its own separate diesel engine.

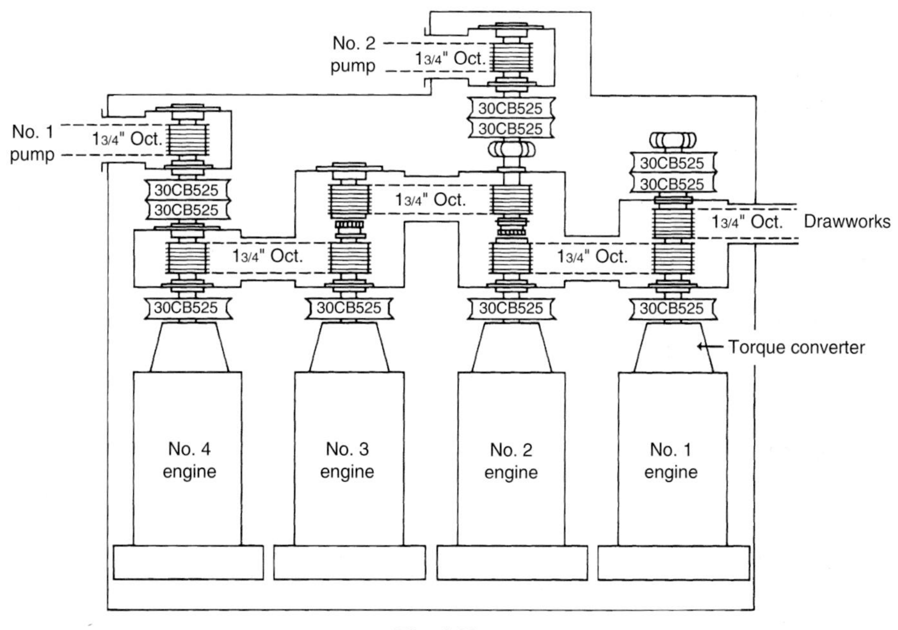

Fig. 4.39

Transmission of diesel power (compound) *(Source: Ideco).*

B. Electric transmission

DC/DC drilling rigs that came on the scene sometime in the 1950s used the Ward-Leonard regulation loop. Direct-current generators driven by diesel engines are connected in a loop with the direct-current drawworks and pump motors. The system was conventional at the time, but had a number of drawbacks, e.g. lack of flexibility in utilization and the need to have one diesel engine for each generator. On the other hand, it was not very complex, fairly simple to use and cheap.

The advent of SCR or silicon controlled rectifiers made it possible to develop the AC/DC system (**Fig. 4.40**). Here electric power is produced by means of a three-phase alternator. Direct-current motors are then powered by SCR rectified current.

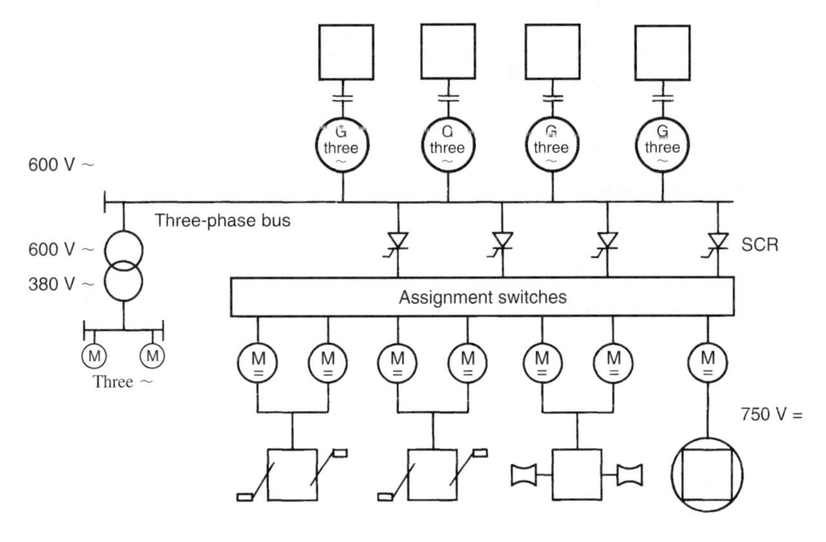

Fig. 4.40

Diagram of an AC/DC electric transmission *(Source: Company Rig Power Transmission).*

C. Hydrostatic transmission

Other than a few prototype exceptions, this type of transmission is found only on lightweight slim-hole rigs or used to power independent components, e.g. power swivel, rotary table.

The mechanical energy supplied by diesel engines is converted into a pressurized oil flow. The hydraulic energy is conveyed by high-pressure hose to hydraulic motors on the drawworks, rotary table, swivel or pumps.

4.3.6 The control panel

All measurement indicators are grouped together on a pressurized explosion-proof panel (**Fig. 4.41**).

The weight indicator showing the load on the hook is the most important instrument, or in any case the one most often checked by the driller. One of the two needles gives the weight hanging from the hook and the other gives the difference between the drill string off bottom and on bottom, i.e. the weight on the bit (WOB).

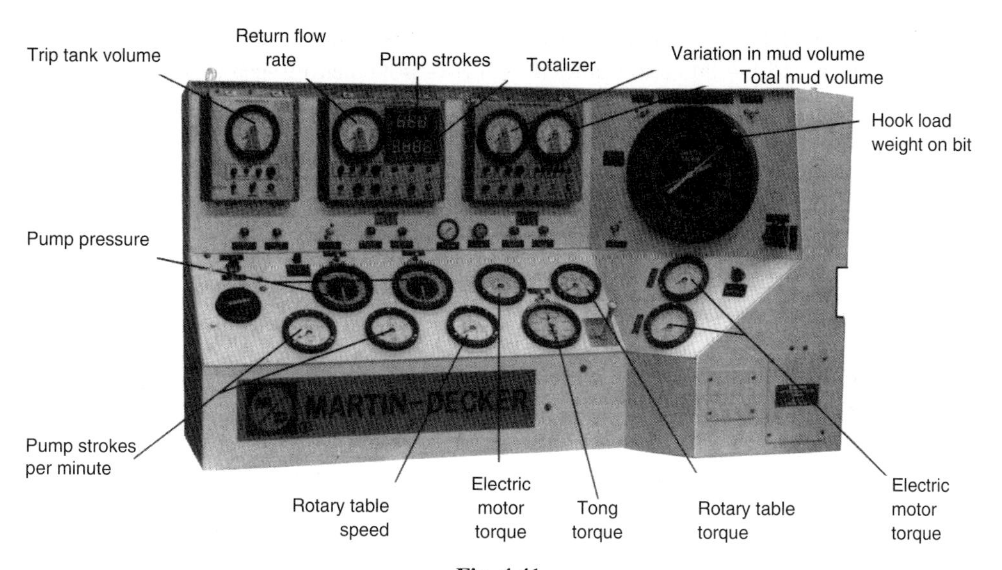

Fig. 4.41

Control panel of a modern heavyweight rig *(Source: Martin Decker).*

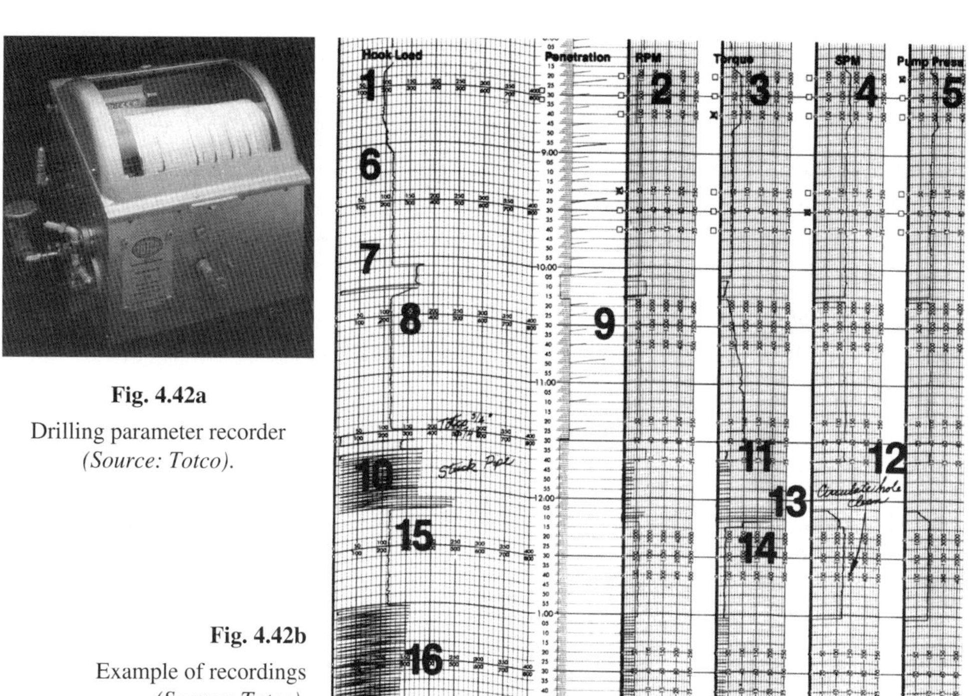

Fig. 4.42a

Drilling parameter recorder *(Source: Totco).*

Fig. 4.42b

Example of recordings *(Source: Totco).*

The other indicators tell the crew the level in the mud pits; mud circulation data, e.g. flow rate, pressure and pump strokes; rotation parameters, e.g. rotary table engine torque and rotational speed; and make up torque measurements on tubulars.

Most of these parameters are recorded in the mud logging office when there is one. In any case the drilling contractor is under contractual obligation to hand over a copy of the drilling parameter recordings to the operator along with the daily drilling report.

Figure 4.42a shows a continuous band recorder for six parameters:

- hook load,
- rate of penetration,
- rotational speed,
- rotary table torque,
- pump strokes per minute,
- pump discharge pressure.

The three measurements that are indispensable to operations are hook load, rotary table rpm and pump discharge pressure.

Chapter 5

DRILLING FLUIDS

The technical breakthrough that resulted from using a fluid in continuous circulation in a borehole was discussed in the introduction. The fluid, as a drilling parameter, has a host of operational features and accordingly fulfills some very important functions. Drilling performance records have made considerable headway due to technical progress in the physicochemical nature of drilling fluids. This is why this chapter on drilling fluids will focus first and foremost on the fluid's functions before dealing with the different types of fluid and how each one is used on a well site.

5.1 FUNCTIONS AND CHARACTERISTICS OF DRILLING FLUIDS

5.1.1 Transporting cuttings to the surface

The fluid circulating and rising in the drillpipe/borehole wall annulus must sweep the cuttings from the working face up to the surface. Three parameters influence the effectiveness of the annular cleaning function.

5.1.1.1 Annular velocity of the fluid

The velocity depends on the fluid flow rate and the annular cross-sectional area:

$$V = \frac{Q}{V_a}$$

V (in m/min) = velocity of the mud,
Q (in l/min) = pumping rate,
V_a (in l/m) = unit volume of the annulus.

The usual range for annular velocity is from 25 to 60 m/min.

5.1.1.2 Density

It is because of the Archimedean principle, i.e. the buoyancy of the cuttings, that the density parameter influences how readily the cuttings are moved up the annulus. However, the density parameter is not altered in order to improve on this function.

5.1.1.3 Viscosity

The cuttings velocity can be considered as the difference between the rising velocity of the drilling fluid in the annulus and the settling velocity of the cuttings. The settling velocity depends on the size, shape and mass of the particles; on the rheology of the fluid and more particularly on its viscosity. A minimum viscosity value is needed to get the best match between the mud velocity and the cuttings velocity.

5.1.2 Suspending the cuttings when circulation is stopped

Drilling fluid circulation must be stopped when a length of drillpipe is to be added. During shutdown, the cuttings rising in the annulus are no longer carried upwards and can settle out. It is the thixotropic property of the drilling fluid that keeps the cuttings suspended by gelling when the mud is no longer moving. Practically all viscous fluids are thixotropic.

5.1.3 Cooling the bit and lessening drill string friction

The drilling bit heats up because of the temperature downhole (geothermal gradient) and due to mechanical friction converted into calories. Circulating drilling fluid serves as a coolant, with the mud pits on the surface as heat exchanger. Additionally, the drilling fluid decreases the friction coefficient between the drill string and the wall of the hole. This function is sometimes enhanced by adding antifriction products such as oil or special additives.

5.1.4 Consolidating the walls of the hole

The liquid phase of the drilling mud filters into permeable formations and deposits a film of colloidal particles on the walls of the hole. The film is termed cake and it is reinforced by specific products called filtrate reducers. This is what plasters and isolates the permeable formations in the borehole, allowing longer stretches of uncased hole by helping to stabilize the formations.

5.1.5 Preventing inflows of formation fluids into the well

The drilling fluid exerts a hydrostatic pressure of $P_h = 9.81\ Zd$ (in kPa). If the pressure P_h remains higher than the pressure of formation fluids, no fluids will enter the borehole.

The drilling mud is considered as the first blowout preventer to control pressures downhole.

5.1.6 Acting as a drilling parameter

(see Chapter 3, Downhole equipment)

The choice of the type and properties of drilling mud govern the instantaneous rate of penetration by the mud's capacity to sweep the bit area clean.

In addition, a downhole pressure produced by the drilling fluid and higher than formational pressure is always detrimental to drilling rate. (Use of air or foam as drilling fluids when possible).

5.1.7 Transmitting power to a downhole motor

In some applications such as directional drilling or diamond bit drilling, a downhole motor (turbine or positive-displacement motor) is incorporated in the drill string to rotate only the bit and nothing else. The motor is driven by the flow rate of the mud pumped down the drill string. The pressure drop due to the operation of the downhole motor is added to the pressure losses in the discharge circuit.

5.1.8 Providing geological information

Because it circulates, the drilling fluid conveys major items of information for the geologist. Some examples are the cuttings that the geologist takes out of the mud return line and the traces of fluids or gases from drilled formations that are detected by sensors on the surface. The physicochemical changes in the fluid (temperature, pH, chloride content, etc.) are also an integral part of mud logging measurements that tell the geologist and the driller how drilling is proceeding (see Chapter 8, Measurements and drilling).

5.1.9 Conclusion

Following this brief discussion of the different functions of drilling fluid, the type of fluid flow will now be dealt with. To determine the type of flow, a mud rheogram will be plotted on the basis of mud measurements and will classify the fluid along the lines of recognized models.

Then characteristics such as viscosity, yield point, N and K numbers and gel strength can be calculated. Another important physical property is of course density.

5.2 MUD MEASUREMENTS

5.2.1 Density

5.2.1.1 Definition

This is the ratio of the weight of a substance to its volume under specific pressure and temperature conditions. It is expressed in N/m^3 or more practically in kg/l. Note that to convert from nonmetric units:

$$10 \text{ lb/gal} = 74.8 \text{ lb/cu ft} = 1.2 \text{ kg/l}.$$

5.2.1.2 Measurement apparatus

Density is measured with a mud balance (**Fig. 5.1**) based on the same principle as a beam balance. Density is very important and must be checked regularly as it must be high enough so that the hydrostatic pressure downhole ($P_h = 9.81\ Zd$ in kPa, Z in m, d density) is high enough to control formation fluids. However, it must not be too high compared to the formation fracturing pressure. There are also continuous density measurement devices that are fitted on the discharge end of the drilling mud pumps. They work on the basis of measuring how the mud attenuates radiation by a radioactive source.

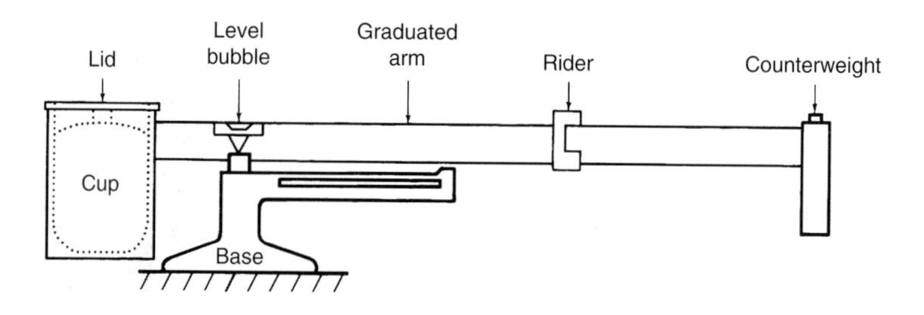

Fig. 5.1

Mud balance *(Source: Manuel du technicien fluides de forage, Milpark CKS)*.

5.2.2 Viscosity

5.2.2.1 Marsh viscosity (Fig. 5.2)

The principle is to measure the time it takes a given volume of fluid to drain out through the calibrated orifice of a funnel.

The funnel is filled with 1500 cm^3 of homogeneous mud, then the time required for a quarter of a gallon (946 cm^3) to drain out is measured. Marsh viscosity is therefore expressed in seconds. The viscosity of pure water at a temperature of 20°C is 26 seconds (946 cm^3).

Marsh viscosity is a practical indication since it can be done quickly on the mud pits, but is gives only a very relative assessment of the characteristics of the mud. It is mainly used to provide a rough but rapid evaluation of any contamination that might dramatically modify the fluid's properties.

5.2.2.2 Fann viscosity (Fig. 5.3)

The Fann viscometer is a device that can help determine the drilling fluid rheogram, i.e. the flow law that is represented by the function:

$$t = f(\gamma),$$

where

t = shear stress,

γ = shear rate. (**Fig. 5.4**)

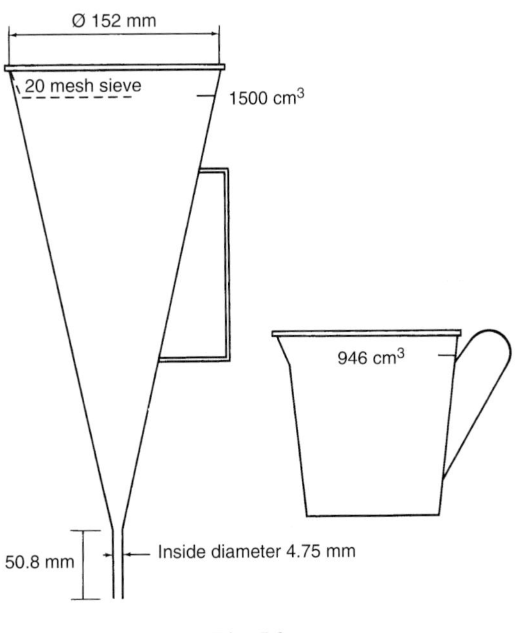

Fig. 5.2
Marsh viscosity meter.

Fig. 5.3
Fann viscometer.

Rheological system	Rheological equation	Flow curve, Cartesian coordinates	Flow curve, Logarithmic coordinates
Newtonian fluid	$\tau = \mu\dot{\gamma}$	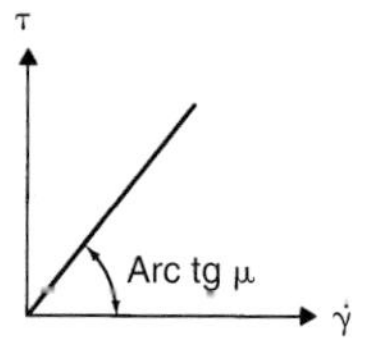 τ; Arc tg μ; $\dot{\gamma}$	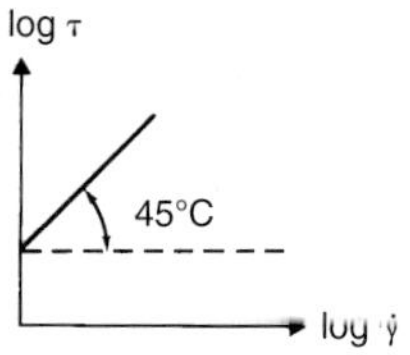 log τ; 45°C; log $\dot{\gamma}$
Bingham plastic fluid	$\tau = \tau_0 + \mu_p\dot{\gamma}$	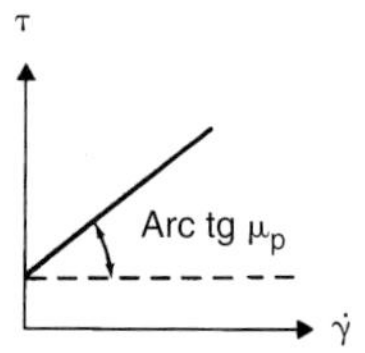 τ; Arc tg μ_p; $\dot{\gamma}$	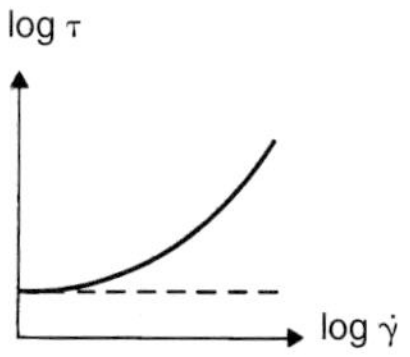 log τ; log $\dot{\gamma}$
Pseudoplastic power law fluid	$\tau = K\dot{\gamma}^n$	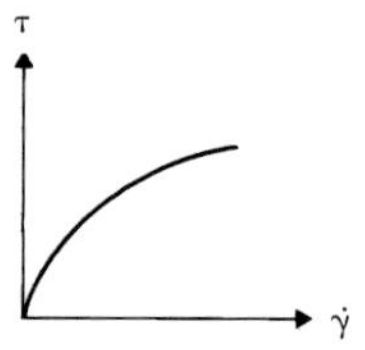 τ; $\dot{\gamma}$	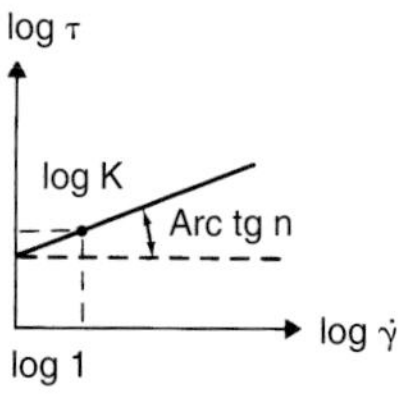 log τ; log K; Arc tg n; log 1; log $\dot{\gamma}$

Fig. 5.4

Rheological systems *(Source: Manuel de rhéologie des fluides de forage et laitiers de ciment, Editions Technip, Paris, 1979).*

The principle of the viscosity meter is two coaxial cylinders with the mud sample contained in the annulus between the inner and outer cup. The outer cylinder (rotor) can rotate at 3, 6, 100, 200, 300 or 600 rpm. For each speed, the fluid's torque on the inner cylinder (stator) is measured. The rheogram can be plotted with the six measurement speeds.

5.2.3 Filtrate

5.2.3.1 Description

The drilling fluid, which is made up of a liquid phase and suspended clayey products, is subjected to hydrostatic pressure while it is in contact with porous and permeable formations.

- If the diameter of the pores is greater than the diameter of the suspended clays, the formation will absorb the whole fluid. The extreme case is lost circulation where the fluid flow is entirely absorbed by the formation and there is no mud return to the surface.

- If the diameter of the pores is smaller than part of the suspended particles, there is filtration, i.e. the clay products will be laid down on the wall of the hole. A filter cake will be formed and the base liquid (filtrate) will invade the formation. The permeability of the cake conditions filtration.

5.2.3.2 API filterpress (Fig. 5.5)

The *API* filtrate is the amount of liquid (in cm^3) collected in 30 minutes under a pressure of 100 psi (7 kg/cm^2).

In practice, the filtrate is often measured after 7.5 minutes, which is multiplied by two to get the *API* filtrate (filtration velocity is proportional to the square root of time: $V = \sqrt{t}$).

The thickness of the cake built up on the filter paper is also noted.

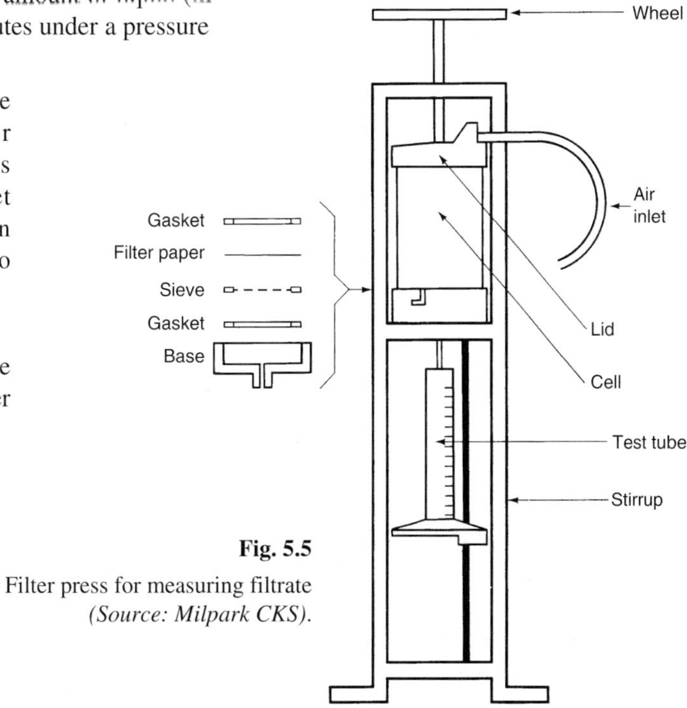

Fig. 5.5
Filter press for measuring filtrate
(Source: Milpark CKS).

5.2.4 Solids content

5.2.4.1 Definition

The drilling fluid is made up of a liquid phase and a solid phase. The solids content is:

$$t = \frac{V_{\text{solids}} \times 100}{V_{\text{mud}}}$$

5.2.4.2 Measurement

The two phases are separated by distillation. Then t can be calculated by measuring the volume of liquid collected:

$$t = 100 \times \left(\frac{1 - V_{\text{solids}}}{V_{\text{mud}}} \right)$$

5.3 DIFFERENT TYPES OF DRILLING MUD AND MAIN COMPONENTS

Drilling muds are commonly classified according to the base fluid, i.e.:
- water-base muds,
- oil-base muds.

Air, foam and aerated muds can also be used as drilling fluids.

First the main products used in making up drilling fluids will be listed, then the most commonly used muds and the areas where they are used will be described. The standard composition will be illustrated with examples in the form of tables.

It should be remembered that these are only standard examples. Whenever there is a problem of maintaining the properties of a fluid, it is necessary to make a specific study and often a more complex formulation and treatment.

5.3.1 The main mud products

5.3.1.1 Viscosifiers

Name	Secondary function	Area of use
Clay for fresh water (bentonite)	Effective filter bed	Fresh-water mud, Cl^- content <25 g/l
Clay for salt water (attapulgite)	Cl^- content salt-water mud	
Biopolymer	Shear-rate thinning	Low-solids, low density mud

5.3.1.2 Filtrate reducers

Name	Secondary function	Area of use
Starch	Viscosifier	Salt-saturated drilling mud, temperature <150°C
Technical CMC Low viscosity High viscosity	 Slight viscosifier Strong viscosifier	 Ca^{++} content <500 g/l and Cl^- content <30 g/l
Refined CMC Low viscosity High viscosity	 Slight viscosifier Strong viscosifier	 Ca^{++} content >500 g/l and Cl^- content >30 g/l
Polyanionic polymer	Viscosifier, stabilizes shales	Seawater mud
Emulsified oil	Lubrication	Emulsified mud

5.3.1.3 Dispersants

Name	Secondary function	Areas of use
Tannin		Fresh-water muds Ca^{++} content <300 mg/l Cl^{-} content <20 g/l
FCL (iron or chrome lignosulfonate)	Filtrate reducer Inhibits swelling at higher concentration	Salt- or fresh-water muds FCL muds pH >9 Temperature <200°C
LC (lignochromates or lignites)	Reinforces FCL action High temperatures	

5.3.1.4 pH control

NaOH (caustic)	Precipitates calcium Extender to increase clay yield	

5.3.1.5 Calcium precipitation

Na_2CO_3	Extender to increase clay yield	

5.3.1.6 Weighting materials

- Barite ($BaSO_4$): average density 4.3.
- Hematite (Fe_2O_3): $4.9 < d < 5.3$.
- Siderite ($FeCO_3$): $3.7 < d < 3.9$. Soluble in hydrochloric acid, so mainly used in completion fluids.
- Galena (PbS): $6.7 < d < 7$. Weighting material for special cases.
- Calcium carbonate ($CaCO_3$): $2.6 < d < 2.8$. For low-density fluids that can be acidified.

5.3.1.7 Lost circulation materials

These materials are used to plug up permeable zones. There are several types:

- granular: hard, calibrated products made of walnut shells, apricot, cherry or olive stones, etc.,
- fibrous: designed to "weave" the granular materials together, made of wood, sugar cane or cellulose fibers, etc.,
- flaky: to cover over the previously listed materials, cellophane waste, mica, etc.

5.3.2 Types of drilling fluids

5.3.2.1 Water-base muds

Straight bentonite mud

Average composition ($/m^3$)	Characteristics	Stability toward contaminants	Area of use
Bentonite: 40 to 60 kg CMC: 0 to 5 kg Caustic for pH: 8.5 to 9	Low initial density: 1.03 to 1.05	Slight	Spud mud Few contamination problems

Bentonite mud with tanning extracts

Average composition ($/m^3$)	Characteristics	Stability toward contaminants	Area of use
Bentonite: 40 to 60 kg Tannin: 2 to 4 kg Caustic: 0.5 to 1 kg CMC: 1 to 5 kg	pH <11 Filtrate: 2 to 4 cm^3	Average Ca^{++} <300 mg/l Cl^- <20 g/l	Depth <3000 m Low-contamination zones (gypsum, anhydrite, shales)

FCL/LC bentonite mud

Average composition ($/m^3$)	Characteristics	Stability toward contaminants	Area of use
Bentonite: 50 to 60 kg FCL: 20 to 40 kg Caustic: 2 to 4 kg CMC: 0.5 kg Possibly with LC: 10 to 20 kg	pH >9 Holds up well to 200°C	Good Cl^- from 50 to 70 g/l	Depth: 5000 to 6000 m Wide area of use: concentrations adjusted according to contamination (gypsum, anhydrite, shales)

Gypsum mud

Average composition ($/m^3$)	Characteristics	Stability toward contaminants	Area of use
Bentonite: 50 to 70 kg FCL: 12 to 15 kg Caustic: 3 to 4 kg Gypsum: 10 to 20 kg CMC: 5 to 10 kg Possibly LC	pH <9 Holds up well to 200°C	Good Cl^- from 60 to 70 g/l	Gypsum or anhydrite sections Shaly sections Slightly salt-bearing sections

Salt-saturated mud with inorganic thinners

Average composition ($/m^3$)	Characteristics	Stability toward contaminants	Area of use
Salt: 300 kg Clay: 50 kg (specific for salt-water mud) Starch: 30 to 40 kg Lime: 0 to 10 kg	d > 1.20 Corrosive Holds up moderately to temperature, 130 to 140°C	Good with gypsum and anhydrite sections Fair with shales	Salt-bearing sections Zones with slightly or moderately dispersing shales

Salt-saturated mud with organic thinners

Average composition (/m^3)	Characteristics	Stability toward contaminants	Area of use
Salt: 350 kg Clay: 50 kg Starch: 20 to 30 kg	$d > 1.20$ Corrosive		Salt-bearing sections Shaly zones

Emulsion mud

Water-base mud + 5 to 10% oil	Same as water-base mud plus: • reduced filtrate • lubricating function • reduced friction	Same as water-base mud	Same as water-base mud plus: • enhanced shale control • reduced risks of stuck pipe

Seawater mud

Bentonite: 75 to 100 kg Clay: 30 to 80 kg (for salt-water mud) Caustic: 5 to 10 kg FCL: 15 to 25 kg LC: 5 to 10 kg CMC: 1 to 5 kg (refined)	$d = 1.10$ to 1.15 pH = 9.5 to 10.5 Holds up well to temperature, 200°C	Good	Offshore for: • spud muds • gypsum or anhydrite sections • shaly sections

Polymer mud

Biopolymer: 4 kg Chromic chloride: 1.2 kg Caustic: 8 to 10 kg Bactericide: 0.3 kg Bentonite: 5 kg Possibly plus FCL and CMC	Very low solids content Minimum density 1.03 Filtrate >12 cm^3 Can be used in seawater	Average	When formation problems are few, used in order to improve penetration rate

5.3.2.2 Oil-base muds

There are two main types: oil-base muds (a few percent water) and invert muds.

A. *Oil-base muds*

Characteristics

- Oil-base muds cause the least damage to pay zones.
- They have the properties required for good drilling conditions.
- The oil does not filter much into formations.

Area of use

- Drilling and coring in pay zones.
- Workover and maintenance of producing wells.
- Drilling sections where a water-base fluid would cause problems (swelling shales, stuck pipe, etc.).

Common composition

- Base oil: 95 to 98% of the volume. Diesel oil to highly asphaltic crude can be used with the following most important properties:
  - specific gravity,
  - flash point,
  - acid number,
  - aniline point.
- Water: 2 to 5%, content must be monitored. Allows general characteristics of oil-base mud to be adjusted. Is emulsified.
- Rheological agents: to control filtration and viscosity, the following products are added:
  - blown asphalt,
  - organophilic clay,
  - flame black, etc.
- Emulsifiers and stabilizers,
- Dispersants,
- Weighting materials:
  - $CaCO_3$,
  - $BaSO_4$,
  - Galena.
- Water neutralizing agents.

Advantages

The advantages of this type of mud are:

- Characteristics are easy to control when no water or crude influxes occur.
- The muds are insensitive to common water-base mud contaminants (NaCl, $CaSO_4$, cement, shales).
- They have excellent static filtration characteristics with temperature and pressure; the cake is very thin.
- Wells can be drilled with a density close to one.
- Drill string drag on the walls of the hole is reduced, thereby decreasing twisting moment and drill string wear.
- Oil-base muds lengthen roller bit lifetime.
- There is no risk of differential-pressure sticking.
- The muds allow better core recovery.
- Gives cores where the value of the content and nature of interstitial water can be better assessed.
- Compared to water-base mud drilling, increased productivity index.
- Less damage to formations.

Drawbacks

However, some drawbacks may be:

- Sensitivity to water and to some crudes.
- Weighting materials may tend to settle out.
- Oil-base muds are dirtier to handle.
- Possible fire hazard.
- Rubbers not specifically designed for oil and gas duty may deteriorate.
- The presence of oil is harder to spot in the cuttings.
- Some mud logging and wireline logging methods can not be applied.
- Cost price per m^3 is higher than for water-base mud.

B. Invert muds

These muds are drilling or completion fluids with a continuous oil phase and a dispersed aqueous phase of at least 50% of the volume.

Characteristics

They have the same characteristics as oil-base muds, except that they perform better where oil-base muds are at a disadvantage.

Area of use

They are used in the same areas as oil-base muds:

- Thick sections of salts or anhydrite.
- Problems with drilling under high-temperature conditions.
- Problems of deviation.
- Drilling under low atmospheric temperature conditions.

Advantages

They have the same positive points as oil-base muds, in addition:

- They are less of a fire hazard.
- Less expensive per m^3.
- Easier to treat on the surface.

This type of mud came into use more recently than oil-base mud and has taken over practically all of the range of applications.

5.3.3 Drilling with air, foam or aerated mud

5.3.3.1 Drilling with air

Compressed air is pumped into the drill string instead of drilling mud and it fulfills all the functions required for drilling. Below are the basic differences between using mud and air as drilling fluids:

- annular velocity is 900 m/min,
- there is very little hydrostatic pressure on the bottom,

- rate of penetration is considerable since pressure is negative on the working face,
- drilled formations are not invaded by the drilling fluid,
- a rotary blowout preventer, or diverter, is required at the wellhead,
- but air can not be used as drilling fluid if there are water influxes in the well.

5.3.3.2 Drilling with foam

What is attempted is to keep the advantages of air drilling while at the same time coping with the problem of water influxes. The circulating foam is the result of mixing air + water + foaming agent.

Advantages

- Foam flow rates are ten times lower than air flow rates (foam has a much greater capacity to clean out the well than air).
- Foam is stable when there are small water influxes.

Drawbacks

Foam is stable and therefore almost impossible to treat continuously on the surface. Applications have been confined to desert regions.

5.3.3.3 Drilling with aerated mud

Some advantages, e.g. faster penetration rate and less wear and tear on bits, can be preserved while at the same time small fluid influxes into the well can be controlled.

5.3.4 Completion and workover fluids

This section is based on *Complétion et reconditionnement des puits, Programmes et modes opératoires,* Chambre syndicale de la Recherche et de la Production du Pétrole et du Gaz Naturel, Comité des Techniciens, Editions Technip, Paris, 1985.

From the time the pay zone containing effluents is drilled until the well is brought on stream, several different fluids will have resided in the borehole. The fluids must be adapted either to the type of operations carried out during all this time or to producing the pay zone.

Likewise, well workover requires the use of a special fluid during the whole operation and before the well is brought back on stream.

The different fluids can be classified into four categories:

- fluid for drilling the pay zone,
- completion fluid,
- control or workover fluid,
- packer fluid.

Later on, it will be seen that in practice — and for economic reasons mainly — the same fluid can be used for various functions.

5.3.4.1 Pay zone drilling fluid

Here the fluid is a drilling mud used when drilling the reservoir that is due to be produced. So that the pay zone is preserved intact in its native state, it is advisable to use a fluid having all the properties of drilling mud, but without its detrimental effects on the reservoir.

In addition to the conventional functions of drilling mud, the pay zone drilling fluid minimizes damage to productive beds. In fact, using a conventional drilling mud might have serious consequences on the reservoir (plugging, incompatibility of mud and effluents, contamination of the reservoir).

These adverse, sometimes irreversible, effects can be remedied only by the use of expensive techniques. To avoid resorting to high-cost techniques, the reservoir drilling fluid must therefore have the same characteristics as the completion fluid or the workover fluid (see below).

In practice, it is often difficult or even impossible —and generally costly to formulate for one single operation— a fluid that is both suited to drilling and compatible with producing the reservoir. Depending on the nature of the well and of the productive beds, the reservoir will be drilled either with the drilling mud that has already been used and possibly adjusted, or it will be replaced by completion fluid.

5.3.4.2 Completion fluid

The completion operation normally begins with drilling the pay zone that is going to be produced. In practice, however, the completion fluid is the fluid used to run in and set well equipment and when any production casing is perforated.

The chief functions of a completion fluid are therefore different from those of a drilling mud. It must be chosen so as to optimize eventual production of the bed while at the same time ensuring safe operation.

The characteristics of the fluid must therefore help consolidate borehole walls (especially when production will be with an open-hole completion system). It must also clean the borehole by keeping particles and cuttings suspended, but mainly avoid plugging up the pay zone.

Plugging is in fact the biggest hazard, since even when slight it can make production drop considerably. It is directly related to the characteristics of the reservoir (porosity, permeability, degree of fracturing) and of the effluent. The effluent can react with the completion fluid and form precipitates that can modify permeability. The filtrate must therefore be adjusted accordingly.

The solid particles contained in the fluid can also plug up the pay zone surface. Though a safety margin must be kept, the fluid's density must not be much higher than the density equivalent to the reservoir pressure gradient. The aim is to help keep particles from invading the reservoir.

In practice, the completion fluid's composition varies widely according to the nature of the reservoir and of its effluent, and the cost of making up the fluid. In many cases, the drilling mud is reused, thereby minimizing costs but increasing plugging risks, however.

Brines are frequently used, especially sodium chloride brines, since their filtrate is well suited to reservoir characteristics.

More complex fluids can also be utilized if permanent reservoir plugging is to be avoided. An acidizing operation can then be performed before the well is brought on stream. Here the completion fluid's composition must be such that it can be broken down by the acid. This is true for calcium carbonate muds.

Oil-base muds are used if their density allows and they generally behave well with the effluents they come in contact with.

5.3.4.3 Workover fluid

This is the fluid used for workover operations. It serves chiefly to keep the well under control. Its properties and functions are similar to those of a completion fluid.

5.3.4.4 Packer fluid

This is the fluid that is placed between the tubing and the production casing, above the packer. It remains there for the well's whole productive life, except possibly during workover operations when it may be replaced by a control fluid. Its functions are specific, it must:

- keep enough hydrostatic pressure on the tubing and on the production casing to keep the first from collapsing and the second from bursting,
- keep enough hydrostatic pressure on the casing packer in order to minimize the risk of leakage at this point and counterbalance formation pressure if possible in the event of a leak,
- retard corrosion phenomena on the casing and the production tubing.

The first two functions are directly related to the fluid's density, however the third is achieved only with a fluid of definite composition. Lastly, the packer fluid must be stable with time and as neutral as possible with respect to the produced effluent to reduce damage in the event of a leak in the annulus. The stability of the fluid may be impaired for three main reasons:

- When a mud with a high suspended-solid content is used as a packer fluid, settling of the particles modifies the value of the hydrostatic pressure throughout the well. In addition, settling can considerably hinder any workover operation requiring well completion equipment to be pulled. Drilling muds should therefore not be used as packer fluids.
- When bacterial growth develops in the packer fluid (a phenomenon that is more frequent at high temperatures and in water-base fluids). Deterioration may cause corrosive gas (H_2S) to be released and damage completion equipment. This drawback can be avoided by using bactericides and by keeping a high pH and salinity.

- When a chemical or organic reaction occurs between the packer fluid and either the produced effluent (packer or tubing leak) or a fluid contained in a shallower geological layer (production casing leak). This is a particularly serious hazard when the effluent is a sour gas.

In practice, several types of packer fluids are used depending on the type of well and produced effluent, and the cost of formulating the fluids. The following types can be distinguished:

- Water-base muds, fairly close to drilling muds, that can be weighted up with barite or carbonates. They are relatively low cost and not very stable with time (they settle out, break down).
- Brines, which are more expensive and often corrosive, contain no solids. They remain stable even at high temperatures. In addition, they are easy to make up and their density can reach a value of 2.
- Oil-base fluids certainly offer the best characteristics, but they are hard to make up and expensive.

In some instances, if conditions allow, the completion fluid can be used as a packer fluid.

5.4 MECHANICAL TREATMENT OF DRILLED SOLIDS

5.4.1 Functions of mechanical treatment

All drilled solids are contaminants in a drilling mud.

The main mud properties that are affected by solids concentration are as follows:

- Density increases variably according to the type of formation drilled. Since limestone is inert, it will be removed more readily than shale, which will disperse to a greater or lesser degree depending on the type of drilling mud.
- Rheology can be affected directly by fine particle load multiplying contact areas and interactive forces. It can also be influenced indirectly making the mud more sensitive to all sorts of contamination. As a result, its properties are more difficult to control.
- The filtrate is affected by solids which throw off the grain-size distribution of the cake. Accordingly, removal of the coarsest solids on the surface will improve filtrate control.

5.4.2 Separation ranges in treatment facilities

Figure 5.6 and **Table 5.1** below help situate the size of solids to be removed in relation to the size of useful solids added to the mud (barite, bentonite, attapulgite). They specify the separation range of each of the components in the treatment unit.

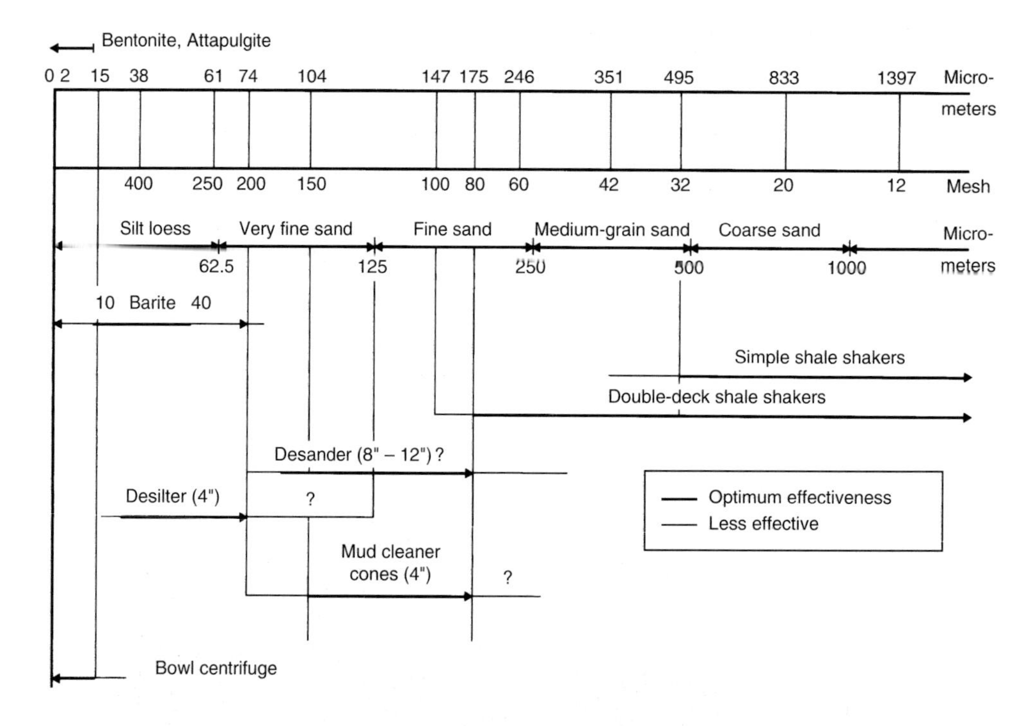

Fig. 5.6

Classification of solids treatment equipment
(Source: Mud Equipment Manual (Handbook 6), IADC).

TABLE 5.1

Mechanical solids control

Separation apparatus	Optimum mesh	Removes particles above (micrometers)
Simple shale shakers	32	500
Complex shale shakers	80	178
8" to 12" desanders		74
4" desilters		15 to 20
Mud cleaners	150 200	104 74
Centrifuges		2

5.4.3 Description of equipment

A. *Shale shakers* (Fig. 5.7)

A shale shaker, or vibrating screen, is made up of two parts: a stationary base and a moving frame that holds the screen cloth. The moving frame is equipped with an electric motor which imparts a circular, elliptical or linear vibrating movement. The vibrating frame is isolated from the base by shock absorbers. Whatever the type of shale shaker, they all have solids removal screens that are defined by the mesh number, i.e. the number of openings per linear inch.

B. *Hydrocyclones*

Hydrocyclones work on the principle of natural cyclones and could be described as a low-pressure center with pressure increasing around it. This gives winds around the edge that converge toward the center and rise.

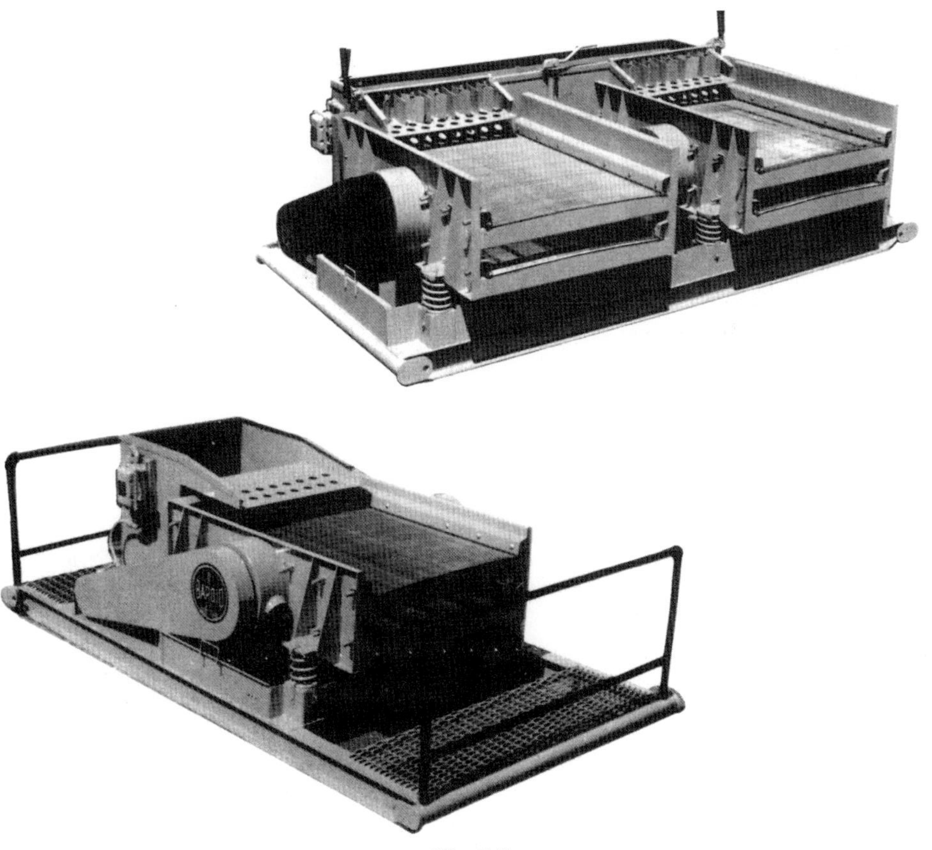

Fig. 5.7
Shale shakers *(Source: N.L. Baroid).*

To produce circumferential velocities, the fluid enters the hydrocyclone (**Fig. 5.8**) and is centrifuged faster and faster since the radius decreases. In the axis of the cone, the fluid rises and exits. Centrifuging deposits solid particles on the wall of the cone and they are removed through the nozzle.

The size of particles removed depends on the dimension of the cones. The separating power of a group of cones is what differentiates between desanders (particle size greater than 74 micrometers) and desilters (particle size greater than 30 micrometers) (**Figs. 5.9** and **5.10**).

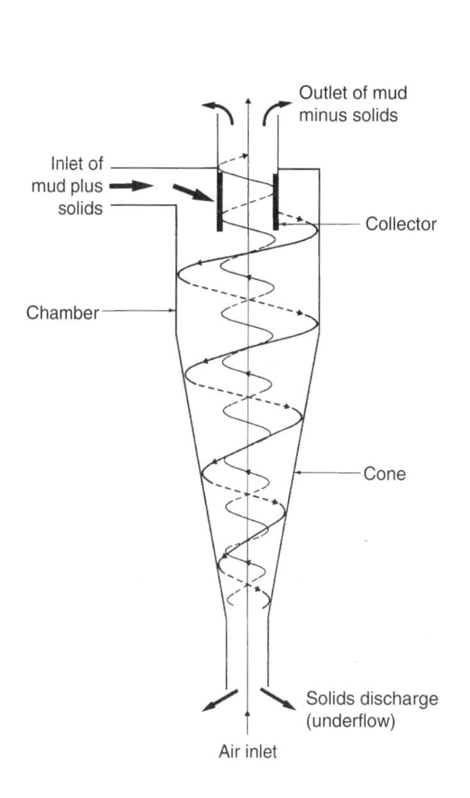

Fig. 5.8

Principle of the hydrocyclone *(Source: Mud Equipment Manual (Handbook 6), IADC).*

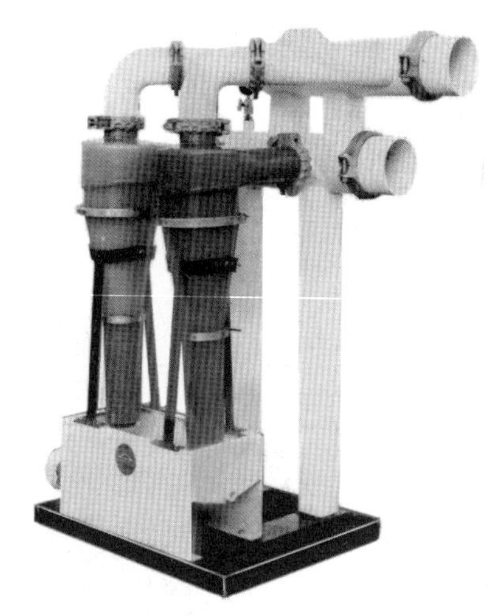

Fig. 5.9

Desander *(Source: Sweco Oilfield Services).*

Fig. 5.10

Desilter *(Source: Sweco Oilfield Services).*

C. *Mud cleaners* (Fig. 5.11)

These devices are simply desilters whose heavy effluent from the nozzle falls onto a shale shaker screen cloth. The fine-mesh cloth (100 to 325 mesh) enhances recovery of the liquid phase and therefore makes the discharged material drier.

D. *The centrifuge* (Fig. 5.12)

This apparatus has a horizontal axis and a conveyor screw and operates continuously. The horizontal bowl rotates at between 1500 and 3000 rpm and the conveyor screw has a slightly lower speed. Since the throughput is generally small compared to the drilling mud flow rate, this type of apparatus can treat only part of the mud returns.

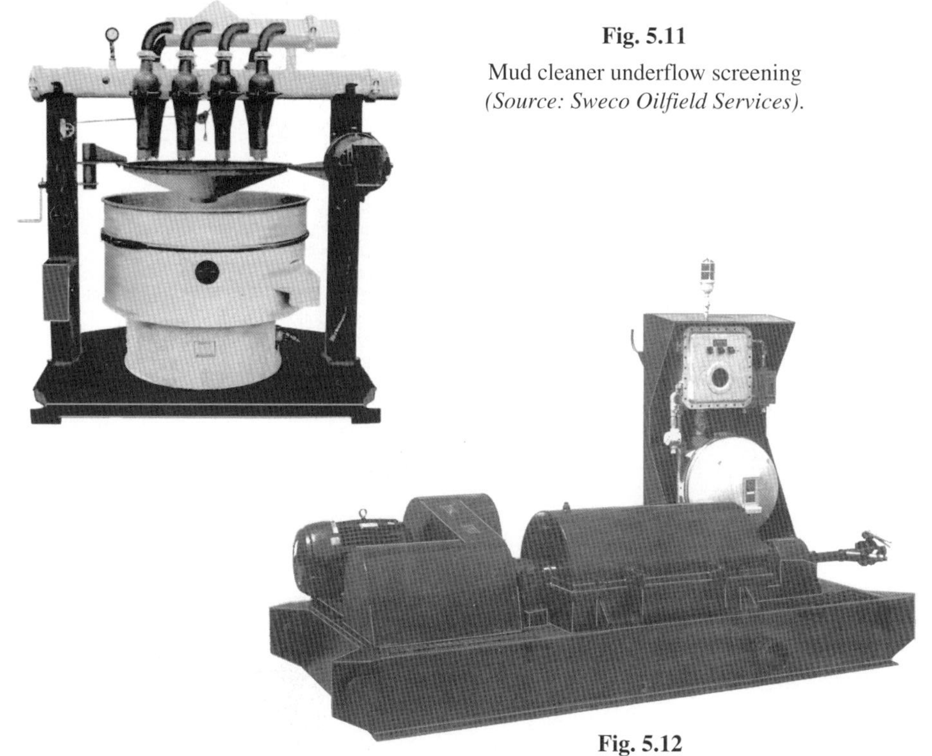

Fig. 5.11
Mud cleaner underflow screening *(Source: Sweco Oilfield Services).*

Fig. 5.12
Horizontal centrifuge *(Source: Sweco Oilfield Services).*

5.4.4 Standard treatment layout

The diagram in **Fig. 5.13** can be considered optimum for a drilling rig for deep wells. The number of mud pits can vary widely. The volume of mud in the surface facilities and on stand by depends on the maximum volume of the objective that is being drilled.

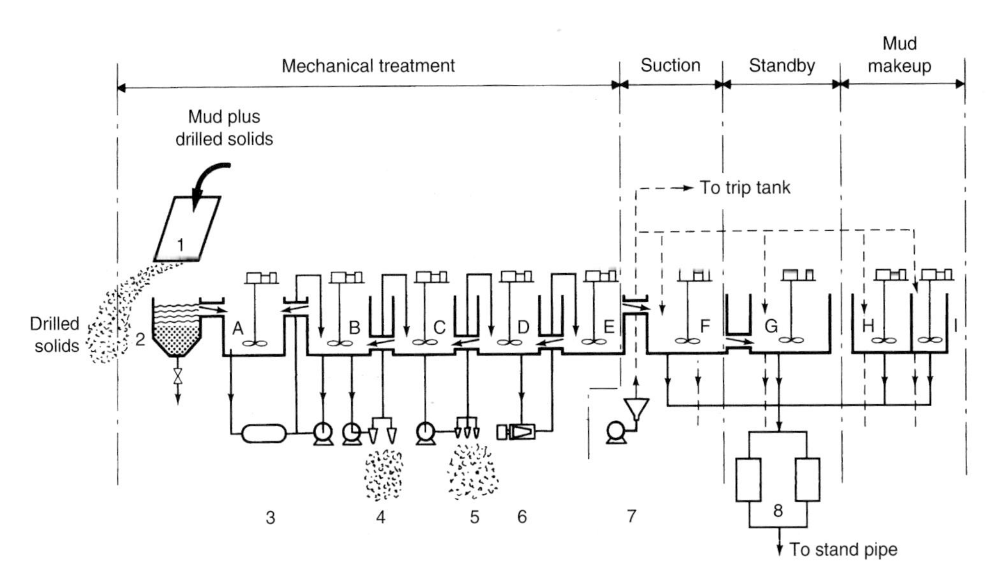

1. Shale shakers
2. Sand pit
3. Degasser + pump
4. Desander + pump
5. Desilter + pump
6. Centrifuge or mud cleaner
7. Mixing pump + hopper (transfer pump)
8. Mud pumps

A. Suction tank for gas-cut mud
B. Suction tank for desander
C. Suction tank for desilter
D. Suction tank for centrifuge or mud cleaner
E. Discharge tank for centrifuge (optional)
F. Main suction tank
G. Standby tank
H. Mud makeup tank
I. Extra tank

Fig. 5.13

Standard diagram of drilling mud treatment layout.

Chapter 6

WELLHEADS

This chapter deals exclusively with the architecture and safety equipment for onshore wells, or more generally wells with "dry" wellheads (above ground/sealevel). The specific features of offshore floating structures will be discussed in the chapter on offshore drilling.

INTRODUCTION

The term wellhead is a general one including all of the surface equipment topping off the architecture of a well. The configurations can be different depending whether the well is in the drilling, completion or production phase. Of course it is the production phase that gives the assembly its final structure. The lower part (often underground) is made up of casing hangers and accessories. Flush with ground level is the equipment for suspending tubing and just above it the hardware called the Christmas tree (**Fig. 6.1**).

During drilling, the wellhead is modified as the casing program progresses. On the surface, each casing string is topped off by equipment to suspend and hold it, to provide seals and to accommodate the blowout preventers as specified in the program.

6.1 CHANGES IN A WELLHEAD DURING DRILLING AND EQUIPMENT TECHNOLOGY

Assembly and test operations will be described based on the following drilling and casing program: drilling 17 1/2", casing 13 3/8"; drilling 12 1/4", casing 9 5/8"; drilling 8 1/2", casing 7"; drilling 6", then completion.

6.1.1 Start-up phase: 17 1/2" drilling

Here the well architecture is nonexistent or consists of the bell nipple (or the conductor pipe). It is only when there is a risk of shallow gas that safety equipment called a diverter must be installed (**Figs. 6.2a and b**). The diverter is an annular type of blowout preventer

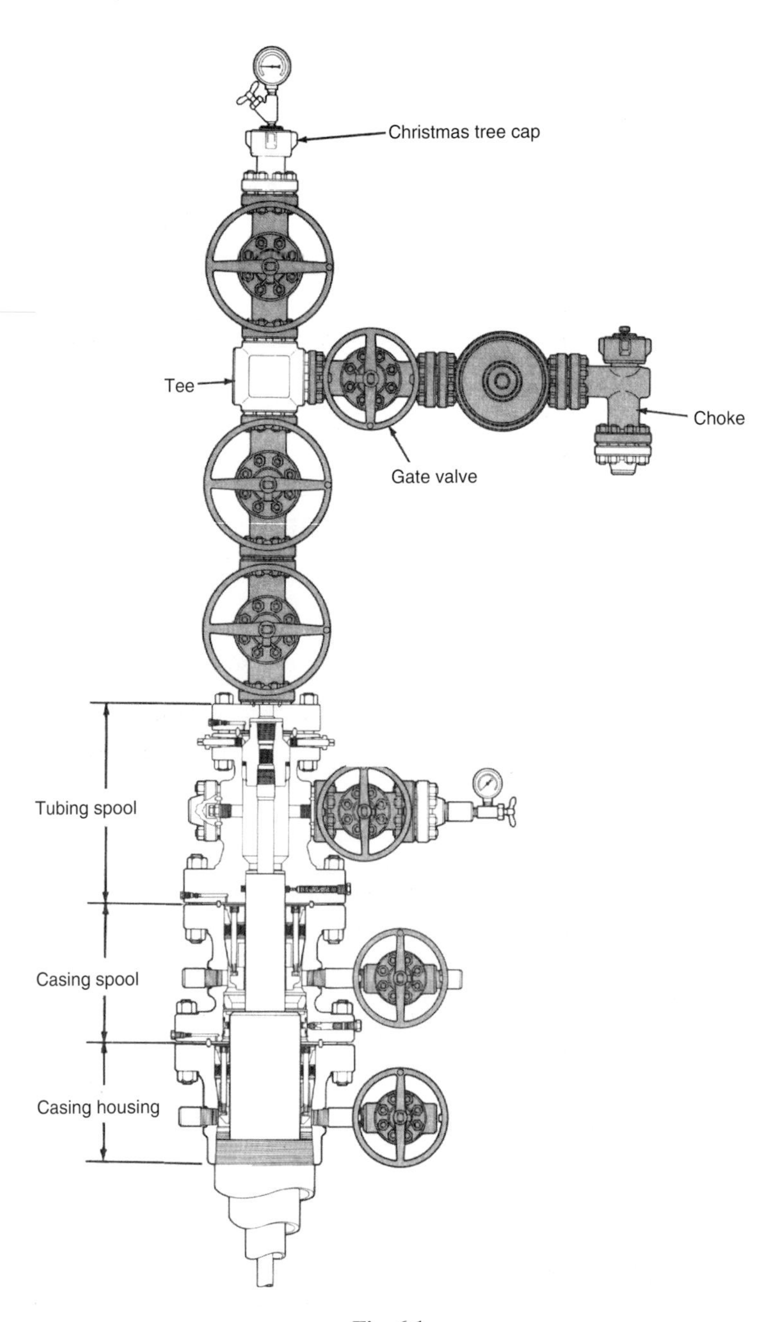

Fig. 6.1
Wellhead for a producing well *(Source: Cameron Iron Works)*.

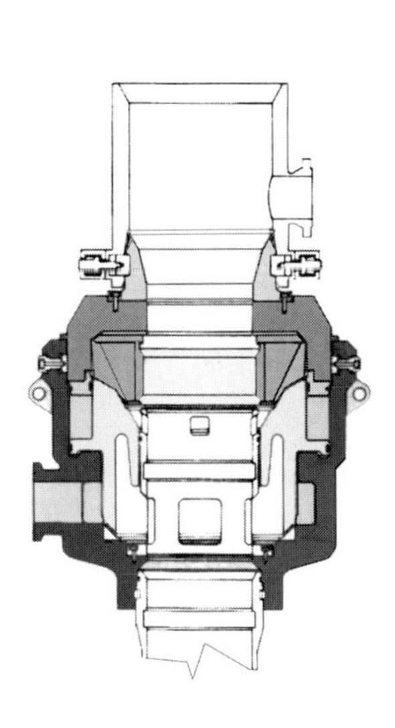

Fig. 6.2a
Diverter *(Source: Hydril)*.

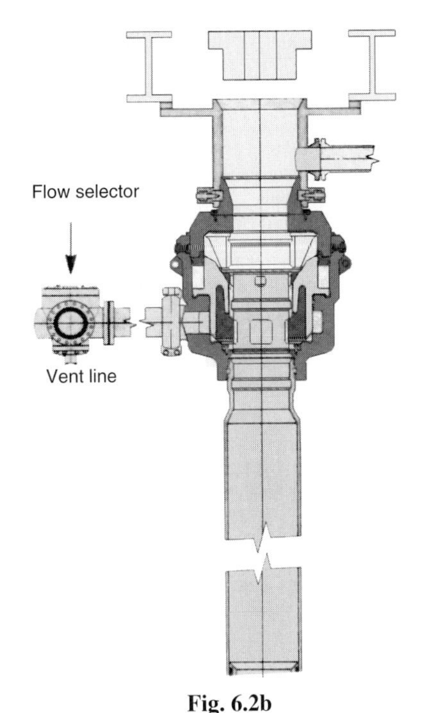

Fig. 6.2b
Diverter during drilling *(Source: Hydril)*.

whose working pressure is generally low (2000 psi). Its function is not to seal off the well. This would not be workable, since without casing the surface soil layer would yield immediately, causing a crater to be formed. Its only function is to divert the blowout to a flare, thereby protecting the well site from fire hazards.

Piping should be designed to cause minimum pressure losses so that the pressure in the well does not rise. This whole assembly is highly vulnerable to the abrasion resulting from the high velocities reached by gas at the surface.

Once the drilling phase has been concluded, the surface casing is run in and cemented right up to the surface. After the required waiting time for the cement to set, the first component of the wellhead, the casinghead housing, can be mounted.

6.1.1.1 Casinghead housings

This is the first component connected to the surface casing. It can be screwed on, by male, or more commonly female, threaded connections. It can also be welded on (slip-on connection) (**Fig. 6.3**) or cold forged, a more recent development (**Fig. 6.4**).

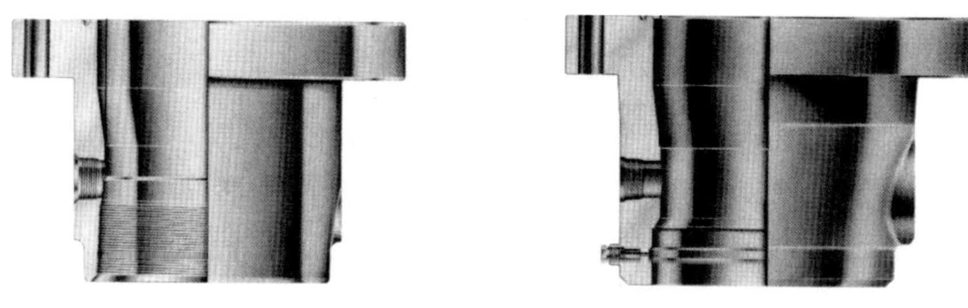

Fig. 6.3

Casinghead housing *(Source: Cameron Iron Works)*.

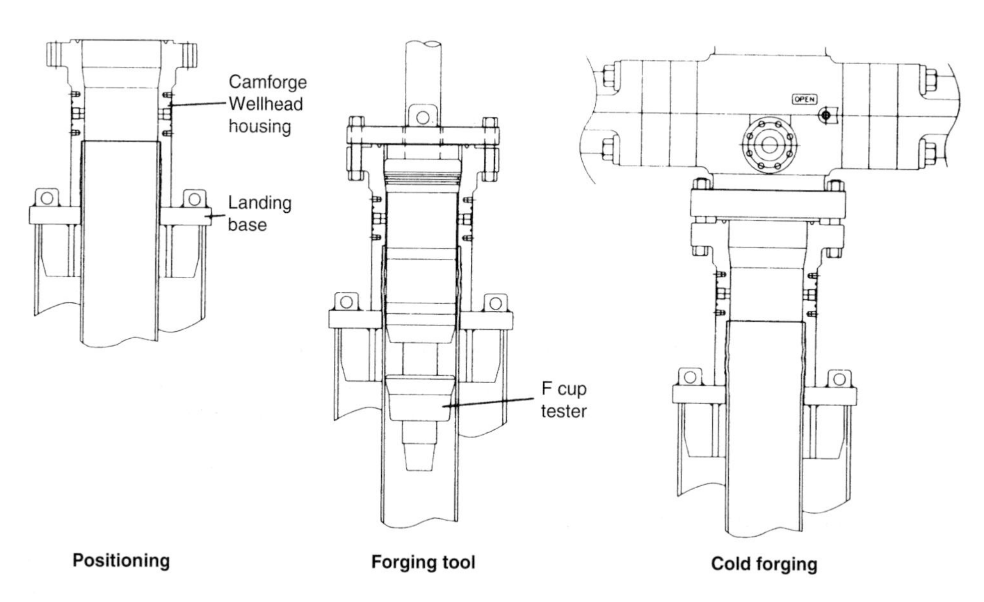

Fig. 6.4

Camforge CIW casinghead *(Source: Cameron Iron Works set FMC)*.

Though the threaded connection is more reliable, the drawback is that the upper casing thread must be at a precise height. This is so there is room enough for the hangers and BOPs to be stacked up under the substructure.

In the event the casing string gets stuck on the way down, the connection can be made only after the pipe has been cut off and then welded.

Connections must be made very carefully keeping the upper flange absolutely horizontal. The inside and outside welds must be tested with a pump providing pressure through a test port located between the two weld seams.

The bore in the upper part of the housing can be cylindrical or tapered (depending on makes and models) to support the casing hangers and the seal assembly for the next casing string. There are two flanged or threaded side outlets to enable inspection of the annulus.

So it is the surface casing and the casinghead housing that support all the casings and BOPs anticipated in the program. In a deep well, the weight of the casing strings (and the tensioning) is quite considerable and requires a casinghead with a circular landing base (**Fig. 6.5**).

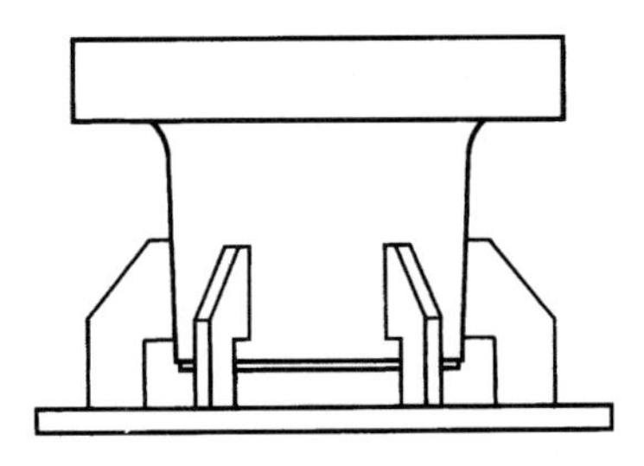

Fig. 6.5

Casinghead housing with landing base *(Source: Cameron Iron Works).*

This type of casinghead provides greater stability for the wellhead and distributes the load more evenly on the bottom of the cellar. This is true provided the cellar is deep enough and the space between the bottom of the cellar and the landing base is filled with concrete. Securing studs are then generally added.

6.1.1.2 Assembling blowout preventers

The wellhead assembly is complete for this drilling phase with the mud cross and the operator's recommended BOP stack. It is advisable to line the inside surface of the casinghead housing bore with a wear bushing to protect it from erosion by the drillpipe. The wear bushing will be locked in place by a special flange called a locking flange and radial threaded rods. The wear bushing must in some cases be removed before the next casing string is run in (**Fig. 6.6**).

6.1.1.3 Pressure tests

Besides BOP operating tests, the wellhead must also be pressure tested:

- whenever anything is assembled,
- whenever anything is dismantled, even partially (when a valve is changed, when gates are opened to change rams, etc.),
- periodically, according to operator's instructions.

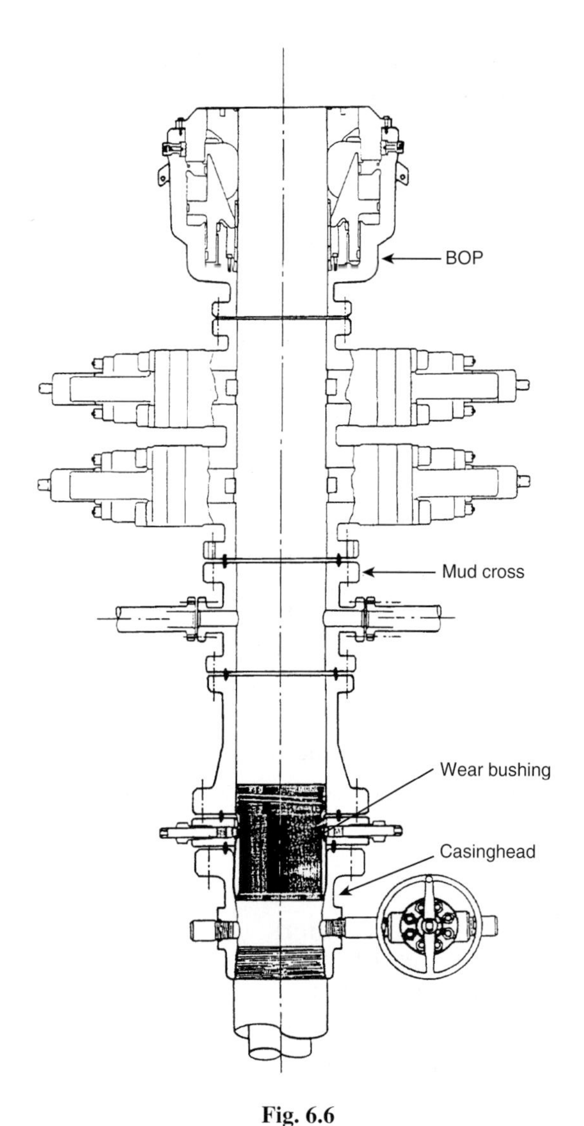

Fig. 6.6

Wellhead for 121/4" drilling phase
(Source: ENSPM Formation Industrie).

Before performing a test, it is necessary to:

- determine which elements will be subjected to pressure to check that the weakest component can withstand the test pressure,
- consider the possibility that test fluid may leak and the consequences it may have (pressure increase in an annulus, etc.).

Prior to each test, clear water should be circulated throughout all the circuits and each component should be tested separately. There are two ways to proceed (**Fig. 6.7**).

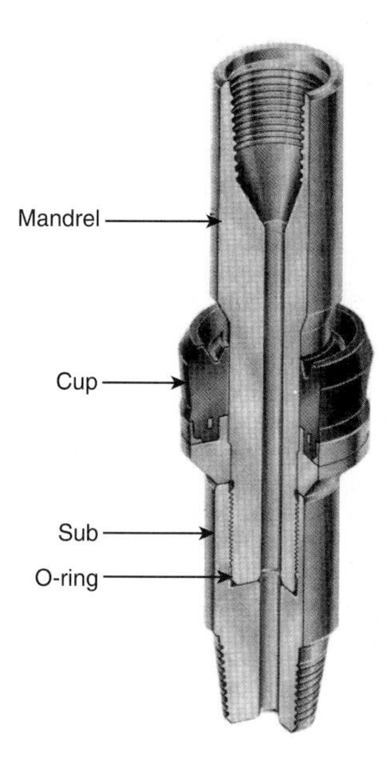

Fig. 6.7

Wellhead test tool. Seal in the casing (tester cup)
(Source: Cameron Iron Works).

A. With a tester cup

This test tool is screwed onto the end of drillpipe and run about 10 to 30 meters into the casing. One or two lengths of pipe may be screwed on below it to serve as a guide and add extra weight to facilitate the lowering operation.

All valves and each pair of pipe rams are tested separately:

- by pumping into the annulus via the kill line, or
- by pulling on the tester cup, if no HP pumping unit is available.

The test pressure should not exceed the working pressure of the wellhead or 60% of the bursting pressure corresponding to the yield strength of the tubulars located at tester cup level.

Three further points require care:

- The tensile strength and collapse strength of the tester pipe itself should be taken into consideration. This means that a tester cup should be run in with 5" heavyweight drillpipe or with higher-grade drillpipe.

- Annular preventers must not be tested at over 50% of their maximum working pressure.
- The inside bore of the tester pipe must not be obstructed. If there is a leak at the cup, the test fluid should be able to flow visibly onto the rig floor without putting pressure on the casing under the cup.

B. With a tester plug (Fig. 6.8)

This is a plug with resilient O-rings that is run in with drillpipe and comes to rest in the casinghead located under the BOPs. The pipe ram preventers, the annular preventer and the accessories can be tested at their working pressure with no risk to the casing.

Blind ram-type BOP closure can be tested by placing a cap on the plug. The drillpipe (not made up) is unscrewed and pulled out after the plug has been positioned.

It is also advisable to open a valve or an outlet below the tester plug so that the casing is not put under pressure in the event of a leak at the O-rings.

In this way, the BOP and casing spool can be tested at the working pressure with this tool.

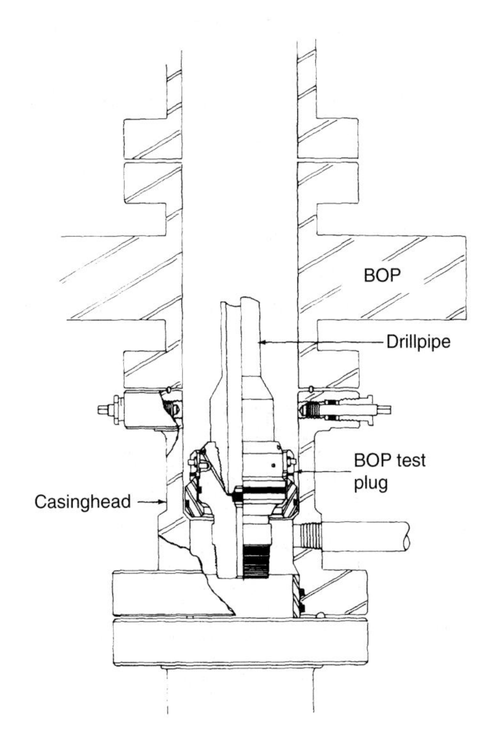

Fig. 6.8

Wellhead test tool. Seal in the casinghead (tester plug) *(Source: FMC).*

6.1.2 121/4" drilling phase

Drilling is resumed in 121/4" diameter through the wellhead that has been assembled. After the 95/8" protection casing has been run in, it is cemented up to a given height from the shoe. Once the cement has set and the wear bushing has been removed from the casinghead housing, the casing can be tensioned and anchored in the housing.

This is done by placing the weight of the 95/8" string on the casing spider and disconnecting the flanges between the casinghead housing and the mud cross (or locking flange). The BOP stack is lifted by the block and tackle in the substructure. Then the casing hangers can be placed between the 95/8" casing and the bore of the casinghead housing.

6.1.2.1 Casing hangers

These are devices used to anchor and hang the casing strings in the casinghead housing. Each one is made up of a set of slips that latch around the casing that is going to be hung, sliding into a tapered part of the casinghead. There is also a rubber ring sealing assembly.

There are a large number of models on the market. It should be noted that for each make, one type of casinghead will accommodate only one type of casing hanger.

Casingheads and hangers are designed for different capacities and are therefore chosen according to the weight they will have to support.

The three casing hanger types below are available in all makes.

A. *Separate packers and hangers* (Fig. 6.9)

When the casing is suspended by the hangers, it can be cut off and the seal assembly is then slipped around the pipe and laid on the hangers. The packer is energized by tightening the cap screws.

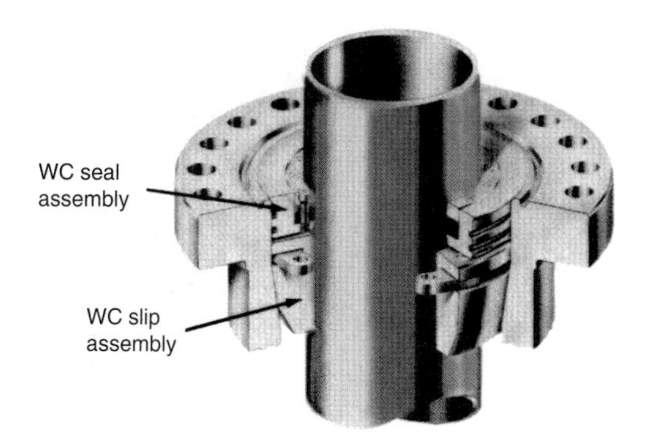

Fig. 6.9

Casing hangers *(Source: Cameron Iron Works).*

B. *Hangers and incorporated packers* (Fig. 6.10)

Here the whole assembly has to open so that it can be placed around the casing. Once it is in place and the casing is suspended, tightening the screws compresses the packing against the body of the casing and against the outside of the casinghead housing. This is the most common equipment.

C. *"Automatic" hangers* (Fig. 6.11)

This hanger system has an integrated packer and a particular feature: the downward pulling force exerted by the weight of the casing compresses the packer and energizes the seal.

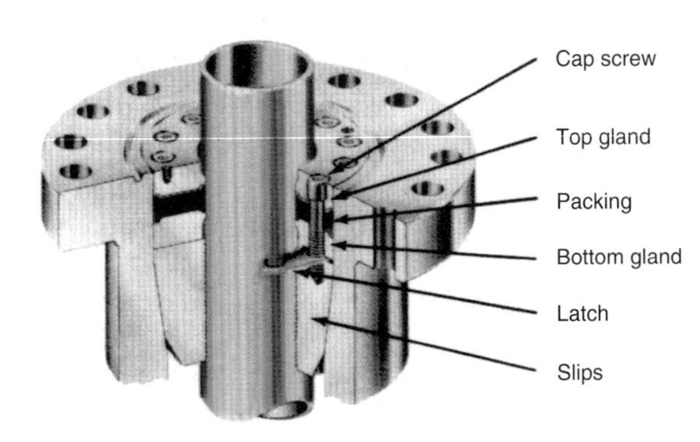

Fig. 6.10
Casing hangers *(Source: Cameron Iron Works).*

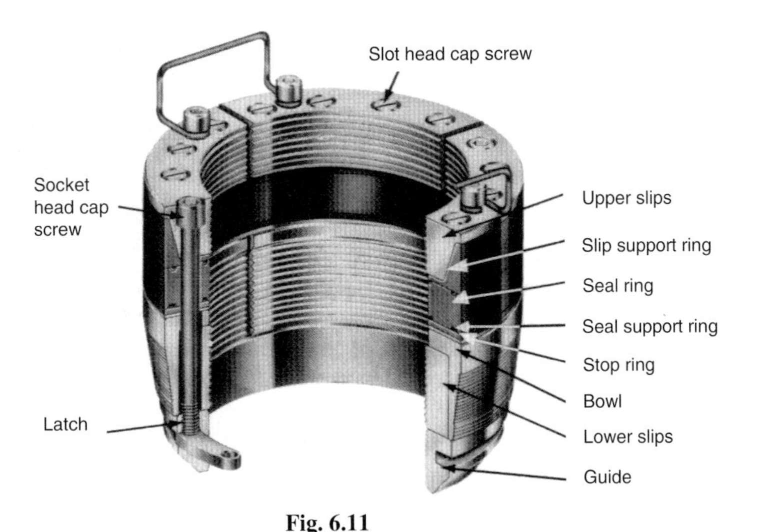

Fig. 6.11
"Automatic" casing hangers *(Source: Cameron Iron Works).*

Once the 95/8" casing string has been hung, the following operations are carried out in this order:

- the casing is cut off according to the distance it penetrates into the base of the casinghead and the distance left between each flange by the ring gasket,
- the 7" casing spool is set in place with the bit pilot and its sealing system (X bushing) at the base,
- these three components are assembled and made up, including the locking flange and the wear bushing lining the bore of the 7" casing spool.

It is only after these components have been made up that the 95/8" casing seal is tested.

6.1.2.2 Casinghead spools

This component (**Fig. 6.12**) also supports a casing string and is composed of:

- Two flanges of different sizes, and different working pressures. The lower flange must be the same dimension and working pressure rating as the casinghead housing flange for the two to be connected.
- The upper part has a cylindrical or tapered bore designed to accommodate the casing hangers for the next string of casing (here 7").
- The lower part has a cylindrical bore designed for a bit pilot and a seal system (bushing).

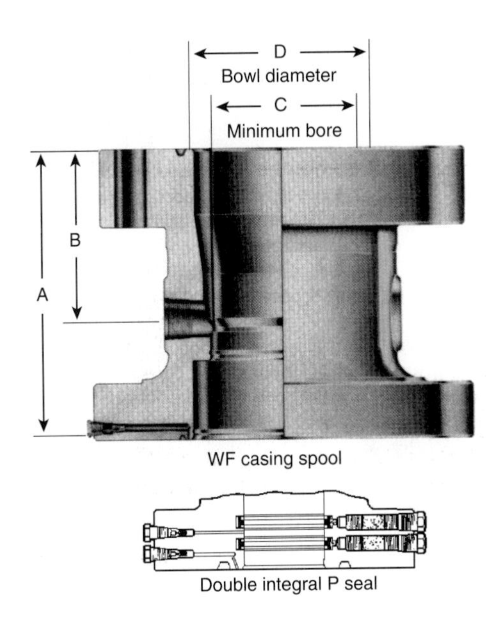

Fig. 6.12

Casinghead spool *(Source: Cameron Iron Works)*.

The annulus between the two casings can be sealed in different ways:

- by compressing a set of packers when the flanges are connected (**Fig. 6.13**),
- by energizing a packer by injecting sticks of plastic grease (*Cameron* type X) (**Fig. 6.14**).

There will therefore be two levels of annular seal on the 95/8": by the casing hanger and by the packer at the base of the casinghead spool.

All manufacturers offer casinghead spools called "all purpose" (**Fig. 6.15**). They may be used for suspending casing or tubing depending on the circumstances of the exploratory well. This is why radial screws are required on the upper flange.

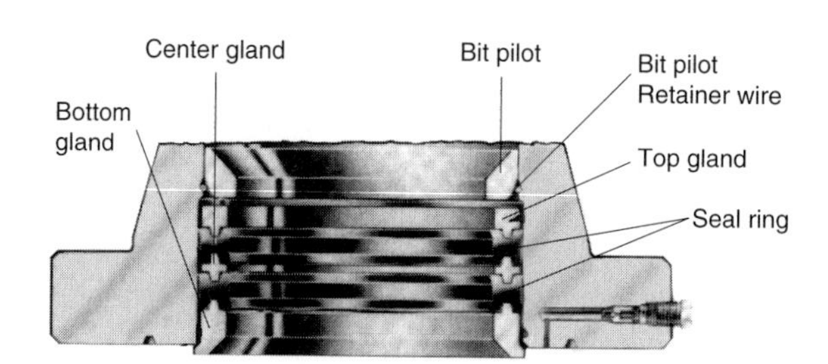

Fig. 6.13

Example of seal on casing
(Source: Cameron Iron Works).

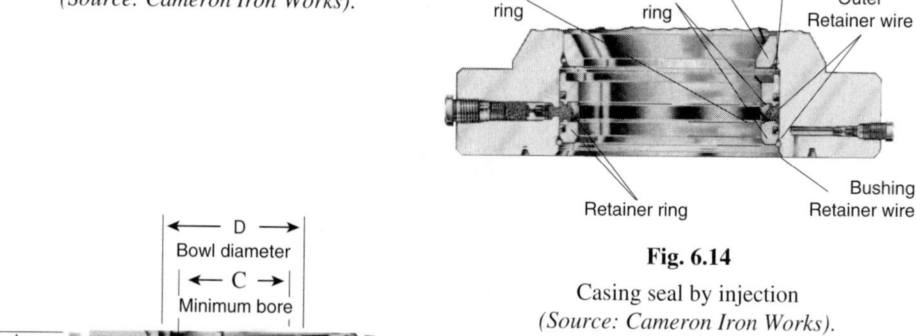

Fig. 6.14

Casing seal by injection
(Source: Cameron Iron Works).

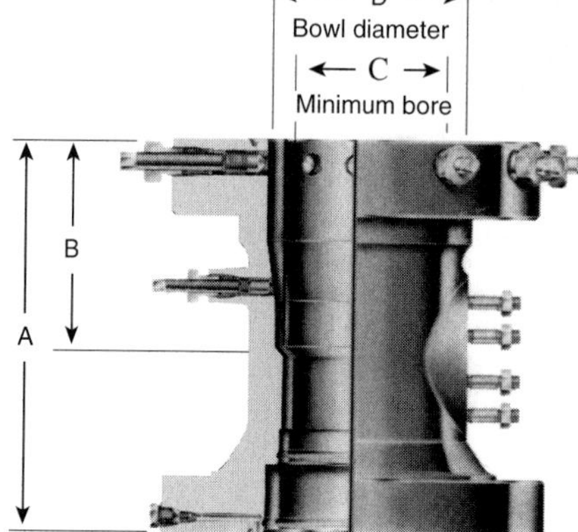

Fig. 6.15

All-purpose spool for casing or tubing
(Source: Cameron Iron Works).

The wellhead assembly will be completed by the mud cross, the kill-line and choke-line connections and assembly of the BOP stack. Their working pressure is generally higher than for the preceding assembly, since the deeper the well gets the greater the risk of high pressures. **Fig. 6.16** shows the wellhead assembly for the following drilling phase.

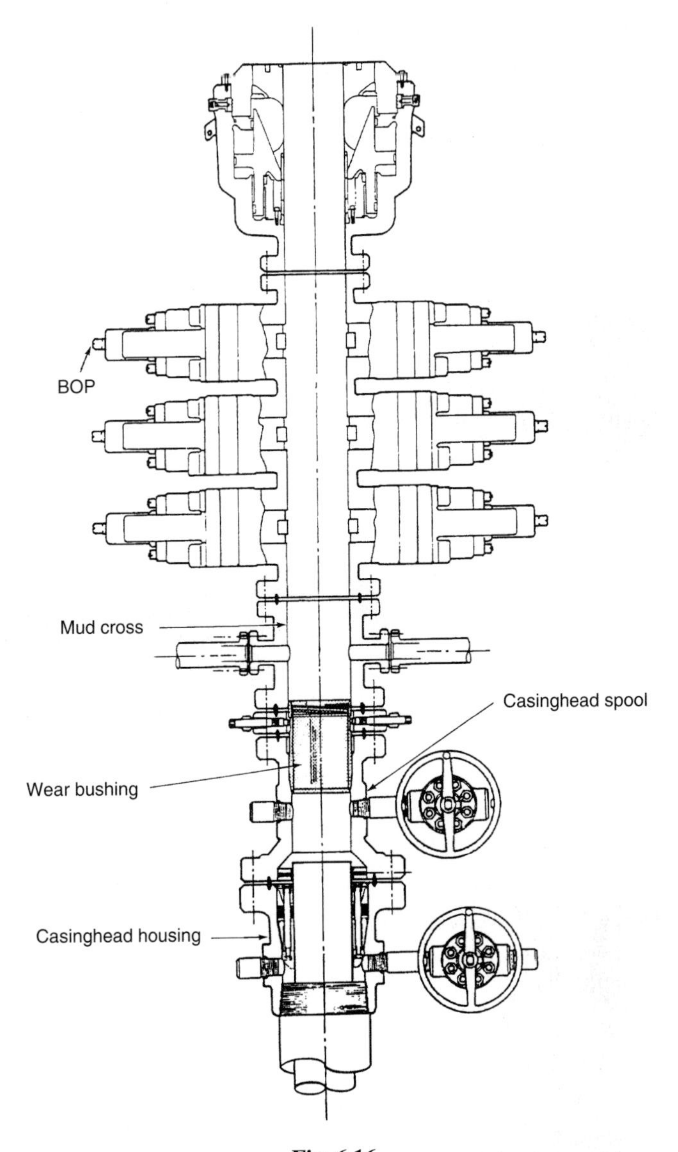

Fig. 6.16

Wellhead for the 81/2" drilling phase *(Source: ENSPM Formation Industrie).*

6.1.3 81/2" drilling phase

This example considers that the casing string that is run in is a 7" production string.

The principle is the same as for the preceding phase:
- the casing string is anchored after the wear bushing has been removed,
- the casing is cut off according to the components that will be assembled on the top of the 7" string,
- a testing or pack-off flange is connected (**Fig. 6.17**).

This flange is connected to the upper side of a casinghead and increases the number of the annular seal on the anchored casing. Test ports allow the seal to be checked.

The test flanges —between two components of the same nominal size and working pressure rating— generally have an extra groove on the upper side for a smaller diameter ring gasket than what is normally used in the pressure range (restricted *API* ring groove). The smaller diameter ring gasket reduces the area exposed to pressure, thereby allowing the next highest working pressure to be used.

- A tubing head spool is installed, which is similar to a casinghead spool. It has a cone-shaped bore to accommodate the tubing hanger.

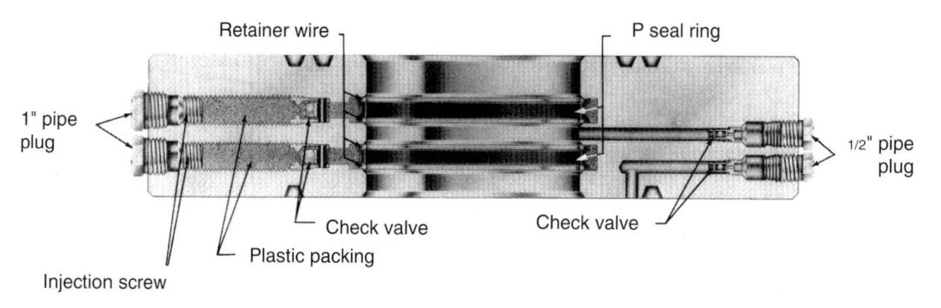

Fig. 6.17

Double pack-off sealing flange *(Source: Cameron Iron Works)*.

6.1.3.1 Tubing head spools

Tubing head spools are identical to double-flanged casinghead spools. The main difference is that the suspended tubular, the tubing, must be easy to pull out of the hole if the need arises (**Fig. 6.18**).

Tubing hanging devices are often very simple. In many cases they consist of only a threaded hanger with a seal assembly on the tapered outside part that presses against the corresponding tapered part of the tubing head. The hanger is held in place by tightened screws in the upper flange. A length of tubing is screwed onto the upper female threaded part of the hanger so that it can be positioned. Two side ports allow the annulus between the last casing string and the tubing to be checked (**Fig. 6.19**).

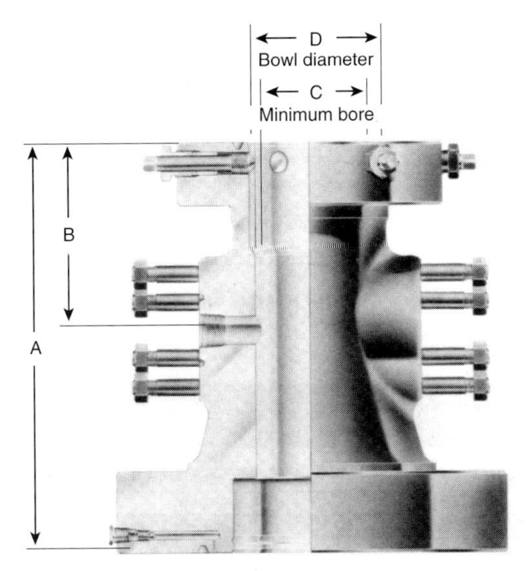

Fig. 6.18

Tubing head spool *(Source: Cameron Iron Works).*

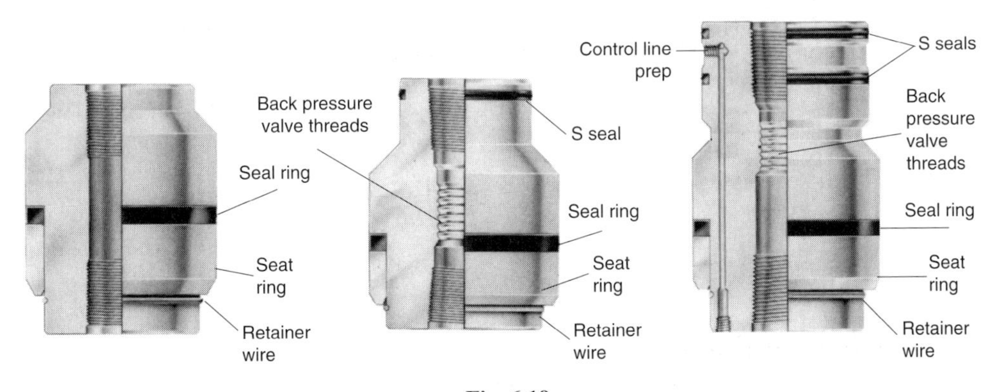

Fig. 6.19

Tubing hangers *(Source: Cameron Iron Works).*

The tubing may also be screwed onto a flange connected to the tubing head. The flange may also serve as an adapter and its upper side is directly connected to the first master valve.

For dual completions (two separate producing reservoirs, i.e. two parallel tubings in the production casing), a special tubing head must be used which allows the tubing hangers to be placed and adjusted in a half-moon shape (**Fig. 6.20**).

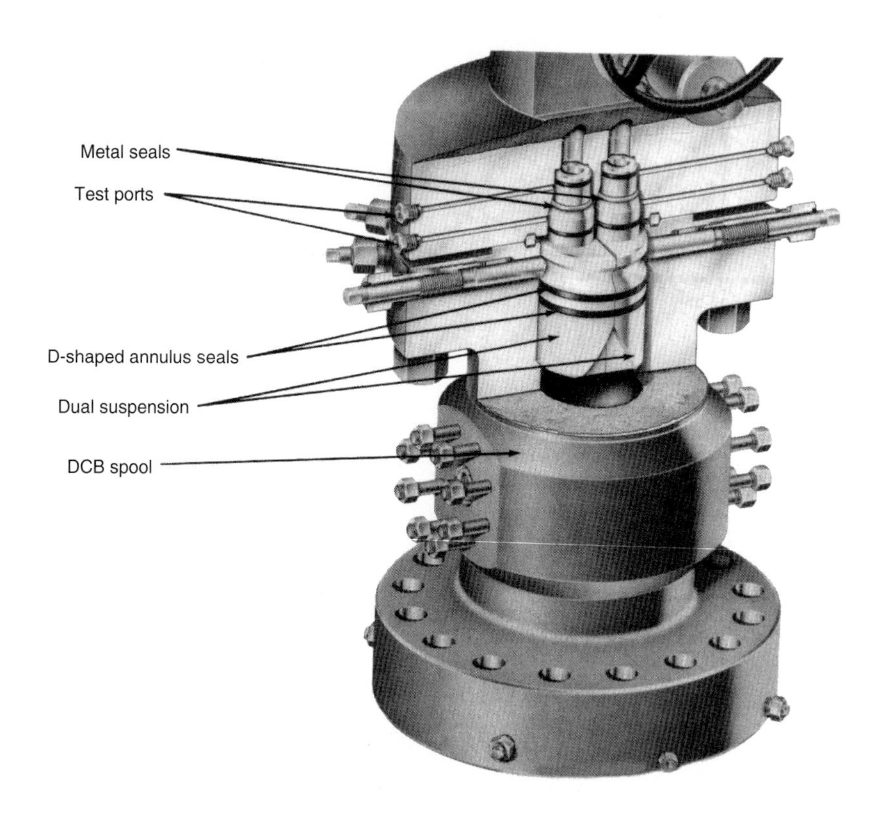

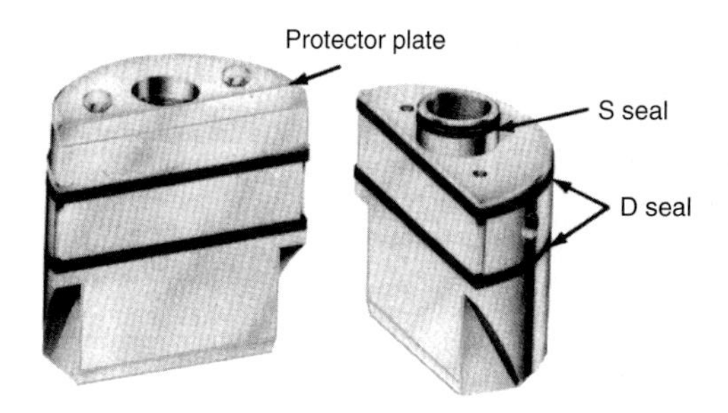

Fig. 6.20
Dual-production tubing hanger and connection *(Source: Cameron Iron Works).*

The BOP stack can then be assembled. Depending on the pressure and the type of fluids contained in the drilled formations, the mud cross may not be needed. The kill and choke lines will be connected to the side outlets of the ram-type preventers (**Fig. 6.21**).

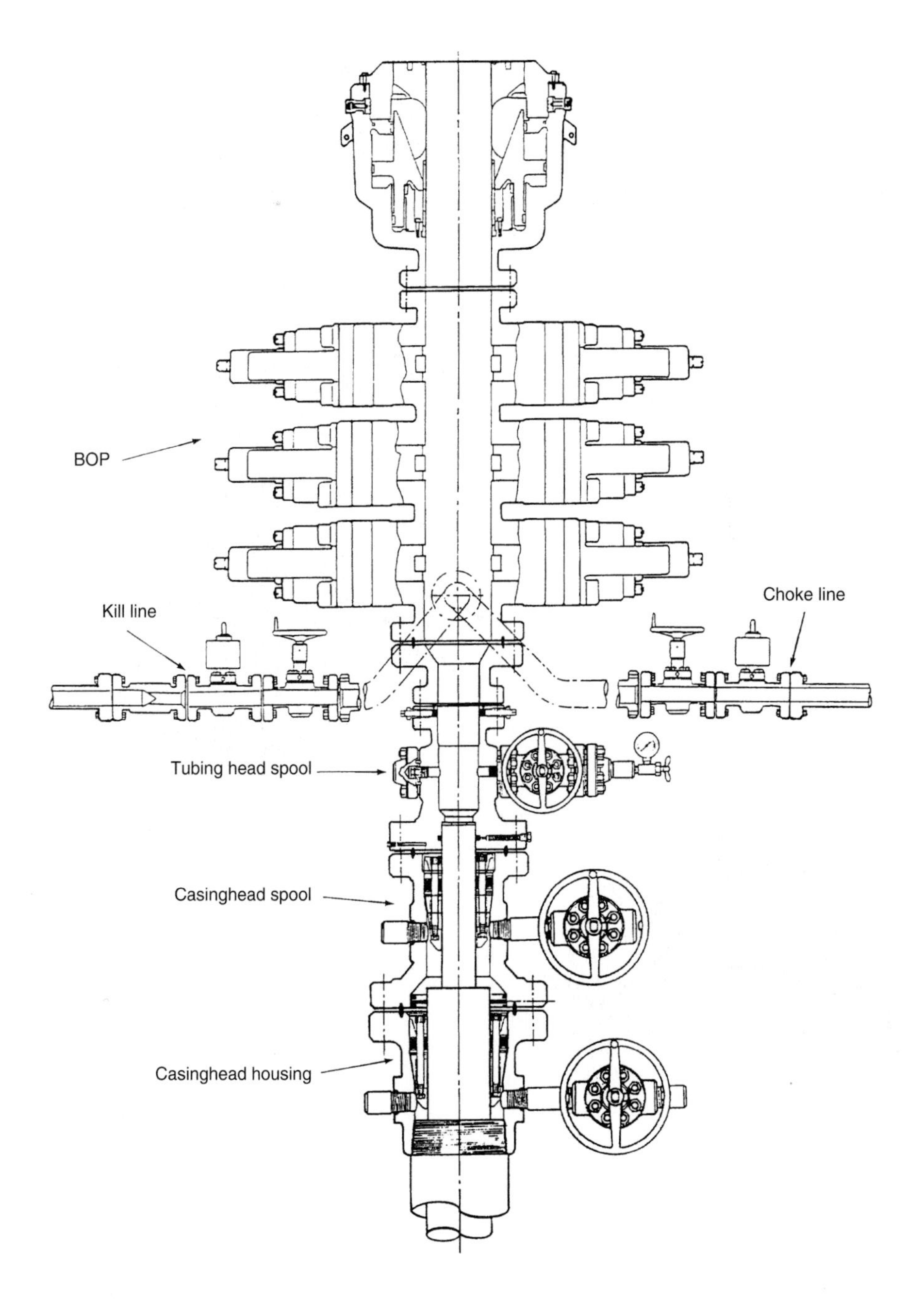

Fig. 6.21

Wellhead for 6" drilling phase *(Source: ENSPM Formation Industrie).*

6.1.4 6" drilling phase

The 7" casing is assumed to have been run in the hole and cemented at the top of the pay zone. The producing layer is drilled and the phase can be completed by running in a 5" liner which is anchored in the 7" casing.

- After the casing has been perforated, the production packer is set.
- The tubing and hanger are run in.
- After the BOPs have been dismantled, an adapter flange fitting on the tubing head and supporting the lower part of the master valves completes the wellhead.

A conventional Christmas tree is shown in **Fig. 6.1**. It has two master valves in the axis of the tubing. A cross channels the effluent laterally through two wing valves. A choke body is set at the branch-off point of the flow line. In the tubing axis above the cross, a swab valve gives access to the inside of the tubing for wireline work or snubbing after a BOP has been screwed to the tree cap.

Dual completions (**Figs. 6.22a and b**) require the same access and safety functions. This makes Christmas tree design more complicated as shown in the figures. There are two examples, one of which is termed solid block as the valves are incorporated in the same block. It is a high-cost solution, but sometimes a necessary one for development wells to gain in height and allow enhanced safety.

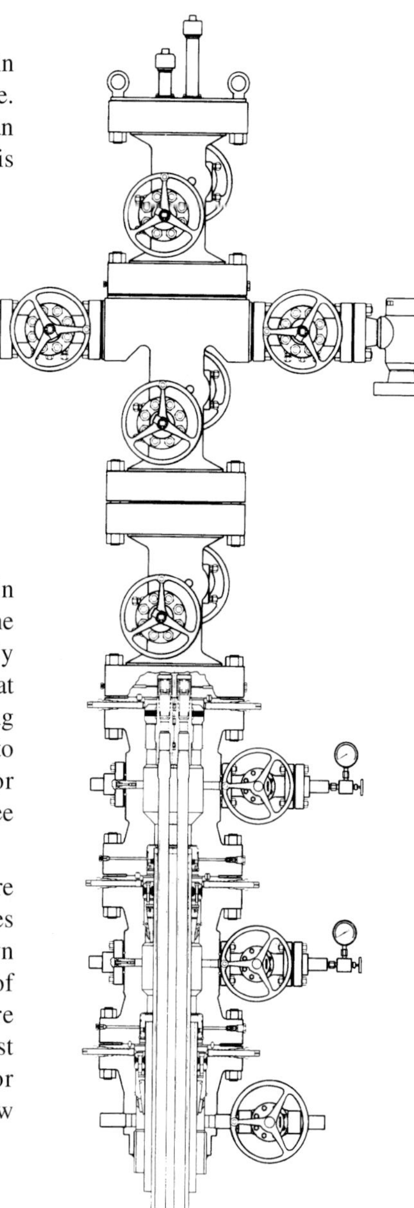

Fig. 6.22a

Dual-completion production wellhead
(Source: Cameron Iron Works).

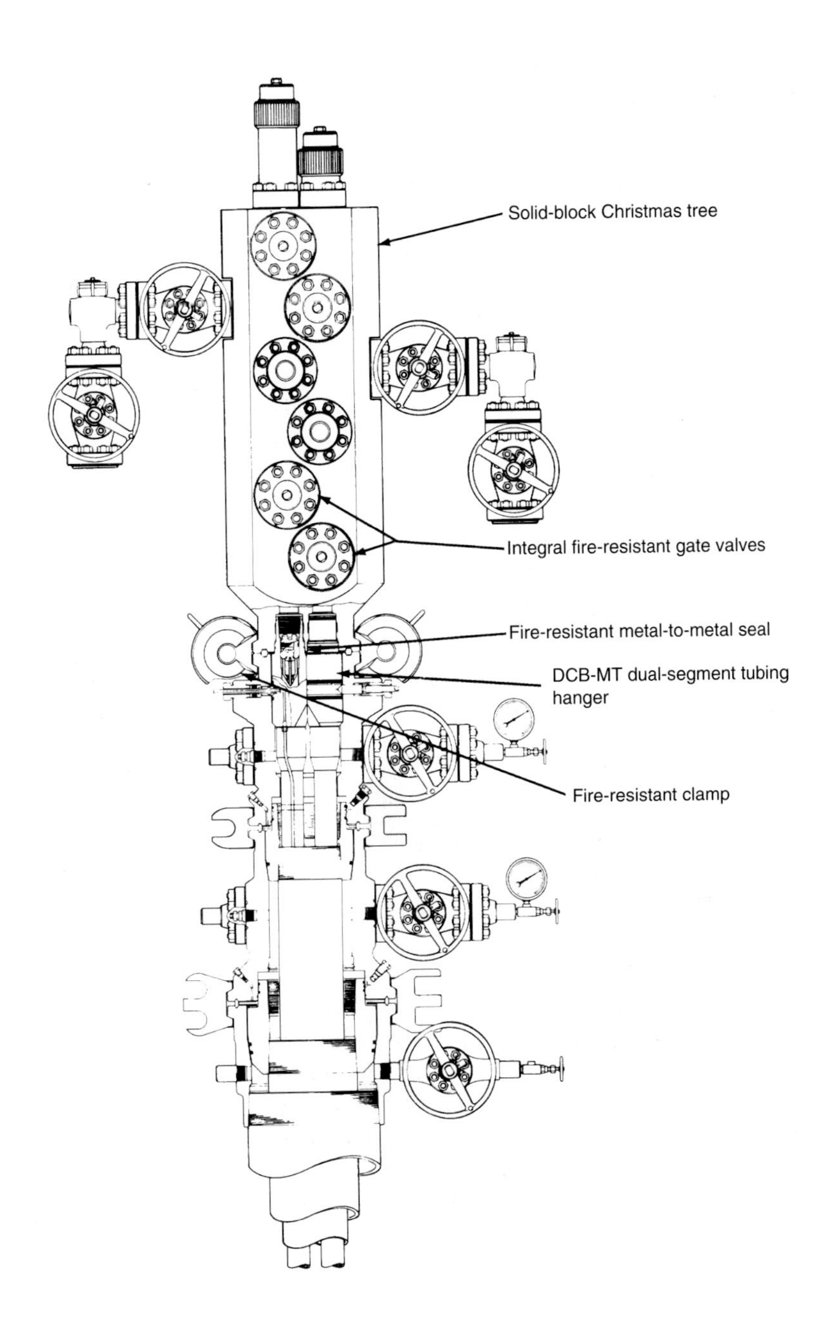

Fig 6.22b

Solid block Christmas tree *(Source: Mac Evoy)*.

6.2 THE SPECIAL CASE OF COMPACT HEADS

All manufacturers propose fairly similar versions of compact heads. The heads are made up of one of two components serving as casinghead housing, casing spool and tubing head. The compact (or three-stage) configuration is screwed or welded onto the surface casing and will then accommodate the tubulars equipped with the corresponding hanger system (*WKM* and *Cameron* examples in **Fig. 6.23**). Compact heads have advantages: height gain and time-saving assembly. They also have a number of drawbacks: precision required in pipe position, sophisticated seals and a large diameter where the highest pressure can be exerted and therefore a more sensitive and higher-cost sealing system.

When there are two stages (**Fig. 6.24**), the casinghead housing is the first stage and the second holds the production string and the tubing. This type of head can first use a low-pressure BOP assembly then a high-pressure assembly but with a smaller nominal dimension, which accordingly decreases the size.

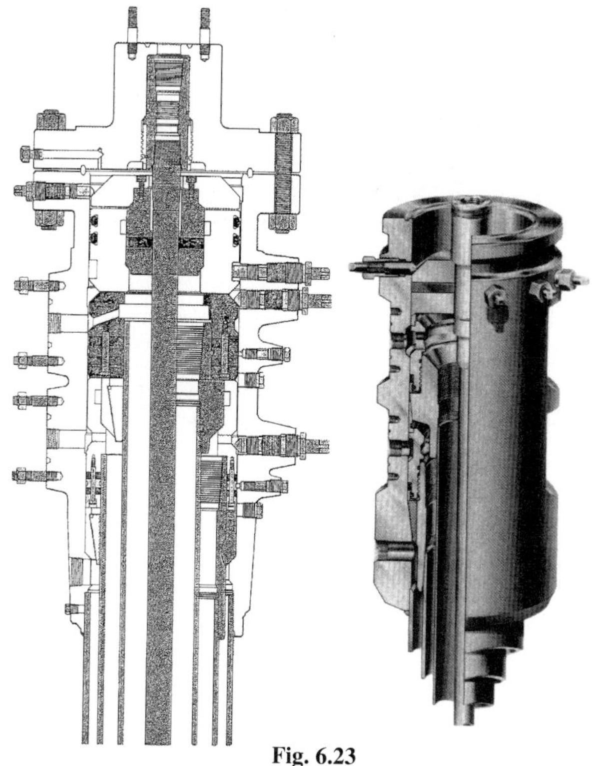

Fig. 6.23

Three-stage compact heads

(Source: Cameron Iron Works and WKM).

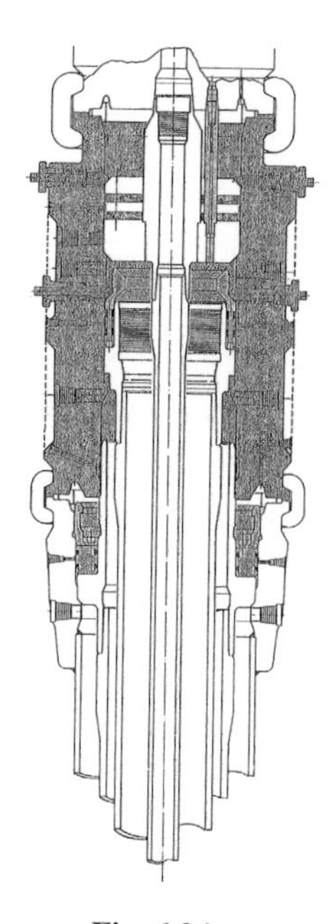

Fig. 6.24

Two-stage compact head

(Source: Vetco Gray).

6.3 BLOWOUT PREVENTERS

6.3.1 General introduction

Blowout preventers and their accessories are designed to:

- Seal off the well when formations are encountered that contain fluids whose pressure is greater than the hydrostatic pressure exerted by the drilling mud.
- Allow circulation so that mud can be treated and its density adjusted according to formation pressure, and so that formation fluids that have entered the wellbore can be circulated out. These operations are carried out under pressure.

A blowout preventer is characterized by:

- the make (the main manufacturers are *Cameron, Shaffer* and *Hydril*),
- the type,
- the nominal size,
- the working pressure.

The last two characteristics give the size of the connecting flanges, or studded ends. The size corresponds to the through-bore diameter of the preventer and to the maximum working pressure.

Chief nominal diameters are: 71/6", 11", 135/8", 163/4", 183/4", 203/4", 211/4", 29" and 30". BOP working pressures have the same names as *API* flanges: 1000; 2000; 3000; 5000; 10,000; 15,000 and 20,000 psi.

The following characteristics are also specified for each preventer:

- The maximum opening diameter or maximum diameter allowing drilling bits to pass through.
- The opening and closing ratios, i.e. the ratio between the pressure prevailing in the well when the preventer is closed (or opened) and the hydraulic pressure required to close (or open) the preventer rams.

For example, the closing ratio of the Cameron U BOP is 7:1, which means that a pressure of 1000 psi has to be exerted on the pistons that operate the rams to close them if the pressure in the well is 7000 psi.

- The volume of fluid required to open or close the BOP.
- The overall dimensions: height, length, width, weight; along with the length or width (depending on the type) when the preventer has been opened to have its rams changed.

6.3.2 Different types of BOPs

6.3.2.1 Ram-type preventers (Figs. 6.25a and b)

- May be of the blind-ram type for full closure. Usually in offshore operations, one of the blind-ram preventers is equipped with blades that can cut through the drill string, called shear rams; or
- Pipe rams or variable rams which close on drillpipe, casing or tubing.

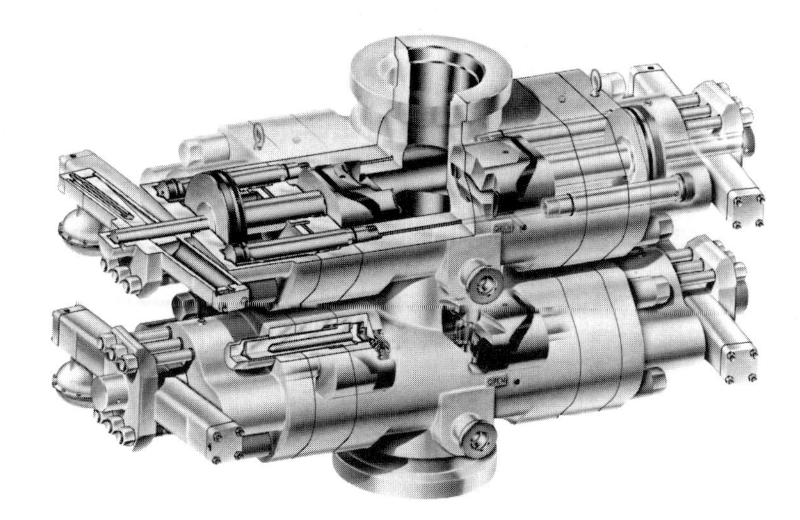

Fig. 6.25a
Double ram-type BOP *(Source: Cameron Iron Works).*

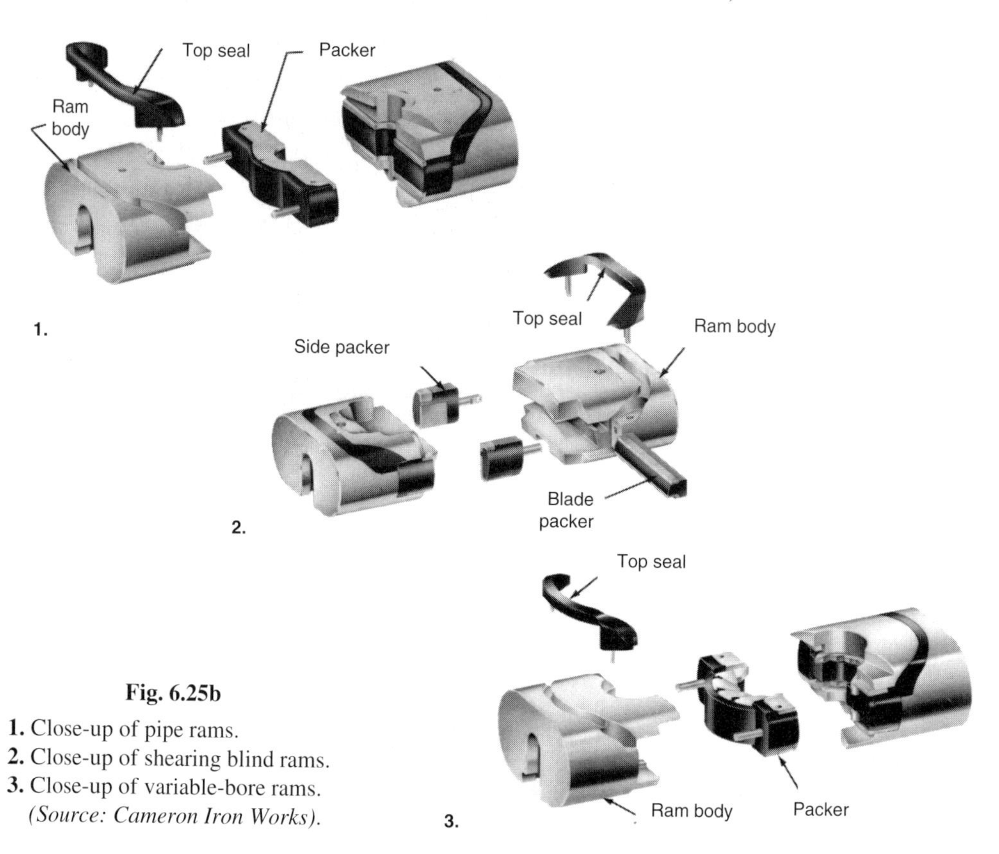

Fig. 6.25b
1. Close-up of pipe rams.
2. Close-up of shearing blind rams.
3. Close-up of variable-bore rams.
(Source: Cameron Iron Works).

6.3.2.2 Bag-type preventers (Fig. 6.26)

These BOPs are also called annular preventers. They can close on any piece of equipment or even on an empty borehole (not recommended). The drill string can be run in or out through the rubber packing element when the well is closed and under pressure (stripping).

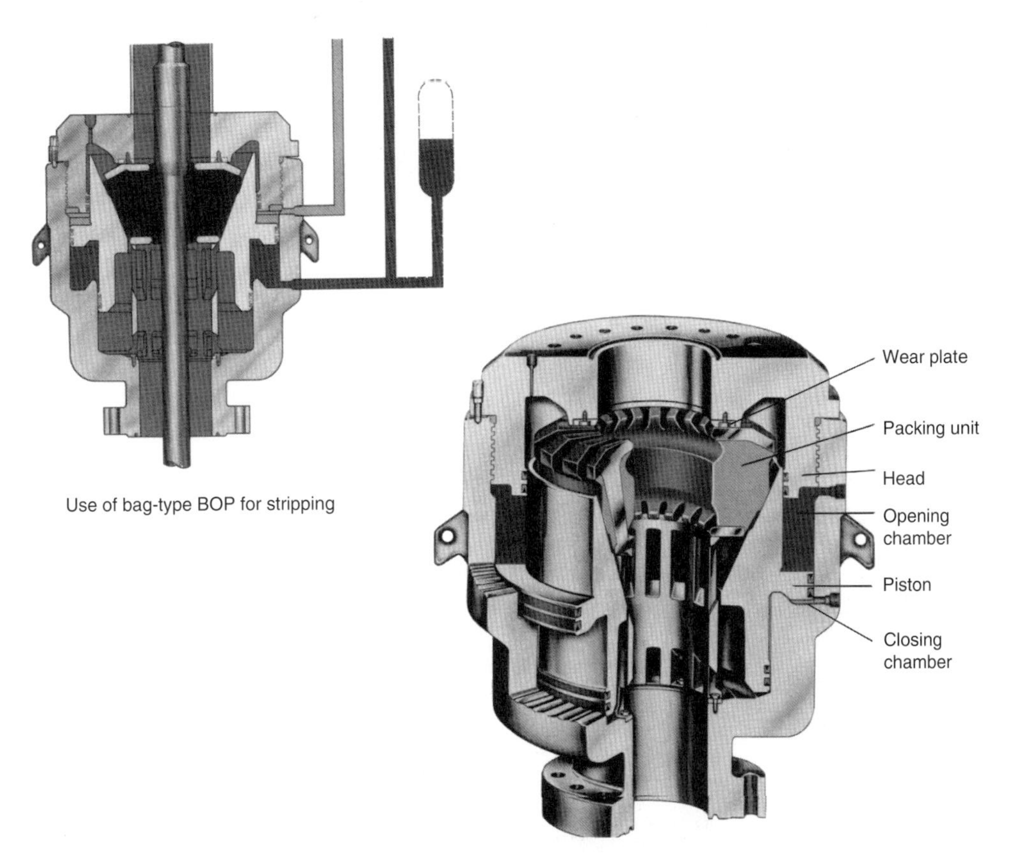

Fig. 6.26
Bag-type BOP *(Source: Hydril)*.

6.3.2.3 Rotating preventers (Fig. 6.27)

This type of preventer allows the drill string to be rotated and run in or out. They are placed above normal preventers and are used for drilling under pressure when low-density mud is required (when increased density would cause lost circulation). They are mainly used for drilling with air or gas as drilling fluid.

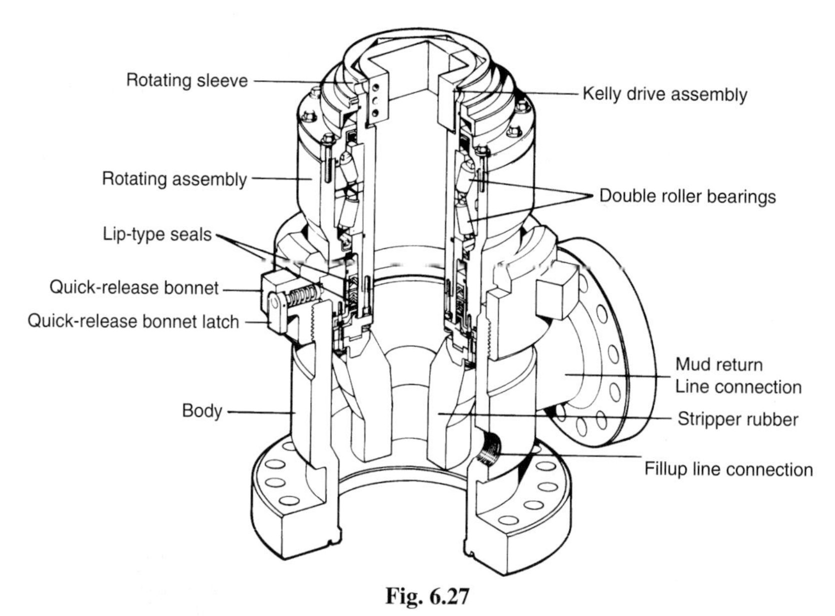

Fig. 6.27
Rotating BOP *(Source: NL Schaffer)*.

6.3.2.4 Inside preventers

These are float valves and back-pressure valves placed inside the drill string:

- Downhole Baker-type valves for drilling with air or with mud in sandy formations (**Fig. 6.28**).

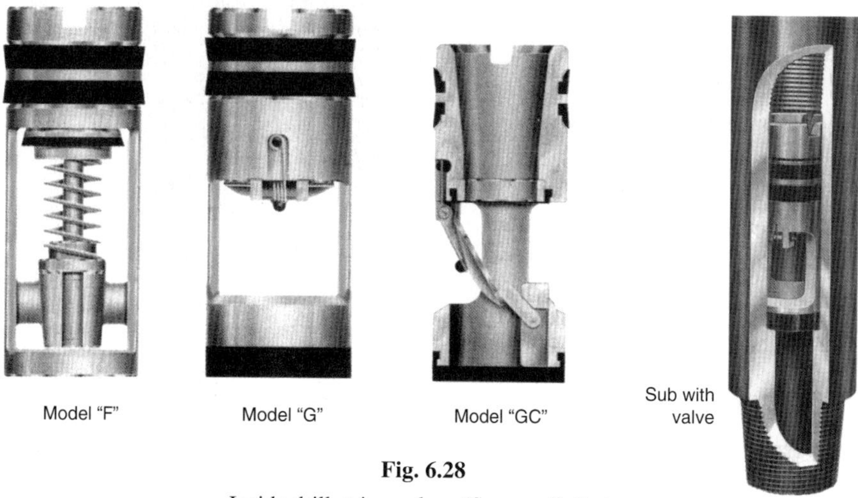

Fig. 6.28
Inside drill string valves *(Source: Baker)*.

- Gray Valve-type valves placed at the top of the drill string in the event of a kick during tripping and allowing the drill string to be run into the closed well under pressure (**Fig. 6.29**).
- Back-pressure valves that are dropped from the surface and pumped to a seat located in a special landing sub placed at the bottom of the drill string (Hydril drop in-type back-pressure valve) (**Fig. 6.30**).

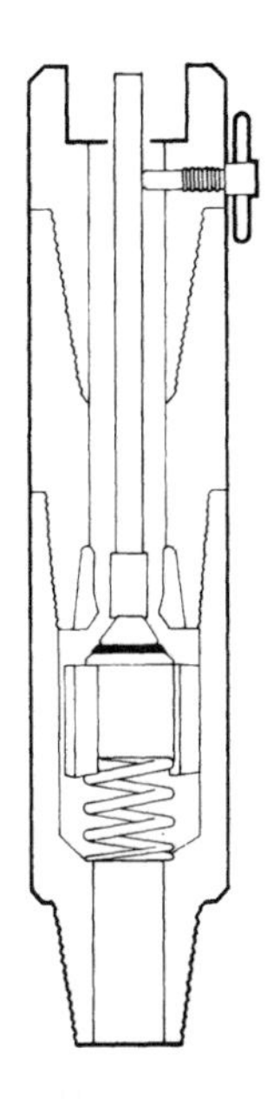

Fig. 6.29
Drill string safety valve.

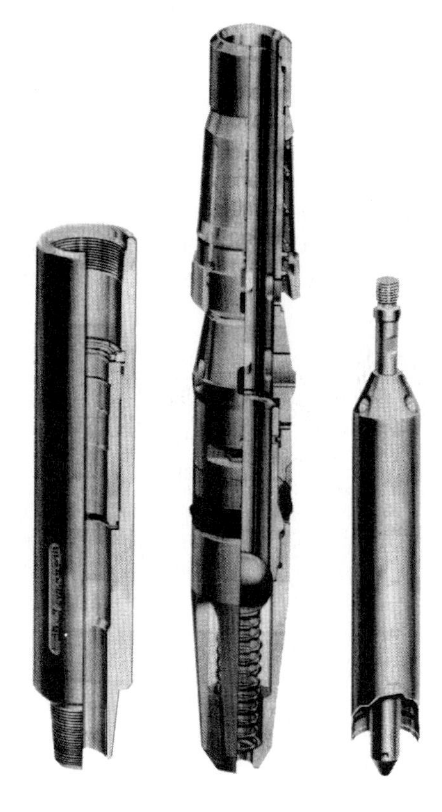

Fig. 6.30
Pumped safety valve *(Source: Hydril).*

6.3.3 Different BOP stack combinations

The *IADC (International Association of Drilling Contractors)* recommends five classes of wellheads depending on the type of duty and the working pressure of the BOPs:

Duty	Working pressure (psi)	Stack
Light	2000	2 ram-type or 1 bag-type (**Fig. 6.31**)
Low pressure Medium pressure	3000 5000	2 ram-type and 1 bag-type (**Fig. 6.32**)
High pressure	10,000	3 ram-type and 1 × 5000 psi bag-type (**Fig. 6.33**)
Very high pressure	15,000	3 ram-type and 1 × 10,000 psi bag-type at least (**Fig. 6.33**)

There is no recommendation on the type and respective position of ram-type BOP internal equipment. The BOP side outlets can be used instead of a mud cross.

A discussion on the position of the different rams is worthwhile to gain some insight into the different configurations and the possibilities available for controlling kicks.

With the annular preventer always placed on the top of the stack, the following describes a medium- and a high-pressure BOP stack consisting of two ram-type BOPs and a spool or mud cross. The ram-type BOPs can be pipe and blind, or shear blind. Combinations can therefore be:

(1)	(2)	(3)	(4)
Blind	Blind	Pipe	Pipe
MC	Pipe	Blind	MC
Pipe	MC	MC	Blind

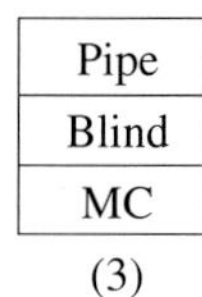

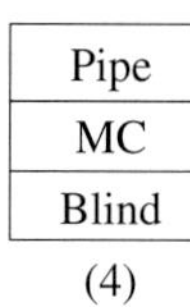

Advantages (in the event of a kick):

- For (1) and (2), the blind rams can be replaced by pipe rams to allow the drill string to be pulled out through the upper rams while the lower rams are kept as back up.
- For (1), if the drillpipe is in the well and a leak occurs at the mud cross (or the kill line or the choke line), the well can be closed with the pipe rams while the repair job is carried out.
- For (2) and (3), when one of the two sets of rams is closed, the MC outlets can be used to control the well.
- For (3) and (4), the well can be closed while the pipe rams are being traded for casing rams.
- For (4), a minimum number of flanges are exposed to pressure when the blind rams are closed.
- For (2), (3) and (4), when the pipe rams are closed, the MC outlets can still be used.

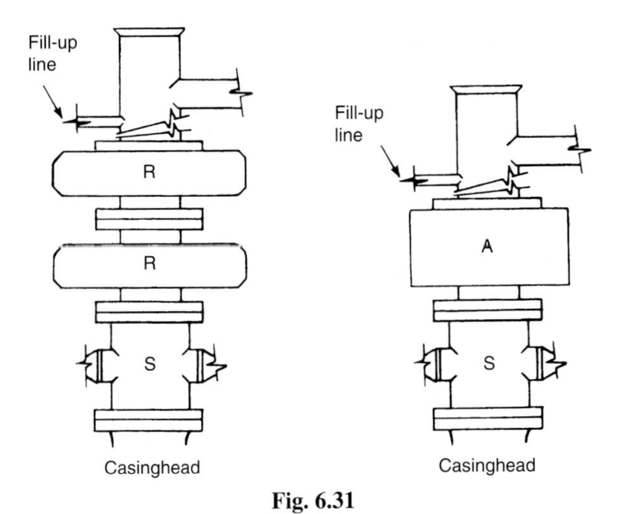

Fig. 6.31

BOP stack for low pressure *(Source: API RP 53).*

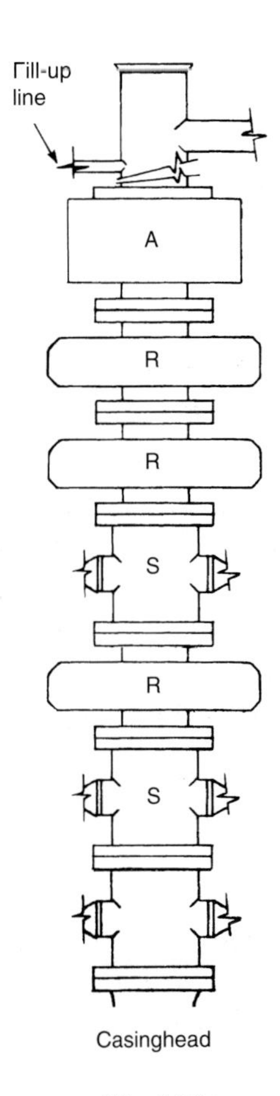

Fig. 6.33

BOP stack for high pressure *(Source: API RP 53).*

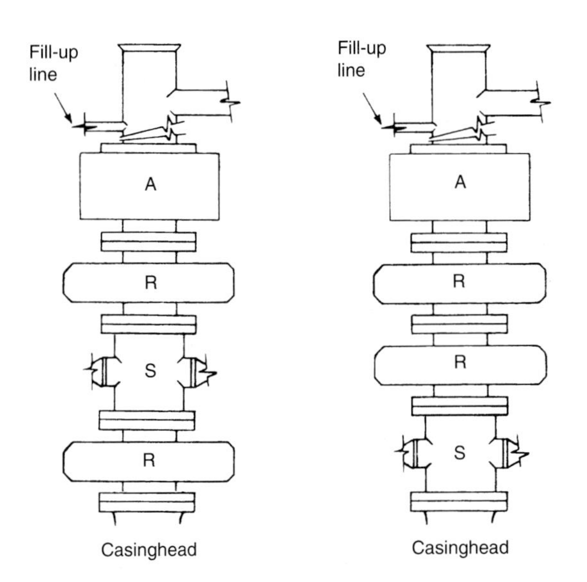

Fig. 6.32

BOP stack for medium pressure *(Source: API RP 53).*

Drawbacks:

- For (1), (2) and (3), if the blind rams are closed, an MC leak can not be controlled.
- For (2) and (3), there are more flanges exposed when the lower rams are closed.
- For (1) and (4), if the lower rams are closed, circulation requires the use of casinghead side outlets.

The most common solutions are the following:

- with two single BOPs:

Blind
Pipe
MC

- with one double BOP and one single BOP:

Pipe		Pipe
Blind	or:	Blind
MC		Pipe
Pipe		MC

Here it can be an advantage to replace the lower pipe rams by variable rams.

If shearing blind rams are used, they are commonly placed above pipe rams where the drill string will hang before it is cut off. Note that the space between the two types of rams must allow a standard tool joint to rest and the cut to be made on the drillpipe body.

6.4 BOP HYDRAULIC CONTROL SYSTEMS

6.4.1 Control principles

All BOPs and the main wellhead valves are hydraulically actuated and operate according to the principle of dual-function hydraulic jacks. Each function requires one line on the opening side and another one on the closing side. Operation is based on having a supply of pressurized fluid available whenever needed for closing or opening the blowout preventers.

The following must be taken into account (for the BOP stack for a given wellhead):

- the volume of fluid required to carry out a given number of emergency open/close functions, with the specifics laid down by the operator,
- the pressure that must be exerted to get a good seal,
- the time required to close all the BOPs.

The system comprises (**Fig. 6.34**):
- several air/oil accumulators,
- a set of hydraulic pumps,
- a direct control manifold,
- one or more remote control manifolds.

Some operators demand that all the wellhead BOPs can be closed (hydraulic pumps stopped) and that there must still be a reserve fluid supply of 50% of the total volume at the residual pressure of 1200 psi.

Others specify a volume of fluid that can open, close and open again all the BOPs. Conditions of course depend on the degree of risk anticipated, whether standard or extreme.

The only accumulators now available on the market have a maximum working pressure of 3000 psi. BOPs operate at 1500 psi except for shear-ram preventers that cut through drillpipe and may require higher pressures depending on pipe strength.

A hydraulic regulator controlled by the driller allows him to adjust the closing pressure of the annular BOP.

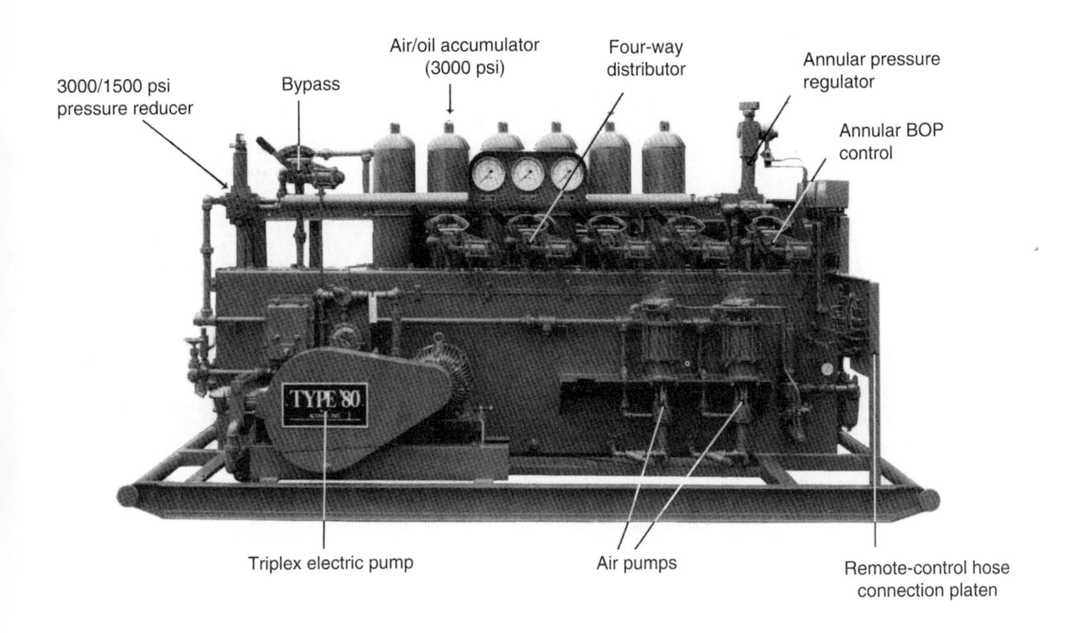

Fig. 6.34

BOP hydraulic control unit *(Source: Koomey Inc.)*.

6.4.2 Hydraulic unit operation

The hydraulic fluid discharged by the pumps to the accumulators compresses a bladder filled with a compressible gas such as nitrogen. There is a pressure equilibrium between gas and the fluid in the accumulator. The gas acts as a spring that expels the oil when the circuit is opened by means of one of the four-way valves that selects the functions. The valves can be operated directly or remote-controlled from the rig floor.

The hydraulic pumps (pneumatic or electric) start and stop automatically according to the accumulator pressure. The main circuit is regulated at 1500 psi, but there is a specialized circuit for the annular BOP where a regulator can adjust the pressure depending on requirements (stripping in particular). The driller can also control this circuit from the rig floor.

Figure 6.35 shows the standard layout for the control unit. It must be located far enough away so that it is outside the first safety perimeter around the wellhead.

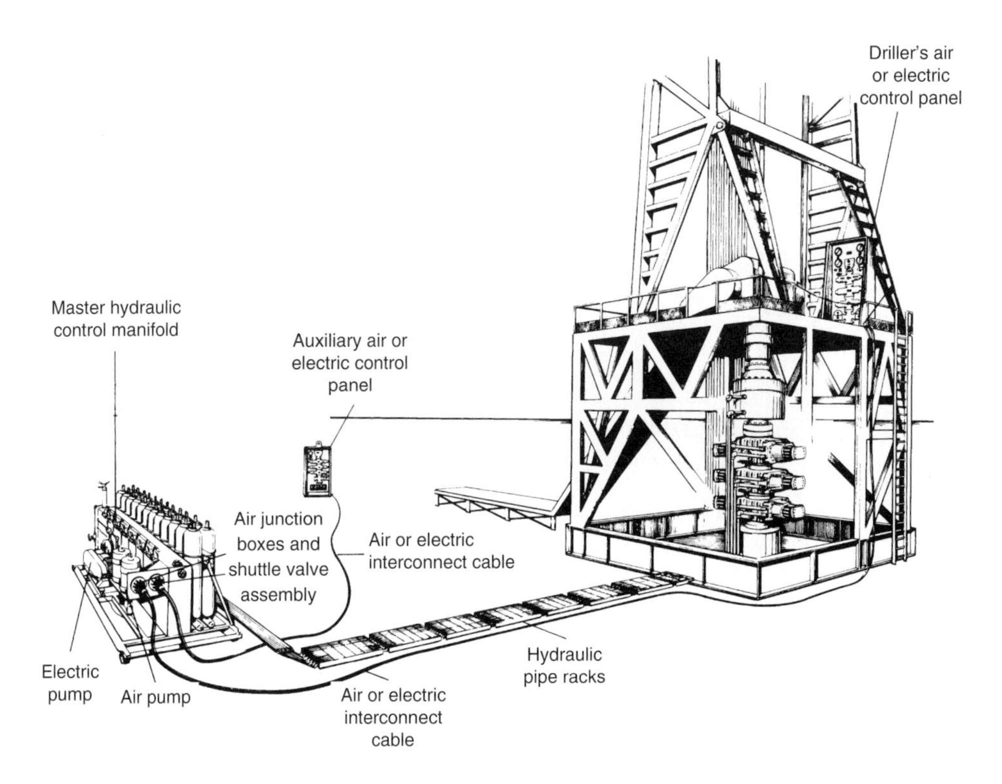

Fig. 6.35

Conventional drilling wellhead control layout *(Source: NL Schaffer)*.

6.5 THE WELL CONTROL CIRCUIT

In order to control a kick, mud of the required density must be added and circulated while back pressure is maintained against the formation. This excess pressure must be slightly higher than the pressure of the fluids contained in the pores of the formation.

There is therefore a need for a line, the choke line, between the annulus and a manifold which directs the effluent to one of the following, depending on the type of fluid involved:

- the mud tanks,
- the degasser,
- the flare,
- the reserve pit (**Fig. 6.36**).

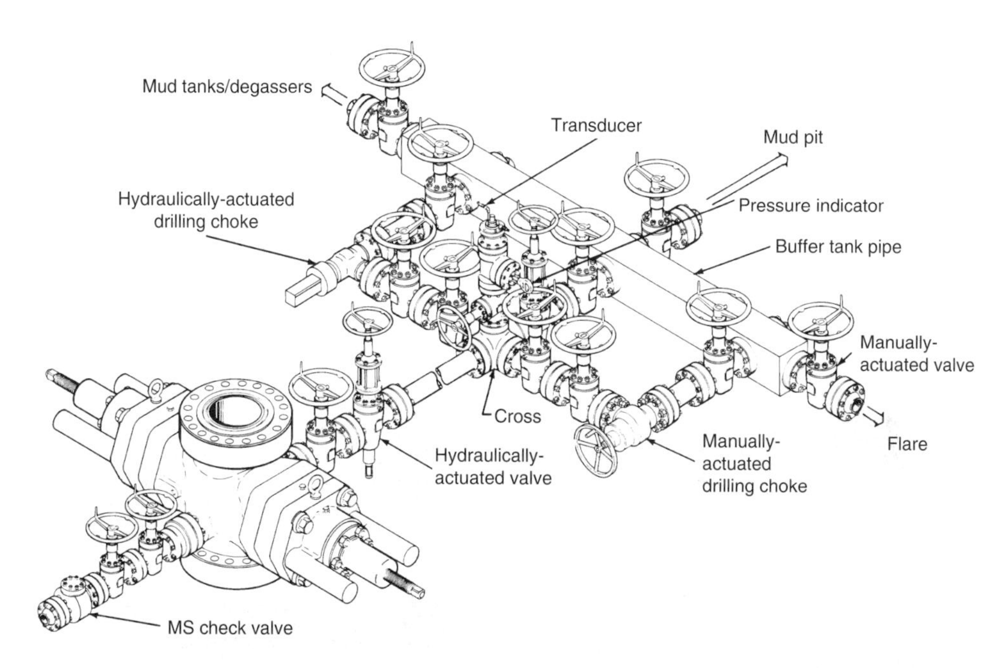

Fig. 6.36

Choke manifold *(Source: Cameron Iron Works)*.

Since flow rates can not be regulated by plug valves, there is a set of chokes at the manifold inlet to adjust the required back pressure. There are several types of chokes:

- the positive choke, a calibrated fixed orifice used in production,
- the adjustable choke, that can be adjusted manually (**Fig. 6.37**),
- the hydraulically-actuated choke (**Fig. 6.38**).

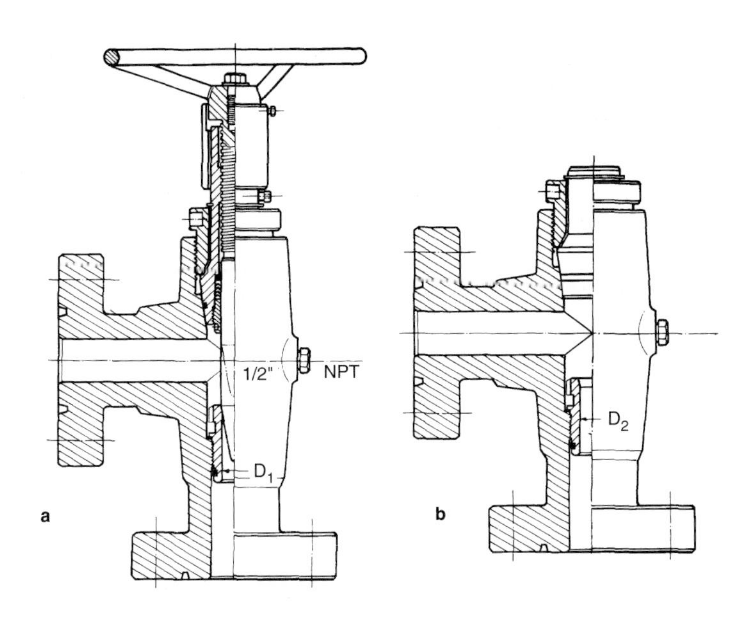

Fig. 6.37
Chokes. **a.** Adjustable choke **b.** Positive choke *(Source: Merip).*

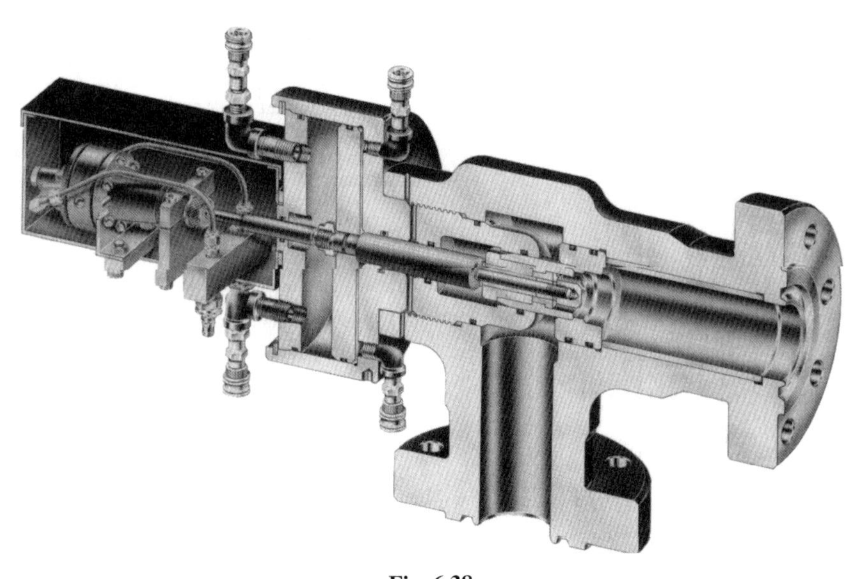

Fig. 6.38
Hydraulically-actuated drilling choke *(Source: Cameron Iron Works).*

It is difficult to control kicks by using the manual choke and the driller is often far from the pressure gages that show the standpipe (discharge) pressure and the casing pressure (in the annulus between the hole and the drillpipe).

There are hydraulically-actuated chokes on the market, remote-controlled from a control panel and console, with a greater or lesser degree of sophistication (**Fig. 6.39**). Conventional systems include:

- a fluid reservoir,
- a pump (with an air-operated motor) supplying hydraulic power,
- one or two control levers for the choke(s),
- two gages giving standpipe and casing pressure,
- a pump stroke counter,
- a pump stroke totalizer,
- a choke position indicator.

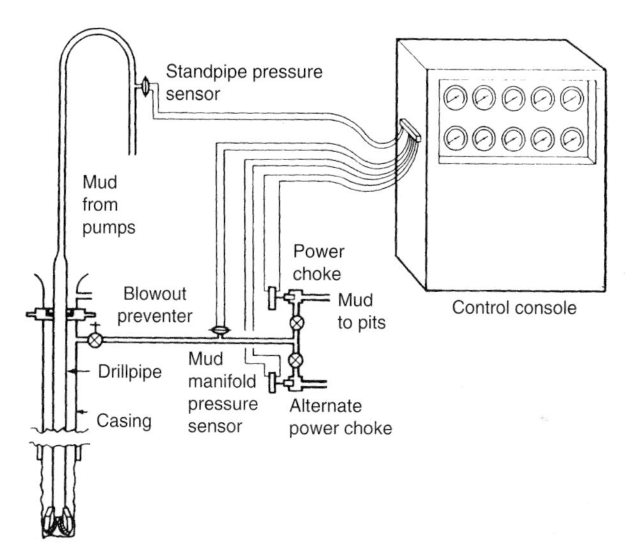

Fig. 6.39

Valve control system by hydraulically-actuated choke *(Source: Cameron Iron Works).*

Chapter 7

CASING AND CEMENTING OPERATIONS

These two operations are indispensable to consolidate a well and are dealt with together in this chapter since the second comes systematically after the first. Generally speaking, no casing is ever run into a borehole without it being cemented in the hole.

Casing and cementing operations have a particular feature in common: any trouble during the operations may have extremely detrimental consequences on further drilling as well as on producing the well later on. This means that the operations must be worked out as carefully as possible in advance. Additionally, considerable resources must often be implemented to cope with even minor trouble immediately.

7.1 THE CASING OPERATION

Casing is the operation of running a string of pipe into a borehole that has been drilled. A special type of tubular —compared to drillpipe— is run in. It is different from drillpipe in its large diameter which leaves only relatively little clearance and its lack of mechanical strength under tensile stresses and torque. It also differs by its equipment, such as scratchers which are centralizers that increase drag. As a result, lowering casing is a tricky operation in many cases, particularly in a directionally drilled borehole. If the casing string gets stuck and does not cover all of the open-hole section, the architecture of the well as a whole is jeopardized along with the program of exploration or production operations.

7.1.1 Preparing the borehole

When the bit has reached the depth anticipated for the casing shoe, the borehole is not usually ready for the casing to be run in.

First, it is advisable to carry out logging operations. Most of them are left up to the geologist, but the drilling engineer can also request measurements that will make his job easier:

- Caliper logging to find out as accurately as possible what the volume is between casing string and borehole wall and work out the amount of cement that will be needed.

The measurement also gives useful information on zones that are calibrated to bit gage diameter where it is preferable to place the centralizers.

- Azimuth and deviation measurements to get a picture of well curvature and spot the depths where sudden changes in hole angle or azimuth may hinder the casing on its way down.
- Maximum temperature measurement needs to be known to determine thickening time for the cement slurry in deep wells.

The logs listed above are automatically run along with certain operations to locate and identify formations because the measurements are needed for interpretation. For instance micrologs and microlaterologs are recorded with a caliper log, and continuous dipmetering with a continuous borehole profile log. Maximum temperature is measured at the same time as the SP-resistivity recording.

Once logging has been completed, the drill string is usually run in before casing is begun. There are two reasons for this:

- to check how well the walls hold up, and
- to circulate and possibly treat the mud.

A borehole that allows the drillpipe, drill collars and bit to pass through freely may cause enormous problems when a string of casing is run in. This is because the pipe is of a larger diameter and much more rigid than the drill string. The borehole is checked out with a rigid string and a one- or two-roller rotary reamer. If there seems to be a lot of reaming work to be done, then it will be necessary to run in several times. The driller must not forget that the more time goes by while these techniques are being applied, the more unstable the walls become.

Simply checking the borehole out with the bit may be enough in the easiest instances, only experience can tell whether this precaution is adequate.

The drilling mud can be treated when the borehole is checked. It is advisable to improve the flow properties of the drilling fluid in the borehole, i.e. decrease the plastic viscosity and the yield value, to run in deep casing strings. This will lessen the surge pressures that may be exerted on the bottom of the hole when casing is lowered and fluid is circulated in a restricted annulus.

7.1.2 Preparing the casing string

Preparation is usually done during the drilling days prior to the casing operation.

The pipe is stored in the pipe yard in layers, one on top of the other (not more than three), so that the normal handling order corresponds to the planned casing string makeup.

As the pipe is brought to the well site, it is identified (grade of steel, thickness, thread) and measured. The pipe is then numbered so that each joint has been unmistakably identified. Characteristics and lengths are written down in a notebook. Then the pipe is sorted and set out in the pipe yard in lowering order. Thread protectors on both ends are removed and the joints are cleaned and greased. Only the protector on the pin end is put back on to protect the thread during handling.

Preparing the casing string also includes fitting the accessories on it (**Fig. 7.1**):

- the shoe and landing collar are screwed on the pipe with a thermosetting resin which sticks the joint together and keeps it from coming unscrewed when the two components are redrilled later on;
- the centralizers are fitted on the bottom of the string, level with the shoe of the previous string and above and below the staged-cementing apparatus;
- the scratchers are fitted on the bottom of the string and possibly a little higher to clean the cake off permeable sections of the borehole.

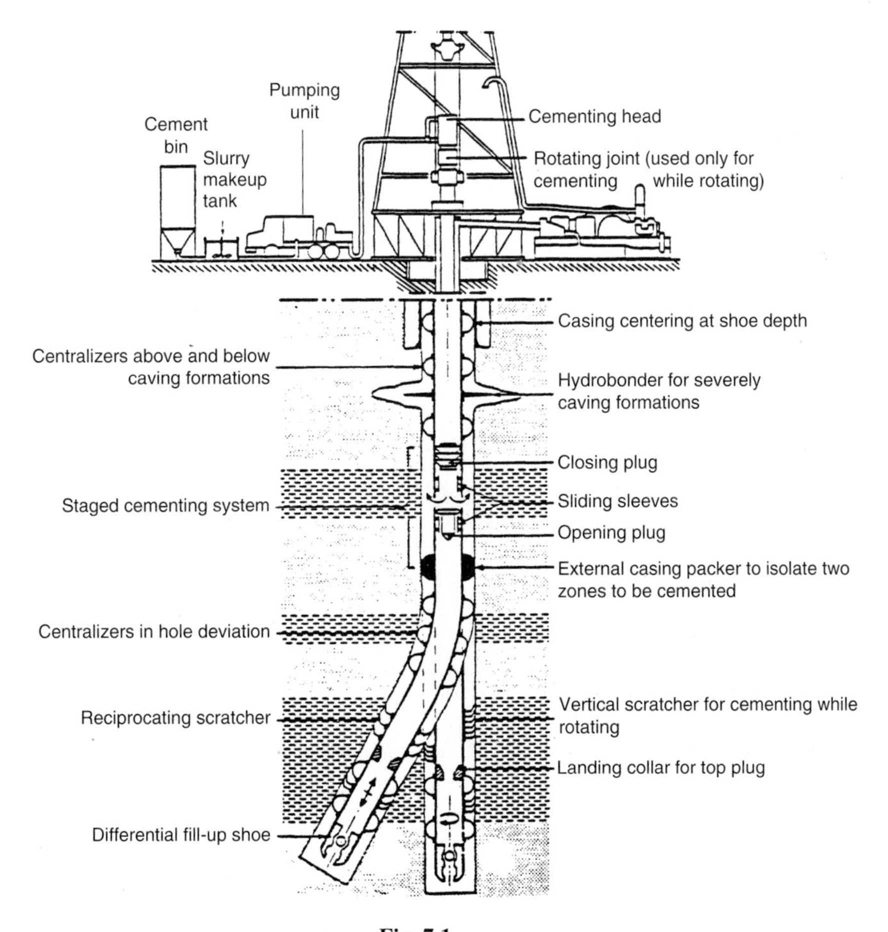

Fig. 7.1
Casing equipment *(Source: Gaz de France)*.

7.1.2.1 Shoes (Fig. 7.2)

Generally rounded, the shoe helps guide and run the casing string in the uncased part of the hole. There are several different types:

- Channel shoes with or without vents that allow the mud to go directly into the casing as it is lowered into the hole.
- Shoes with built-in check valves:
  - advantages: keep the cement slurry from coming back up once it has been flushed out and prevent any blowouts from occurring through the inside of the string while the casing is being lowered into the hole;
  - drawbacks: the casing string must be lowered slowly into the hole to minimize surge pressure on the formations, in addition the string must be filled from the top down, which causes wasted time.
- Shoes with adjustable check valves that allow the string to be filled from the bottom up as long as the check valve system has not been activated by dropping a ball down or by pumping at a given flow rate. These shoes are usually of the automatic fill-up type or the differential fill-up type.

7.1.2.2 Landing Collars (Fig. 7.3)

The purpose of the landing collar is to serve as a seat for the cementing plug(s). Depending on the context or according to the technique of the company, a landing collar with a check valve is chosen when the shoe has none. Most landing collars and shoes both have check valve systems.

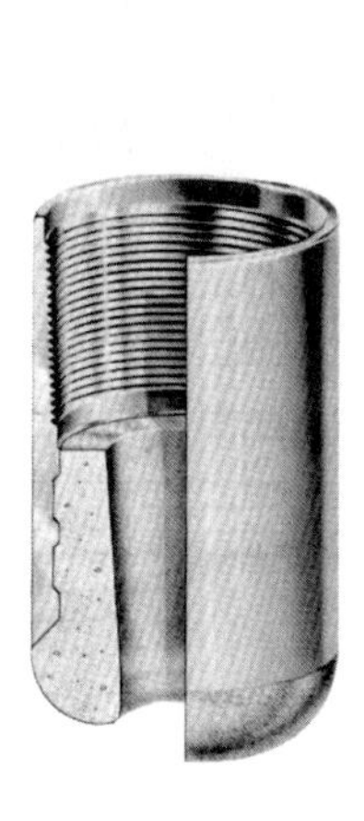

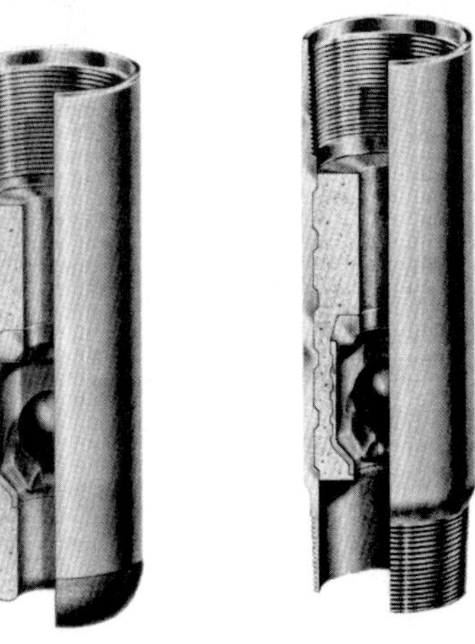

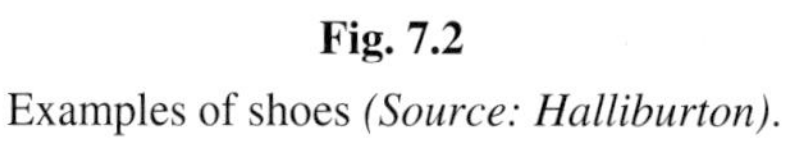

Fig. 7.2
Examples of shoes *(Source: Halliburton)*.

Fig. 7.3
Landing collar
(Source: Halliburton).

The landing collar is always run in two or three joints above the shoe so that the inside volume of the string between the shoe and the landing collar can accommodate the slurry that has been polluted by the upper cementing plug scraping down through the pipe.

7.1.2.3 Centralizers (Fig. 7.4)

Centering the casing string is one of the parameters that governs the success of a cement job. There are centering rules for vertical and directional wells.

According to whether the centralizers are located inside casing or in the uncased part of the hole, they will be rigid or flexible:

- rigid or positive centralizers (with U-shaped blades) are designed for casing-casing annular spaces,
- flexible or spring-bow centralizers are used to center casing strings in the uncased section of the hole. There are two sorts of flexible centralizers: straight ones and spiraled ones.

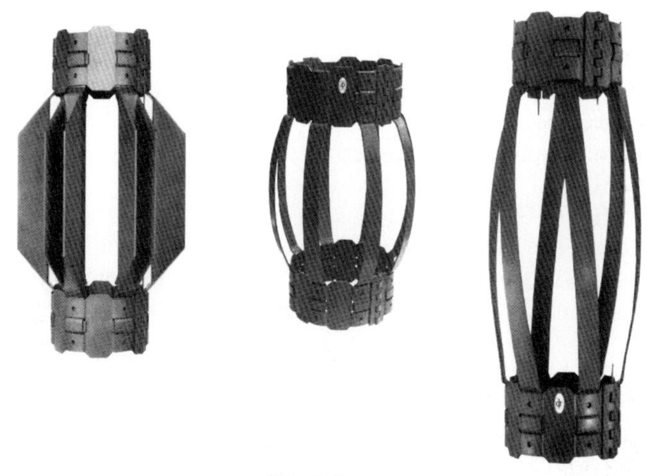

Fig. 7.4
Casing centralizers *(Source: Weatherford)*.

7.1.2.4 Scratchers (Fig. 7.5)

Scratchers are designed to break down the mud cake mechanically and promote a better bond between the cement and the formation. They are particularly recommended on permeability barriers when pay zones need to be sealed off. On porous formations with fairly high permeability, getting rid of the cake with scratchers does not automatically guarantee a good cement job. Other factors can have a detrimental effect on the slurry quality (filtration, dehydration, etc.).

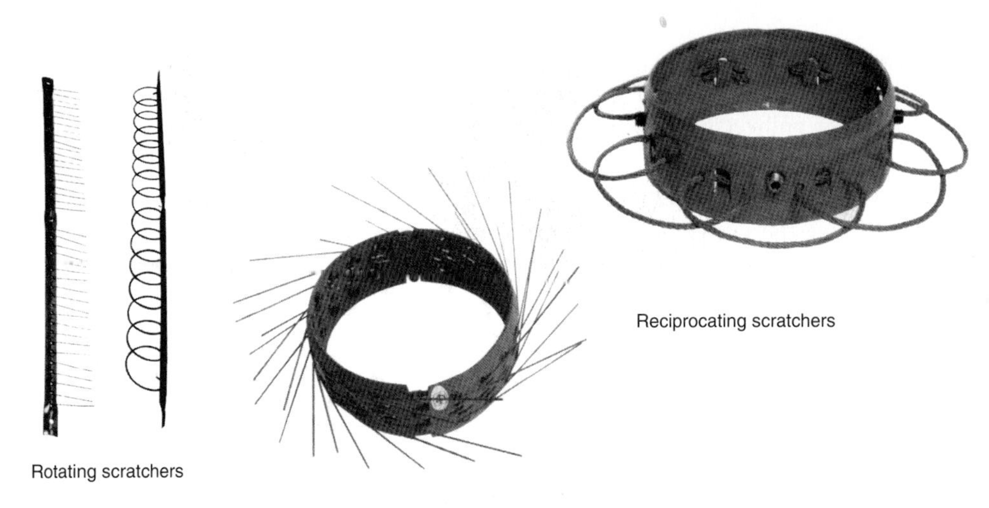

Fig. 7.5
Scratchers *(Source: Halliburton).*

Scratchers are chosen according to the way the casing string will be moved during cementing:

- rotation: rotating scratcher,
- up and down: reciprocating scratcher.

A. *Rotating scratchers*

These scratchers are longitudinal and attached along the axis of the casing. They are metal bars covered with steel spines (scratchers) or equipped with a wire that forms spirals (wipers). They are either welded (when the grade of metal allows) or clamped between two stop rings.

B. *Reciprocating scratchers*

The reciprocating version is ring-shaped and is attached around the circumference of the casing either with a self-locking mechanism or between stop rings.

7.1.2.5 Hydrobonders (Fig. 7.6)

By modifying the fluid flow path, hydrobonders improve the displacement efficiency of the drilling fluid by the cement slurry and promote a better bond between the cement and the formation. If the borehole has caving walls, hydrobonders can be placed in the casing at the depth of the enlarged part of the hole. They consist of rubber disks with a diameter larger than the nominal wellbore diameter. The disks have spiraling grooves that force the slurry to circulate toward the back of the enlarged sections of the hole and fill up the cavities better. It is recommended to reverse the direction of the spirals of two subsequent hydrobonder collars if they are close to each other to get maximum alteration in the fluid flow.

Fig. 7.6
Hydrobonder
(Source: Halliburton).

Fig. 7.7
Cementing basket
(Source: Halliburton).

7.1.2.6 Cementing baskets (Fig. 7.7)

This equipment is generally used to minimize losses of cement slurry in weak zones. The baskets are used only in shallow to medium depths, since they act in a purely mechanical way. They do not prevent pressure transmission but do curb fluid passage considerably. There are two types:

- the conventional model in the shape of an umbrella which is not recommended for the uncased part of the hole,
- the centralizer-basket in a one-piece apparatus, with a receptacle protected by an outside centralizer, that can be used in the uncased part of the hole as well as to move the casing during cementing.

7.1.2.7 Cementing heads (Fig. 7.8)

A cementing head is screwed onto the top of the casing and holds the two cementing plugs. It allows the drilling mud to circulate, the cement slurry to be pumped in after the bottom plug has been released and the top plug to be pumped down with more drilling mud.

7.1.2.8 Cementing plugs (Fig. 7.9)

The plugs are mainly designed to separate the different fluids (mud, intermediate fluid, slurry) physically while they are being pumped through the casing. The plugs keep the fluids from mixing for as long as possible and lessen contamination risks. There are two types of plugs:

- The bottom plug has a membrane that bursts when a slight overpressure is exerted. Though optional, it is strongly recommended. Besides separating the fluids, the bottom plug scrapes the walls as it moves down through the casing. As a result, it helps keep the slurry behind it from becoming overly contaminated.
- The top plug is designed to provide a seal and to withstand high pressures. It is mandatory and pumped in on the tail end of the slurry. When the slurry has been flushed out the plug comes to rest on the bottom plug (or on the landing collar) and then allows pressure testing to be performed on the casing string.

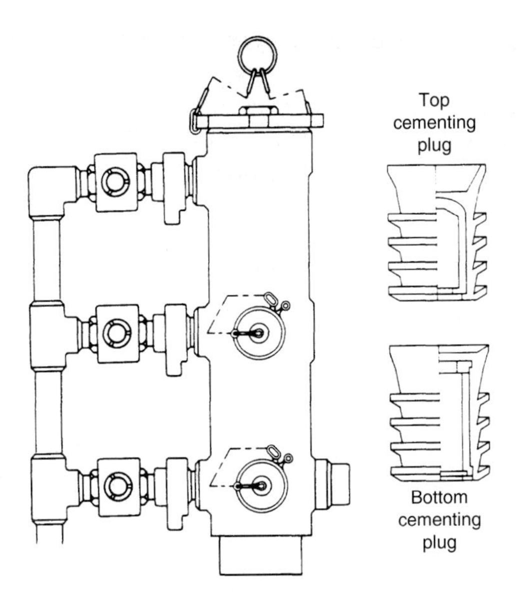

Fig. 7.8
Double-plug cementing head
(Source: Dowell Schlumberger).

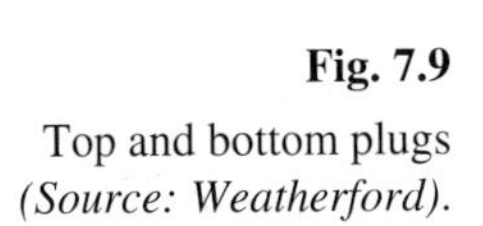

Fig. 7.9
Top and bottom plugs
(Source: Weatherford).

7.1.2.9 Primary cementing equipment with a subsea wellhead (Fig. 7.10)

The subsea plug assembly is connected onto the running tool of the string in the subsea wellhead. The plugs are released by a ball that is dropped down the drillpipe to seal the bottom plug or by another top plug pumped through the drillpipe to the top plug.

7.1.2.10 Specific equipment for two-stage cementing (Fig. 7.11)

Primary cementing equipment is standard, but for second-stage cementing, a stage float collar or DV (diverter valve) must also be incorporated in the string. The DV serves as a bypass between the inside of the casing and the annulus so that the cement can be circulated and flushed through the annulus at the selected depth.

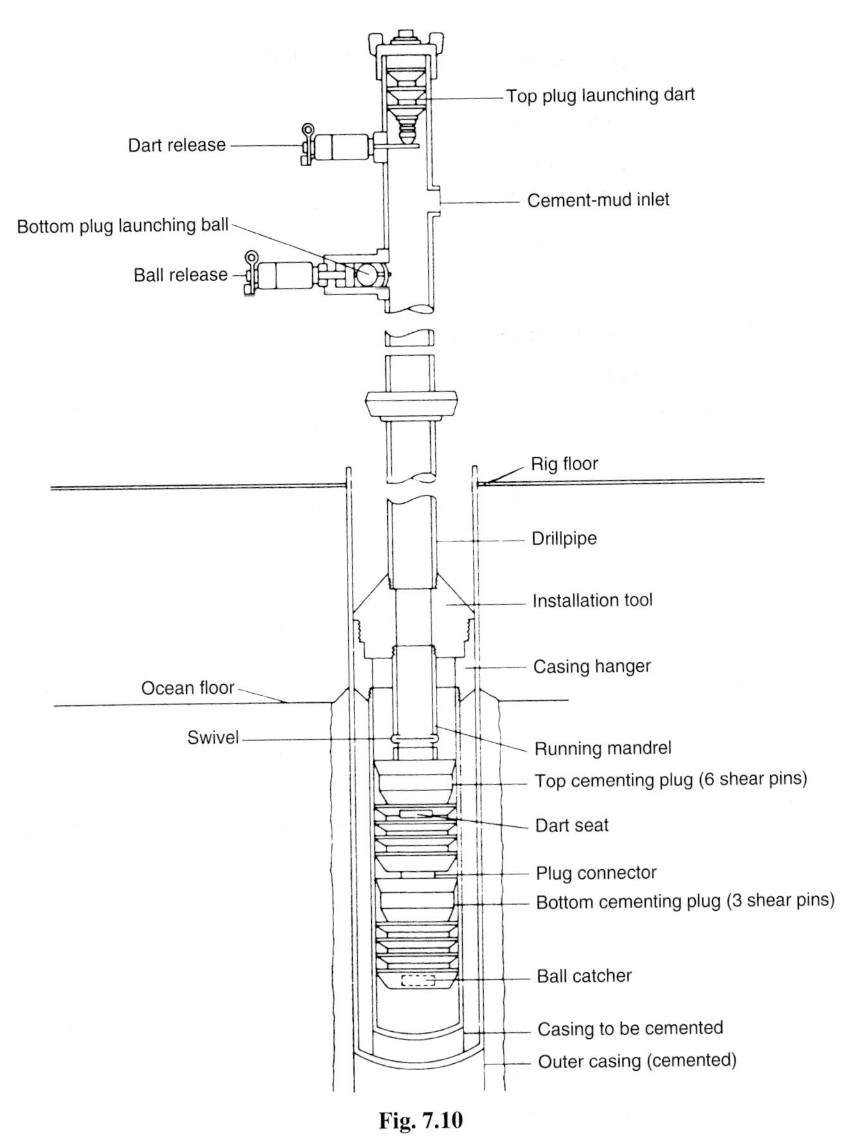

Fig. 7.10

Principle of cementing with a subsea wellhead *(Source: BJ Hughes).*

When primary cementing has been completed, the DV is opened by pressure applied on the bomb that plugs off the lower sleeve. The second stage can then be cemented. Then a closing plug is released to move the upper sleeve downward at the end of the flushing phase, that closes the ports.

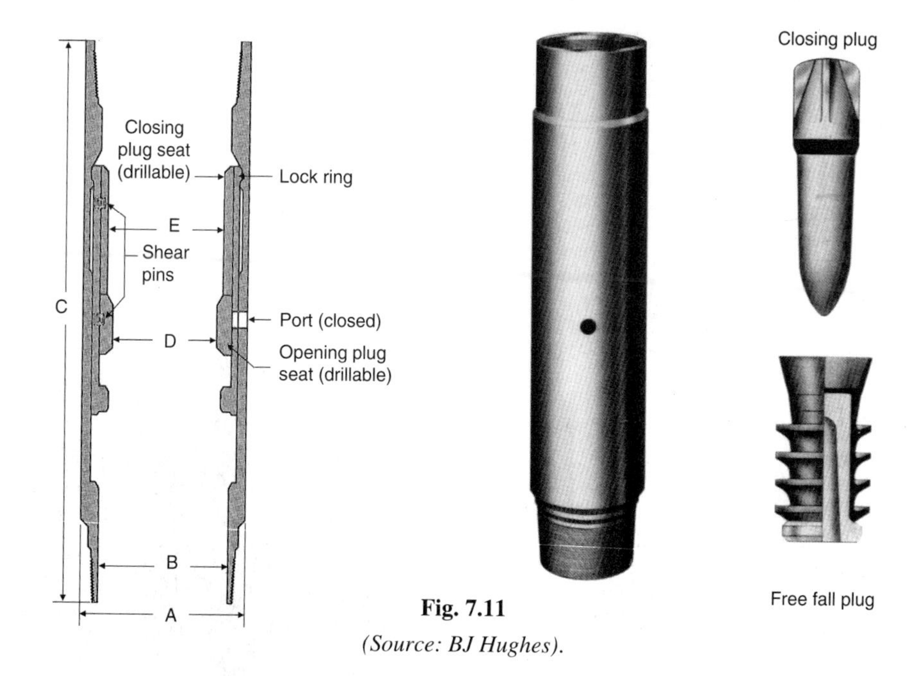

Fig. 7.11

(Source: BJ Hughes).

7.1.2.11 Liner cementing equipment (Fig. 7.12)

Liners are equipped the same as for primary cementing with a subsea wellhead, but they are hung in the preceding string by what is called a liner hanger. There are two types of liner hangers:

- mechanical: (J slot) a longitudinal rotating movement unlocks the locking wedges and makes them grip inside the casing,
- hydraulic: releasing a ball and exerting hydraulic pressure move a sleeve that activates the gripping process.

7.1.3 Lowering the casing

The casing must be lowered as fast as possible, since the time spent is nonproductive the same as any running in or pulling out operation. The casing string lowering speed must be controlled according to the surge pressures generated on the bottom and the walls of the hole. The work needs to be well organized, since any stopping on the way down because trouble has arisen may well cause the casing to get stuck at that depth.

The handling principles are the same for casing and drillpipe, but the equipment is adapted to casing diameters and to the lower collapse strength. A spider and a casing elevator are often used (**Fig. 7.13**).

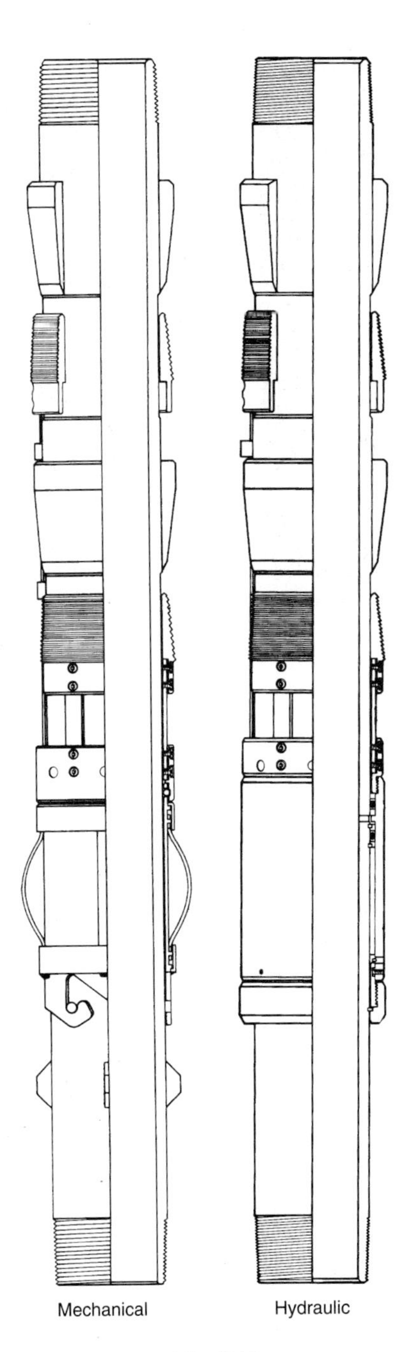

Fig. 7.12
Liner hangers *(Source: Baker)*.

Fig. 7.13
Casing table and hoisting spider *(Source: BJ Hughes)*.

The lengths of casing are screwed together by means of hydraulic tongs (**Fig. 7.14**). A service company is often used for this difficult operation.

The driller monitors the casing string as it is being filled if need be and keeps track of the hydrostatic level in the well. Once the casing has been run in, the mud can be reconditioned and circulated while the string is being maneuvered to put the scratchers to work. Circulation will be stopped only when:

- the mud returns no longer contain cuttings,
- the gas content is low and constant,
- there are no lost returns or kicks,
- the total volume of circulating mud is homogeneous.

As the casing string is equipped with the cementing head and plugs, the cementing operation itself can begin.

Fig. 7.14
Casing tongs *(Source: Weatherford).*

7.2 THE CEMENTING OPERATION

Cementing operations consist in placing an appropriate cement slurry in the annulus between the walls of the hole and the casing that has been run in.

There are several types of cementing jobs and each one meets a particular need:

- Cementing casing strings for a number of purposes:
  - isolating a producing formation from adjacent beds,
  - securing the casing mechanically to the borehole walls,
  - protecting casing from corrosion by fluids contained in the beds that have been drilled,
  - providing a leak-proof base for safety and control equipment that is installed on the wellhead.
- Cementing under pressure, called squeeze cementing, in cased and perforated boreholes, also for a number of purposes:
  - injecting extra cement through the perforations in the casing to consolidate or repair the primary cementing job,
  - sealing off a depleted productive layer,

  - isolating a bed from adjacent zones to reduce the per cent of water or gas in oil production.
- Placing cement plugs in an open hole during drilling in order to:
  - seal off water influxes,
  - plug up lost circulation zones,
  - serve as the basis of a side track,
  - comply with well abandonment procedures.

7.2.1 Principles of cementing methods

The aim is to force a cement slurry into the annulus between the outside of the casing that has been landed in the well and the walls of the borehole. The slurry is pumped directly down the casing that is going to be cemented or through drillpipe. It is then pushed up in the annulus between the casing and the hole to a predetermined height.

The slurry is usually mixed continuously on the surface by two pressurized water jet mixers that wet the cement in powder form and convey it to a small surge tank where the slurry density is checked constantly. It is then sucked up by high-pressure piston pumps and pumped into the well.

The slurry density is adjusted by varying the flow rate of the water that comes into the system downstream from the point where the dry cement and the mixing water meet.

Dry cement is fed into the system by gravity from a silo, the cement bin. Modern-day equipment has forced-air feed with the cement piped to where it comes into contact with the mixing water. This setup is typical offshore.

7.2.1.1 Primary cementing

Cement slurry is pumped into the well, flows through the shoe and then starts flowing up through the annulus. The landing collar, as its name indicates, serves as a shoulder for the bottom and top wiper plugs that respectively precede and follow the slug of slurry in the casing (**Fig. 7.15**).

A pressure surge ruptures the bottom plug and allows the slurry to pass through and circulate up into the annulus. It is the slurry that displaces the drilling mud directly while at the same time “washing” the borehole wall and the outside wall of the casing as it flows by. When all the slurry has been pumped down, the top plug is released and is pushed along down by the drilling mud. This operation is called flushing and the flush volume is the volume of drilling mud between the landing collar and the cementing head. When the flush has been completed, a pressure increase is seen as the top plug bumps against the landing collar.

The overpressure can be held steady for a few minutes in order to test the seal on the casing string at the same time.

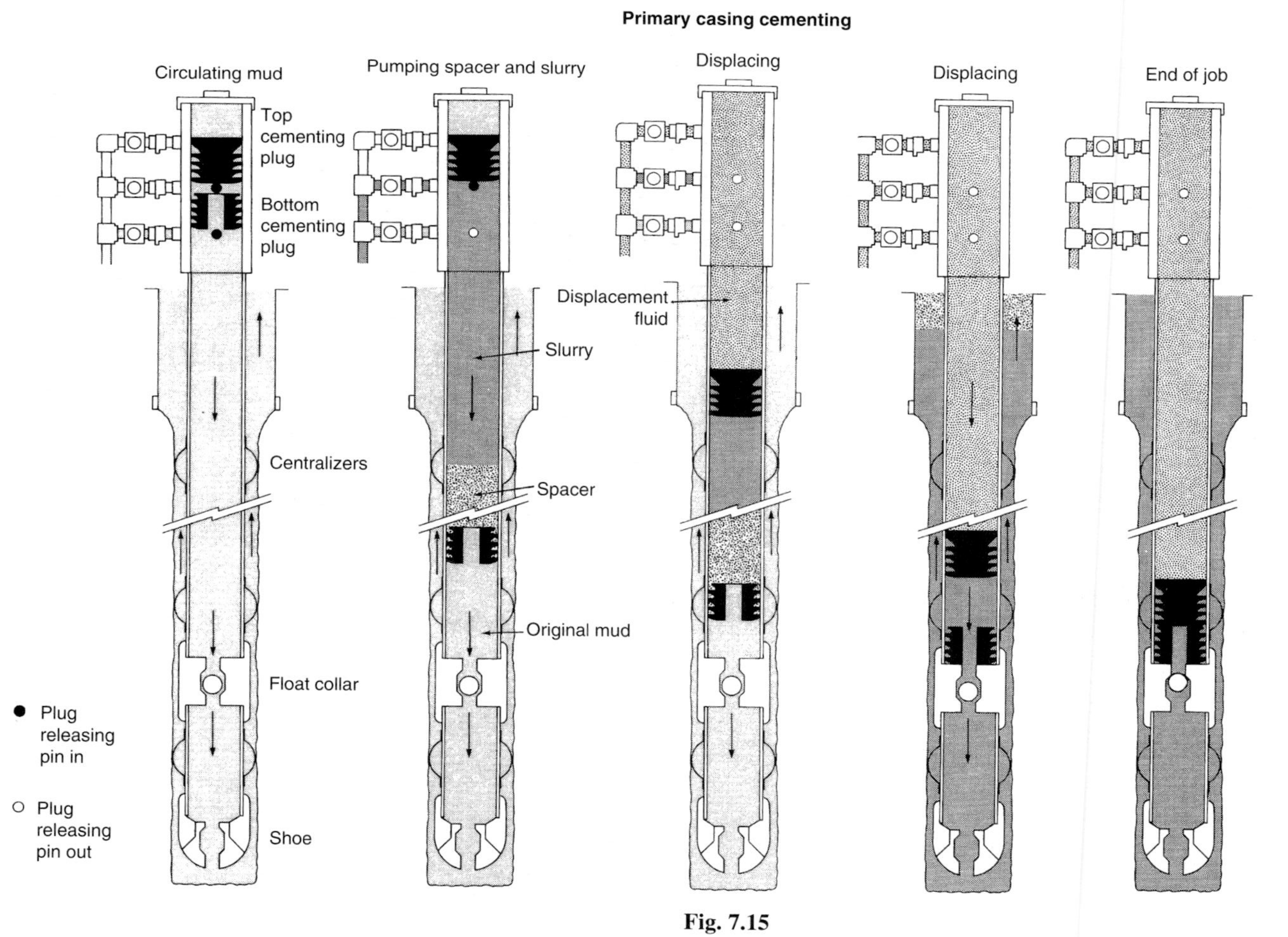

Fig. 7.15
Principles of primary cementing procedure (*Source: Dowell Schlumberger*).

7.2.1.2 Stage cementing (Fig. 7.16)

The string is equipped with a DV at the predetermined depth. A primary cement job is done in the routine way with, however, the diameter of the cementing plugs selected so they can pass through the cross-section narrowed down by the DV. After the pressure surge, the bomb is dropped (50 to 60 m/min depending on the deviation). The opening pressure (around 10 MPa) shears pins and moves the sleeve so that the slurry can then be pumped in, but no bottom plug is used. Once the slurry has been pumped in, the top plug is released and flushed down to the DV. The top plug then closes the DV by moving a second sleeve.

7.2.1.3 Cementing liners (Fig. 7.17)

A liner goes up to a limited height inside the preceding casing, not all the way up to the wellhead. It is cemented with an overlap inside the last casing.

A liner is cemented for two reasons:

- to bond it mechanically to the preceding string,
- to ensure a good seal between the liner and the string.

The best odds of getting a good cement job for the top of the liner dictate an overlap of 80 to 150 meters depending on the diameters. In addition, the liner is often the last tubular set in the well. It is cemented at the depth of producing formations and then perforated to bring the well on stream selectively. It is therefore important to have a high-quality cement job so that the different levels of the reservoir can be isolated and unwanted influxes (water, gas) can be minimized.

A production casing may be connected to the top of the liner by a tie-back. Liners fulfill the functions of conventional casing, but are used instead of a complete string:

- to reduce costs of: pipe, running time, wellhead components, small-diameter drill string (combination string),
- to use a larger tubing diameter,
- when the wellhead does not allow a given-diameter string to be anchored.

Depending on how liners are set, a distinction is made between two types:

- **Liner set on the bottom.** This method is used less and less as it affords little safety. It means there is a limit on the length of the string that is set under compressive stress.
- **Suspended liner.** The trend was to run in longer and longer casing with increasingly large diameters. This led to the use of liners hung in the casing already in place in the well.

Circulation equipment includes only one plug attached to the top of the liner on the extension of the running tool. An initial plug adapted to the drillpipe diameter displaces the slurry in the drillpipe. When this plug bumps the liner plug and fits into it, the two together flush the slurry in the liner to the landing collar.

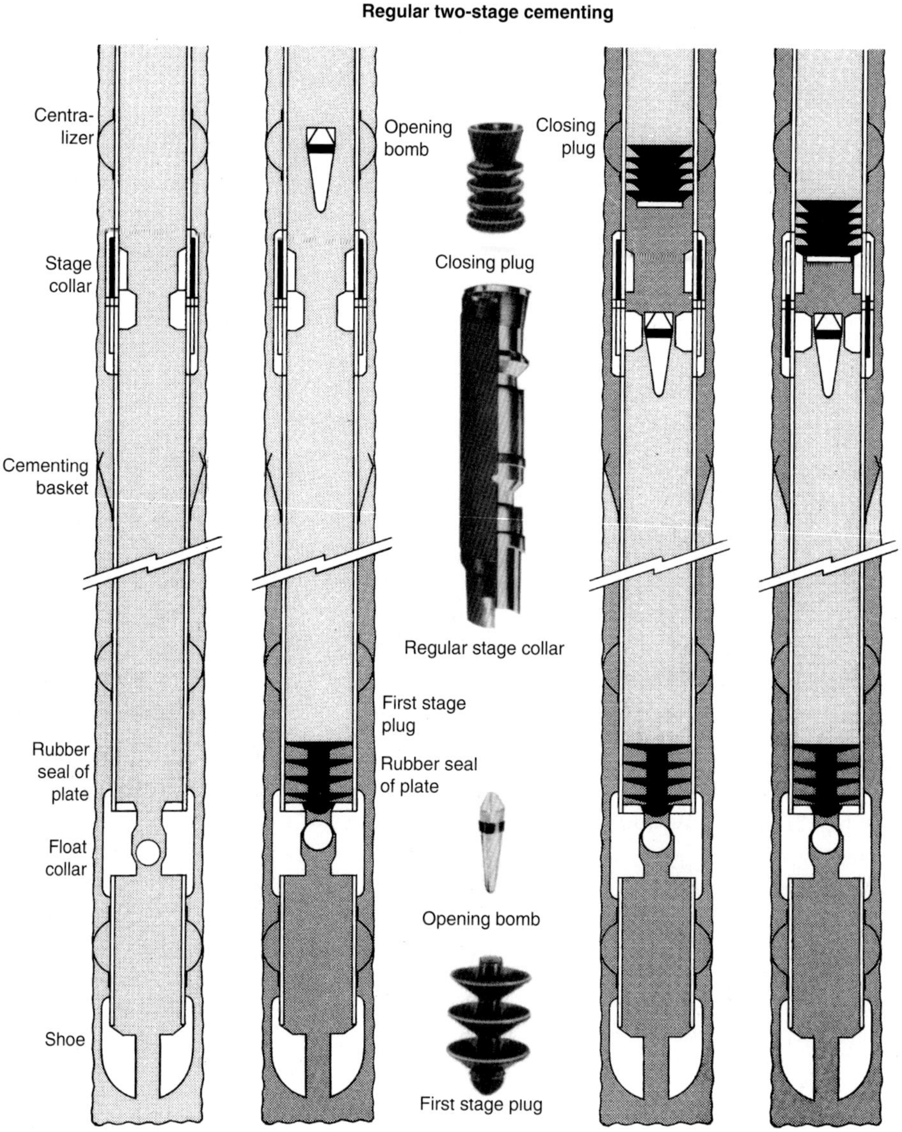

Fig. 7.16
Principle of two-stage cementing *(Source: Dowell Schlumberger).*

The driller then unscrews the running tool by turning it to the right and proceeds to reverse circulation to clean the excess slurry from the upper part of the liner hanger.

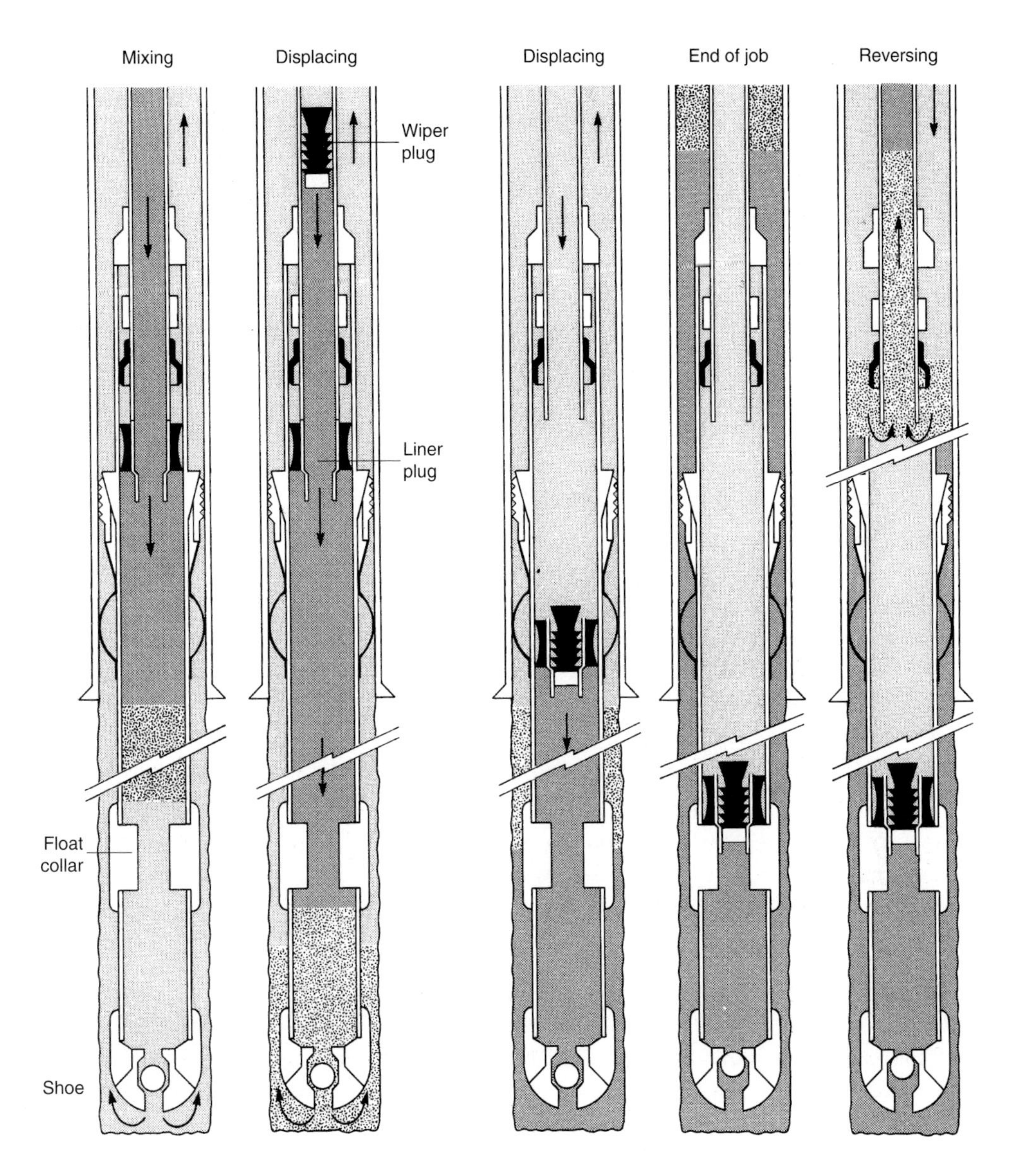

Fig. 7.17

Principle of the liner cementing procedure *(Source: Dowell Schlumberger).*

Reverse circulation is achieved by closing a BOP on the drillpipe and using the kill-line to inject mud into the system. The mud returns up through the drillpipe.

7.2.1.4 Squeeze cementing

This operation involves placing a cement slurry under pressure in a given location in the well. The purpose is to repair a defective seal or produce a new seal (to plug up perforations of a depleted bed). Pressure is exerted on a permeable formation so that the slurry loses water gradually and forms a cement cake. The cake sets and fills in any gaps in the seal or plugs up lost circulation zones.

The slurry is usually injected by the "hesitation squeeze" method. This involves stop and start pumping of batches of a few hundred liters of slurry, with pumps cut off for a few minutes and then on again and pressure increasing gradually until the final squeeze pressure is reached.

7.2.1.5 Positioning plugs of cement

Here a given amount of cement slurry is placed in the uncased part of the hole or in a string with the drill string by simple circulation.

There are a number of reasons for this type of operation:

- to abandon a well by isolating permeable formations and final closing procedure with a cement plug at the casing shoe,
- to close off an aquifer,
- to make a seat for a well to be resumed with directional drilling or a seat for a test drill stem,
- to plug up lost circulation zones,
- to abandon a depleted reservoir before bringing a shallower pay zone on stream.

The drill string or tubing string is used as the common way to spot the plug. The amount of slurry required to fill up the planned height is made up and injected, then flushed out until the column of cement inside the drill string balances the column outside. The drill string is then pulled out.

The same as for other types of cement jobs, some specific precautions must be taken.

The height to be covered depends on the required effect. If the aim is a seat for a sidetrack, the plug usually has enough height (about a hundred meters) so that several attempts can be made one after the other. In small-diameter wells, the height can be quite considerable, because little height would mean such a small volume of slurry that it would be extremely difficult to spot the plug properly.

Either ordinary or retarded cement is used depending on the plug's final depth and all the conventional additives may be added. However, bentonite is not recommended, as it decreases the slurry's compressive strength. For operations at shallow depths where setting time is very long for straight cement, calcium chloride is used as a cement setting accelerator. Sand, added at 5 to 20% by weight of straight cement, reinforces the cement's mechanical strength when the aim is to build a seat for a sidetrack. It also increases the density of the cement slurry.

The plugs are preferably placed where the borehole is well calibrated or at the permeable zones they are designed to seal off.

When a sidetrack is being prepared, the upper part of the plug will be placed where there is a relatively soft formation to make the job easier.

The drill string or tubing string is first run in to the required depth for the bottom of the plug. Circulation is begun and continued until the mud density is the same all throughout the system.

The slurry must be made up fairly slowly and care must be taken to use a minimum amount of water and get a good homogeneous mix. Since a small amount of cement is usually needed, cement in bags is used.

The volume of slurry to be pumped must be calculated and measured carefully. The principle is to wind up with equal heights of cement inside and outside the drill string at the end of the pumping phase. If a caliper log has been run, accurate calculations can be made. If not, the actual diameter of the hole — always a little larger than the bit — must be estimated.

Immediately after the volume of slurry has been pumped down, the drill string is pulled out slowly. When the upper part of the plug must be located at a given depth, a volume of slurry is used that is greater than the volume of the height that is to be covered. Then drilling mud is circulated, usually in reverse circulation, at the predetermined depth to flush out the excess cement slurry.

Setting time is chosen according to downhole conditions and to the final mechanical resistance required for the cement plug.

The plugs used to prepare a well for abandonment are checked after approximately 24 hours have elapsed. The first few meters, always contaminated by drilling mud, are redrilled and the rate of penetration is recorded. When the drilling rate falls between 5 and 10 m/h, the cement's strength can be considered satisfactory. The final check consists in landing between 10 and 15 tons on the bottom with circulation.

It is always a difficult operation to place a strong, leak-proof plug of cement. Factors influencing its success are: careful choice of the zone to be covered, exact calculation of the volume to be pumped into the well and accurate on-site measurements, use of wall scratchers to help the cement bond, choice of additives for straight cement (sand, cement accelerator), and slow, steady handling of the drill string once the cement is in place.

7.2.2 Choosing the cement slurry

The choice is governed by:

- the static temperature downhole, which conditions setting time and therefore the thickening time,
- the circulating downhole temperature while the slurry is being pumped down as it modifies the setting time and therefore the thickening time in a positive sense,

- the slurry density required by the hydrostatic pressure limitations of some of the drilled formations,
- the plastic viscosity of the slurry and its filtration characteristics,
- the rheological parameters of the slurry,
- the setting time and the time it takes the slurry to develop compressive strength,
- the way the cement withstands a number of agents likely to deteriorate it, e.g. some types of corrosive water and high downhole temperatures.

Commonly used slurries are made up chiefly of cement and water with one or more additives, each with a given function. However, some additives counteract the action of others, creating compatibility problems among them. Slurries are laboratory-tested under conditions that simulate actual on-site conditions and downhole environments.

There are a variety of cement categories, defined by API standards, according to the depth and mainly the downhole temperature of the well. Another determinant is the likelihood of contact with corrosive formation waters. The cement types are as follows:

API Spec 10

Class	**Type**
A	For use from surface to 1830 m (6000 ft) depth when special properties are not required. Ordinary type.
B	For use from surface to 1830 m (6000 ft) depth when conditions require moderate to high sulfate resistance.
C	For use from surface to 1830 m (6000 ft) depth when conditions require high early compressive strength. Available in low, moderate and high sulfate-resistant types.
D	For use from 1830 m (6000 ft) to 3050 m (10,000 ft) depth under conditions of moderately high temperatures and pressures. Available in moderate and high sulfate-resistant types.
E	For use from 3050 m (10,000 ft) to 4270 m (14,000 ft) depth under conditions of high temperatures and pressures. Available in moderate and high sulfate-resistant types.
F	For use from 3050 m (10,000 ft) to 4880 m (16,000 ft) depth under conditions of extremely high temperatures and pressures. Available in moderate and high sulfate-resistant types.
G	For use from surface to 2440 m (8000 ft) depth as manufactured, or can be used with accelerators and retarders to cover a wide range of well depths and temperatures. Available in moderate and high sulfate-resistant types.
H	For use from surface to 2440 m (8000 ft) depth as manufactured, or can be used with accelerators and retarders to cover a wide range of well depths and temperatures. Available only in moderate sulfate-resistant type.
J	For use from 3660 to 4880 m (12,000 to 16,000 ft) depth under conditions of extremely high temperatures and pressures. Available only in sulfate-resistant type.

7.2.2.1 Cement characteristics

Manufacture

Artificial Portland cement is produced by calcining at over 1400°C a mixture of calcarious and argillaceous materials mined in quarries. The two basic materials are ground and fed into a rotating furnace. The combustion product, called clinker, is cooled and stored before being pulverized. It is pulverized with a number of added materials such as gypsum, but also natural or artificial pozzolan and blast furnace slag, etc. The pulverized, screened, ready-to use product is delivered to well sites in bags or in bulk.

The resulting cement is a complex mixture whose basic components are silica, calcium, aluminum and iron. Its mineralogical composition varies within a given range and is defined as a state of equilibrium among different theoretical compounds. It is obtained by calculations based on a mineralogical analysis. The compounds are as follows:

- tricalcium silicate C_3S,
- dicalcium silicate C_2S,
- tricalcium aluminate C_3A,
- tetracalcium ferroaluminate C_4AF.

7.2.2.2 Cement hydrating mechanisms

When cement is mixed with water it becomes hydrated starting with the less stable anhydrous components. They are dissolved to saturation more readily than the more stable hydrated components. When there is insufficient water, the solution is supersaturated with respect to hydrated components which crystallize. This allows the anhydrous components to dissolve once again. Gradually, the crystalline needles will become bonded together until the whole system hardens.

The first component that reacts is the C_3A which gives a calcium sulfoaluminate with the gypsum. This is what will determine how quickly the cement sets according to the composition of the aqueous phase, i.e. the equilibrium conditions of the soluble salts, hydroxides and other products that are normally or accidentally found in solution.

The C_3A that has not reacted with the gypsum is hydrated singly and subsists in this form in the hardened cement. However, if sulfates are present later on in water infiltrations, a C_2 sulfoaluminate is formed with an increase in volume. The reaction breaks up the crystalline network and destroys the continuous structure of the cement. This preliminary hydration phase is very important in oilwell cement jobs since it corresponds to the thickening time of the slurry when it can still be pumped.

A number of factors influence cement hydration.

A. Temperature

Temperature has a great influence on cement hydration speed. Higher temperatures will shorten setting time by accelerating hydration speed.

B. Pressure

By modifying the equilibrium conditions in the liquid phase, increasing pressure speeds up hydration.

C. Contamination

Contamination is a phenomenon with random effects, it may be due to the water when the cement is being mixed or to fluids encountered in the well when the cement is being pumped into place.

Any modification in the equilibrium of the aqueous phase by unwanted additions of soluble or insoluble materials affects cement hydration. As a result, the thickening time and final strength, etc. are also affected.

Sodium hydroxide, sodium carbonate and sodium silicate cause acceleration. However, it should be noted that the acceleration is totally unpredictable and often has disastrous consequences on the final cement quality.

Sodium hexametaphosphate and acid sodium pyrophosphate (drilling mud additives) have even more erratic effects, either accelerating or retarding depending on the concentration. Sodium chloride also has the same accelerating or retarding effect depending on the concentration.

Other contaminants act at the liquid phase-solid phase contact area level. Examples are starches, tannins, cellulose derivatives, lignosulfonate acids and carboxylic acids. These products in solution are adsorbed on the cement particles and form a leak-proof barrier that retards hydration. It may even keep the cement from setting at all when contamination is substantial.

Using oil-base mud also causes contamination, though the nonmiscibility of the two phases reduces the risks of mixing.

Using spacers helps overcome the drawbacks of contamination.

7.2.2.3 Cement additives (Table, Fig. 7.18)

Accelerators

These products speed up cement setting at low temperature or offset the retarding effects of other additives. They help shorten waiting-on-cement time before drilling operations can be resumed.

Retarders

These additives slow down cement setting and thereby lengthen thickening time for pumping the cement into place. They are used when high downhole temperature or the accelerating effect of another additive might dangerously reduce the time available for pumping the cement.

Cement Characteristics	Cement Additives / Effect	Bentonite	Perlite	Diatomaceous earth	Pozzolan	Sand	Barite	Hematite	Calcium chloride	Sodium chloride	Lignosulfonate	CMHEC ([1])	Diesel oil	Water loss additive	Lost circulation material
Density	Decreased	•	•	•	•										
	Increased					•	•	•	x	x	x				
Water required	Decreased										•				
	Increased	•	x	•	x	x	x	x							x
Viscosity	Decreased								x		•				
	Increased	x	x	x	x	x	x	x							
Thickening time	Accelerated	x					x	x	•	•					
	Retarded			x						x	•	•	x	x	
Setting time	Accelerated						x	x	•	•					
	Retarded	x	x	x	x						•	•		x	
Early strength	Decreased	x	x	x	x		x	x			•	•		x	x
	Increased								•	•					
Final strength	Decreased	x	x	•	x		x					x		x	x
	Increased														
Duration	Decreased	x	x	x									x		x
	Increased				•										
Water loss	Decreased	•									x	•	x	•	x
	Increased		x	x											

x Denotes minor effect
• Denotes major effect and/or purpose of additive
([1]) Carboxymethyl hydroxyethyl cellulose

Fig. 7.18
Effects of some additives on cement properties (*From Dowell Schlumberger Engineer's Handbook*)
(Source: Drilling Data Handbook, Editions Technip).

Extenders

These are lightweight, inert materials mixed with cement and designed to reduce both slurry density and costs. However, most lightweight additives have an effect on cement setting time and compressive strength. Special additives are often necessary to offset this tendency.

Fluid-loss additives

These agents stop the slurry from losing water by filtration into permeable formations. Water loss may trigger either unwanted setting or no setting at all since not enough water is available for hydrolysis and crystallization of the cement components.

Heavyweight additives

These products serve to increase slurry density. They are inert and mixed in with the dry cement.

7.2.2.4 Miscellaneous agents

- **Dispersants** offset an overly-high viscosity and some slurries' tendency to gel. They also help establish turbulent flow when it is needed.
- **Antifoam agents** prevent the excessive foam that is produced when some cements are mixed. Foaming can interfere with smooth pump operation.
- **Gelling agents** modify the thixotropic properties of some slurries.
- **Liquid additives** can be added to the mixing water instead of being incorporated dry into the cement before it is hydrated.

7.2.2.5 Spacers

The spacer fluid on the leading edge of the slurry is designed to:

- make it easier to move the mud through the annulus,
- isolate the cement slurry from any contact with the mud so that the mud does not gel,
- make it easier to get rid of mud that has gelled on the walls of the casing.

The calcium present in cement causes the argillaceous particles in drilling mud to flocculate out and this in turn causes slugs of highly viscous mud to be formed. The slugs may be channeled by the cement, resulting in a poor cement job. Spacers act mainly as a "cushion" to prevent contamination of the mud by the cement.

7.2.2.6 Choosing the type of flow

Laboratory results and experience show that slurry moving in turbulent flow achieves a 95% displacement of the mud. Displacement by slurry in streamline flow is much less effective.

Research into cement slurries focuses particularly on two characteristics: thickening time and rheological properties.

A. *Thickening time*

This is an empirical criterion that makes it possible to determine how long the cement slurry will be pumpable by the equipment on a well site under real conditions of use. Determining the thickening time involves a viscosity measurement defined by an arbitrary scale graduated in units of consistency (UC) from 0 to 100. By definition, the limit on utilization is 100 UC. The measurement is made in an apparatus that simulates real pressure and temperature conditions. Testing is done at a temperature, termed circulating temperature, that is determined on the basis of the static temperature. Static temperature is the temperature of the well when it is in thermal equilibrium with the environment.

B. *Type of flow and rheology*

It is necessary to consider the characteristics (density, rheology) of the drilling mud already in the well before the cement job is done. The fluid that is most readily displaced by the cement slurry is water. Any increase in the density or rheology of the drilling mud will have a negative effect on displacement by the slurry. So that displacement defects can be minimized, the density and rheology of the slurry need to be properly adjusted taking the mud characteristics into account.

To prevent any contamination of the slurry that would be detrimental to the cement job, a spacer fluid is run in between the mud and the slurry (see paragraph 7.2.2.5). The spacer will also have density and rheology characteristics such that it can:

- displace the drilling mud,
- be displaced by the cement slurry.

Adjustments involving density, rheology and annular velocity are made so that one fluid can be more readily displaced by another in an annulus.

Regulating the density

In an initial approach, the following rules can be used as a basis:

The drilling mud has a density governed by drilling conditions and equal to d_{mud}:

- the density d_{spa} of the spacer fluid will be such that:

$$d_{spa} = d_{mud} + 0.1$$

- the density d_s of the slurry will be such that:

$$d_s = \text{at least } d_{mud} + 0.4$$

Adjusting the rheology and the annular velocity

These two parameters are closely related. The fluids involved in cementing (drilling fluid, spacer fluid, cement slurry) are usually non-Newtonian. They can be characterized by Bingham plastic or power law rheological models. Depending on the geometry of the annulus and the flow rate, each one of these types of fluid moves respectively in the following type of flow: plug, laminar and turbulent.

The best displacement of one fluid by another occurs when the second fluid has a flat velocity profile. The aim for the cement slurry is therefore either plug or turbulent flow in the annulus. In turbulent flow, the energy supplied by the eddies can only be beneficial in displacing the fluid preceding the slurry.

7.2.3 Implementation

Cement is delivered to the user in bulk or bags, with bulk delivery preferred. But in the oil industry, distribution in bags has been the rule for so long that the bag has become a unit of measurement. Here the bag is standard at 94 lbs with a volume of one cubic foot.

However, the bag usually goes no farther than the supply base. It is in bulk that the cement is taken out to the well site after the different dry additives have been mixed in.

The cementing operation itself is performed by a cementing unit that includes all the equipment and materials required to emplace the cement (**Fig. 7.19**):

- one or more tanks to ensure an extra supply of water is on hand for mixing,
- one or more centrifugal pumps to supply the tanks, mix the fluids and boost the high-pressure pumps,
- a mixing apparatus for the cement,
- one or two high-pressure pumps,
- monitoring and measurement equipment.

Fig. 7.19

Truck-mounted cementing unit *(Source: Halliburton).*

Chapter 8

MEASUREMENTS AND DRILLING

This chapter is based on École Nationale Supérieure du Pétrole et des Moteurs (ENSPM) course material, Center for Exploration, with the kind assistance of Sylvain Boyer and Bernard Michaut.

INTRODUCTION

A distinction should be made between two kinds of measurements. Both are used in a drilling operation, whether for exploration or development and whether offshore or onshore.

They are as follows:

- Mud logging, or geological control of drilling, which consists of measurements collected from all over the drilling site while drilling is under way. There are a variety of very different measurements:
  - drilling parameters,
  - mud characteristics,
  - downhole measurements during drilling,
  - geological monitoring by sample-taking.
- Wireline logging, which involves local techniques of study based on one well. Logs provide more or less continuous information on the formations surrounding the borehole.

The oil industry has used the second type of measurements increasingly. They date back to 1927, when the first electrical measurement was made by two Frenchmen, the Schlumberger brothers, in the Pechelbronn field in France.

Wireline logging, or well logging, is a basic implement among the major exploration techniques and is used to discover, quantify and produce the world's oil and gas resources.

8.1 MUD LOGGING

Mud logging chiefly involves measurements while drilling is in progress and includes:

- drilling parameters: weight on the hook (on the bit), rate of penetration, rotational speed and drilling fluid pumping rate;
- operational variables: rotary table torque, mud pump discharge pressure, drill string progress and levels in mud tanks;
- geological variables which depend on the formations that are drilled: mud temperature and resistivity (inlet and outlet), gas content and mud density (inlet/outlet).

Data are acquired by sensors laid out all over the well site and are collected in the geological supervision unit. Data can be acquired and processed in real time by means of a computerized system connected directly to the sensors (solution recommended for offshore operations or complex sites). An example of this is the *Geoservices* on-line type system.

In less complex operations, only interpretation is done by a microprocessor. This means sensor data must be input by a user (*Geoservices* off-line type system).

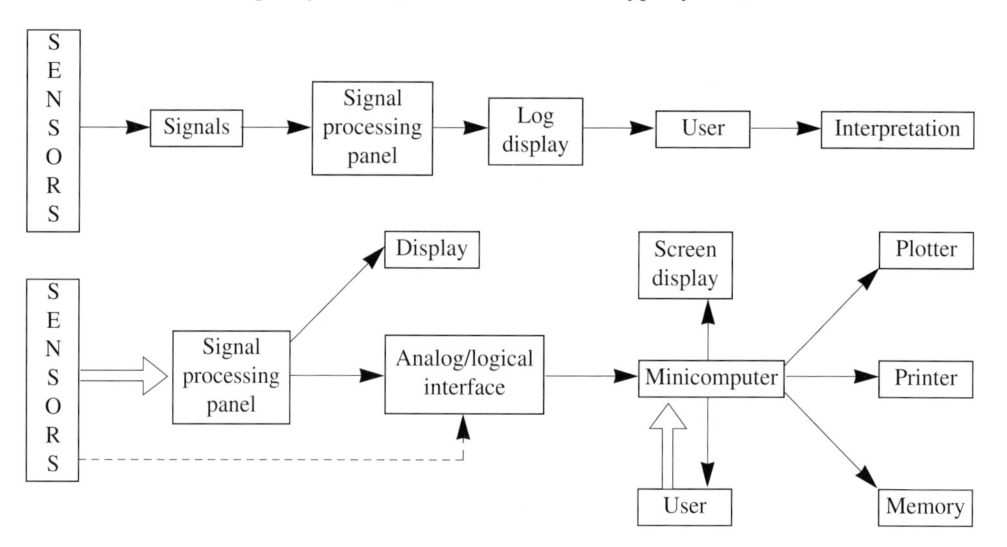

There are three main objectives for these measurements:

- ensure the safety of the whole site in general, the men and the well;
- establish a resource document base, the major aim in exploratory drilling; or gain further information in development drilling;
- serve as a drilling aid with simultaneous analysis of several variables to diagnose present, past or even future events.

The measurements therefore involve all the senior personnel on a well site, whether on the company side or the contractor side.

The following discussion will not deal with technical considerations of measurement instruments, though the technical angle is of prime importance. This is because any interpretation is governed by the system's field of application and the validity of the information. The reader should consult other specialized literature for further information if need be.

8.1.1 Measuring the rate of penetration

8.1.1.1 Aim of the measurement

The rate of penetration measurement is of great importance for the geologist. It provides him with an immediate indication that is sensitive to the variations in lithology and can be correlated with wireline logs. As a result:

- correlations can be made with neighboring wells,
- reservoirs can be located, thereby helping decide when coring should be done,
- a core sample can be depth-recalibrated when part of it is missing,
- abnormal pressures can be detected (rate of penetration increases in undercompacted zones and this is one of the parameters in the *d* exponent formula).

Integrating drill string progress while drilling also gives the depth reached at any given time and the position of the bit in the hole during tripping. Continuous recording of the drilling rate makes it possible to reproduce all mud logs versus depth by using computerized systems. This is in addition to recordings versus time.

The driller is also interested in rate of penetration measurements since they help him:

- check how well bits are working (balling up, junk in the bottom of the well, wear and tear on teeth),
- optimize drilling parameters (find the best trade-off among drilling rate, weight and rotation),
- supervise how well orders are being carried out.

The apparatus that measures drilling rate can even (depending on the type) monitor traveling block speeds during tripping so as to:

- keep surging or swabbing problems from occurring,
- safeguard well stability.

8.1.1.2 Principle of the measurement

The rate of penetration of the bit is likened to the displacement or rotation of a mechanical part, assuming a direct relationship between the bit and the movement seen on the surface. The relationship is provided by the drill stem, which can be considered as having a constant length as long as the stress measurements on it do not vary. Only measurements made over substantial depth intervals (where stresses vary little) are significant. The depth intervals selected are generally 25 cm, 50 cm and 100 cm.

Choosing the mechanical part

- Recording a displacement (**Figs. 8.1 and 8.1a**):
  - movement of the kelly or the swivel (no monitoring during tripping),
  - movement of the traveling block.
- Recording a rotation:
  - movement of a pulley on the crown block (slow pulley to avoid slipping),
  - rotation of the drawworks drum (**Fig. 8.1b**).

8.1.2 Weight on the hook. Weight on the bit

The weight hanging on the hook represents the pulling force exerted by the drill string suspended in the drilling mud.

In actual fact, the parameter that is monitored is the weight on the bit. This is the difference between the weight hanging on the hook with the bit off bottom and the weight with the bit on the bottom (drill stem rotating and mud circulating).

Though the difference in pulling force can be considered to represent the actual weight on the bit in a vertical well, the same is not true for a deviated well. Here drag, which can be substantial, should not be disregarded.

8.1.2.1 Aim of the measurement

The weight on the bit is an essential measurement for the driller to check on how well the drilling bit is working and to detect abnormal drag during tripping (weight on the hook/overpull).

It is also a useful measurement for the geologist, since the weight on the bit governs the rate of penetration.

8.1.2.2 Principle of the measurement (Fig. 8.2)

The weight hanging on the hook is measured from deadline tension measurements by a hydraulic pressure cell. The geological supervision sensor is usually connected directly to the driller's measurement circuit. The pulling force exerted on the wire rope is converted into pressure in a hydraulic circuit. The sensor, a strain gage incorporated in the circuit, gives an electric signal that can be weight-calibrated.

8.1.3 Rotational speed

8.1.3.1 Aim of the measurement

This parameter is required by the driller, since the rotational speed influences bit lifetime (particularly bearing lifetime). It helps choose (along with the weight on the bit) optimum drilling progress parameters.

Fig. 8.1

Displacement sensor
(Source: Geoservices).

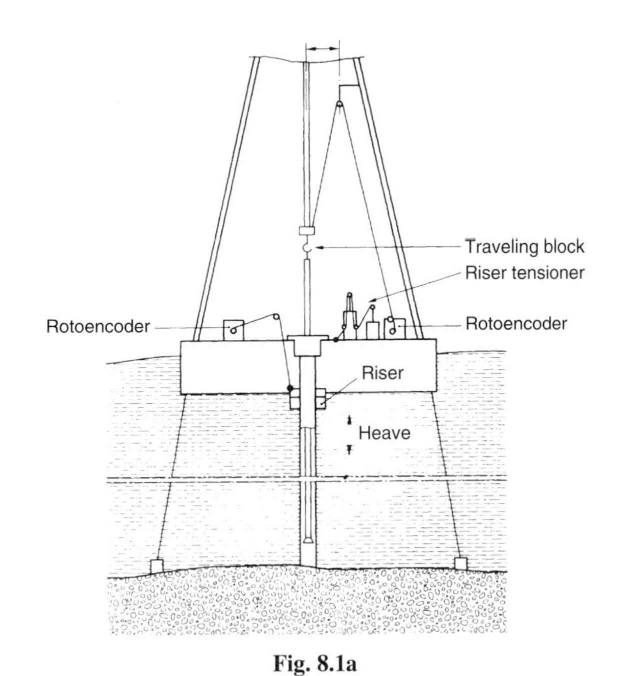

Fig. 8.1a

Sensor layout on floating rigs *(Source: Les mesures en cours de forage, Editions Technip, Paris, 1982).*

Fig. 8.1b

Displacement sensor on the drawworks drum
(Source: Les mesures en cours de forage, Editions Technip, Paris, 1982).

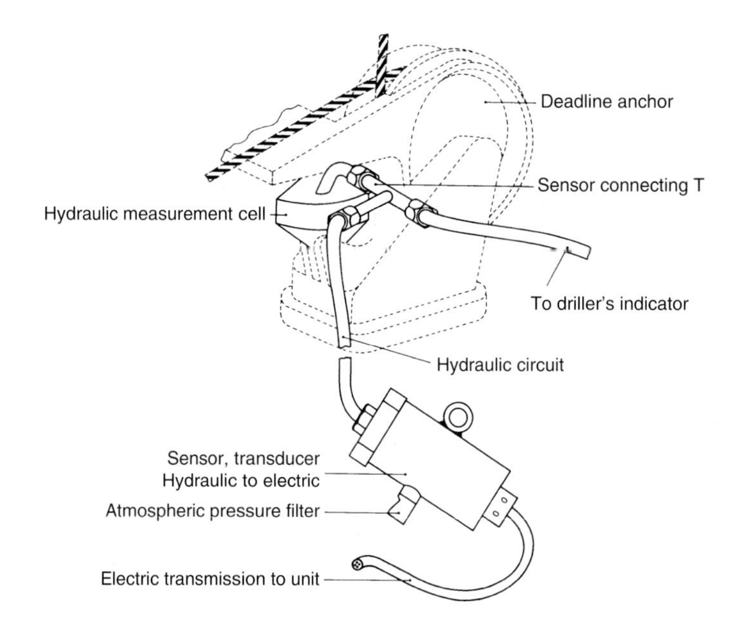

Fig. 8.2

Hook weight sensor

(Source: Les mesures en cours de forage, Editions Technip, Paris, 1982).

Here the geologist's interest is the same as for the weight on the bit: rotation is related to penetration rate.

8.1.3.2 Principle of the measurement (Fig. 8.3)

An electric pulse is generated each time the rotary table turns by means of a contactor on the table and a proximity detector. Integrating the number of electric pulses in a given lapse of time indicates the rotational speed. Another possibility is to use infrared telemetering (*Anadrill*).

8.1.4 Torque

The torque measured on the surface is the force required to keep the drill string rotating.

A number of forces tend to counter drill string rotation. Some are related to mechanical parameters (weight on the bit, rotational speed, diameter and type of bit), while others are connected to the rock formation (resistance to pulling). What technicians are interested in is torque at the bit, but transmission of torque from the rotary table to the bit is far from efficient because of drill string inertia and drag (particularly at stabilizer depth).The problem is even more serious in deviated and horizontal wells.

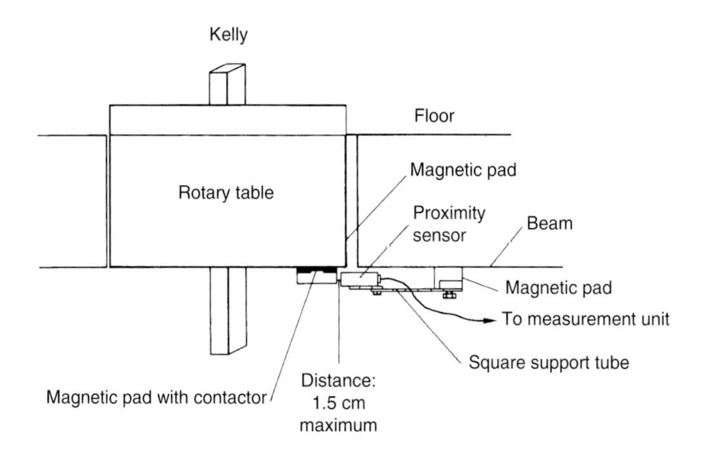

Fig. 8.3

Rotary table rotational speed sensor
(Source: Les mesures en cours de forage, Editions Technip, Paris, 1982).

The information gathered is significant only if drill string friction torque can be disregarded. It is the relative variations in torque which are in fact useful.

8.1.4.1 Aim of the measurement

The changes in the average torque value are what tell the driller about the status of the bit and especially the bearings.

Frequency variations may mean:
- problems with a stuck bit cone
- formation caved in on top of the bit.

The geologist is also interested in torque variations to:
- detect changes in lithology,
- spot abnormal compaction zones,
- locate possible fractured formations when torque variations are sudden.

8.1.4.2 Principle of the measurement

On rotary drilling rigs where power is transmitted by a chain, the measurement is made by means of a hydraulic system. Pressure in the system varies according to the degree of tension on the chain. A strain gage converts the pressures into electric signals (**Fig. 8.4**).

When the drive system is electric, a Hall effect sensor is placed around the current input wire (measuring electric current consumption by the rotary table motor) (**Fig. 8.4a**).

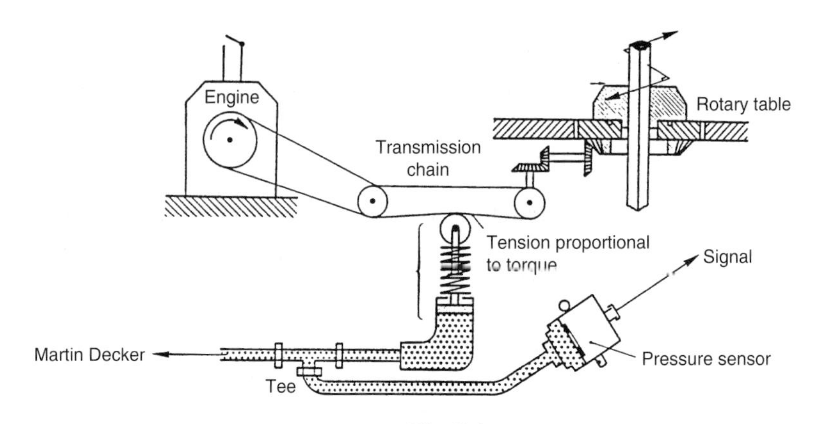

Fig. 8.4

Rotary table torque sensor

(Source: Les mesures en cours de forage, Editions Technip, Paris, 1982).

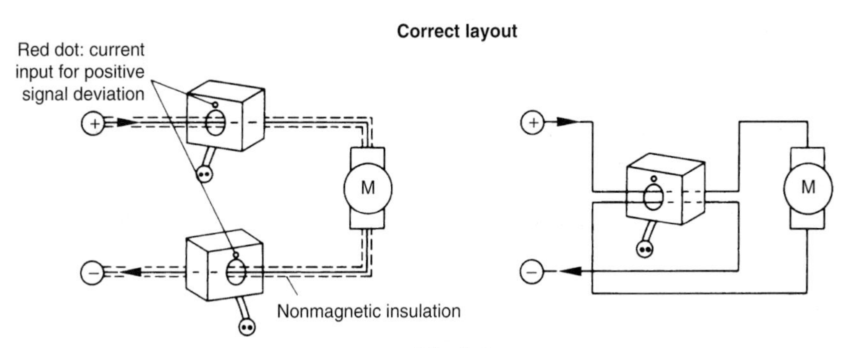

Fig. 8.4a

Hall effect torque sensor on an electric rotary table

(Source: Les mesures en cours de forage, Editions Technip, Paris, 1982).

8.1.5 Pumping pressure. Casing pressure

8.1.5.1 Aim of the measurements

A. Pumping pressure

The driller needs pumping pressure measurements to:

- check on operating conditions (problems with nozzles, drill string or pumps),
- monitor the movement of drilling mud slugs in the well,
- detect wall stability problems.

The geologist is interested in them because of the impact of pressure on rate of penetration. Penetration will be less when there is an increase in the differential pressure exerted on the bottom of the well.

B. Casing pressure

This measurement is not warranted in normal drilling when mud returns are at atmospheric pressure. It is indispensable for the driller during choke line circulation when a fluid intrusion has occurred in the well.

8.1.5.2 Principle of the measurements

Pressure gages are connected to the standpipe to measure pumping pressure and to the choke manifold to measure casing pressure. They are usually strain gages that convert pressure into an electric signal.

8.1.6 Density measurement

8.1.6.1 Aim of the measurement

Drilling mud density is a value that conditions well balance. The density must be checked at the inlet and at the outlet of the well. It is crucial for the driller and should not be disregarded by the geologist (detector sensitivity to gas and lag time can be checked).

A. Inlet mud density

Drilling mud density is fixed and will not vary if the mud is homogeneous on the surface.

B. Outlet mud density

Monitoring outlet mud density can help identify any annular pressure imbalances due to:

- an increase in solids concentration which could cause lost circulation,
- water or oil influxes that cause a decrease in density,
- gas influxes coming from the formation, with the resulting substantial decrease in mud density well-site degassers should immediately be started up,
- air-cut mud after a round trip.

8.1.6.2 Principle of the measurement

A. By pressure sensor(s) **(Fig. 8.5)**

The principle is to measure the hydrostatic pressure of a column of mud with a constant height. The mud is tapped as close as possible to the bell nipple and circulates continuously through a column.

This system gives good results, but requires regular maintenance (cuttings settle out in the column). Another way to assess the density of the mud is to measure the difference in pressure between two sensors.

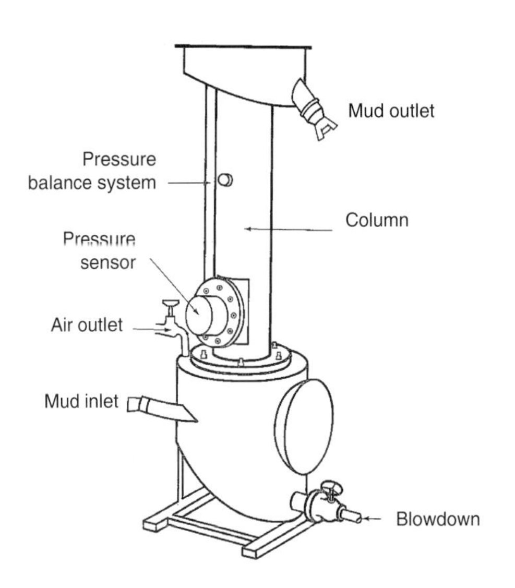

Fig. 8.5
Column-type density meter *(Geoservices type)*.

Fig. 8.5a
Gamma-ray density meter on the standpipe.

B. By gamma-ray density meter **(Fig. 8.5a)**

What is measured is the attenuation of a beam of gamma rays emitted by a natural radioactive source (100 mCi cesium 137). The degree of attenuation depends on the electron density of the medium that is penetrated. The electron density is in turn closely related to the average density of the medium.

Gamma-ray density meters can measure the density of inlet mud under pressure, thereby eliminating the gas-cutting effect. The meter is generally installed on the standpipe.

8.1.7 Variations in mud tank volume

When the hydraulic system is well balanced, the volume of mud available to the driller does not vary. With the mud either in the well or in the tanks, the only acceptable variation is a decrease in the amount of mud in the tanks because the well has been deepened. The reduction in volume can be calculated.

In actual fact, the volume of mud can also decrease because of losses due to mud treatment on the surface: desanders, desilters, mud cleaners and shale shakers. The losses are generally regular.

Likewise, a regular increase in the volume of mud in the tanks can come from new mud makeup or additive inputs. Outside of these variations, any modification in the volume of the mud in the tanks is a sign of well imbalance. Though it is easy to spot sudden, severe events, detecting slow variations is much harder.

8.1.7.1 Aim of the measurements

Mud levels in the tanks are constantly monitored:

- by the driller to:
  - check on the available mud in each tank,
  - supervise well stability,
  - monitor well filling during a round trip when there is no trip tank;
- by the geologist to:
  - check for partial mud losses that might indicate fractured formations.

It is recommended that geologists and drillers carefully monitor and compare mud levels in the tanks before and after pipe is added to the drill string, during tripping to change bits and when the casing is being lowered.

8.1.7.2 Principle of the measurements

A. *Float system* (Fig. 8.6)

A float shows the level in the tank and actuates a potentiometer that delivers a signal. Since the dimensions (area) of the tank are fixed, level is converted to volume.

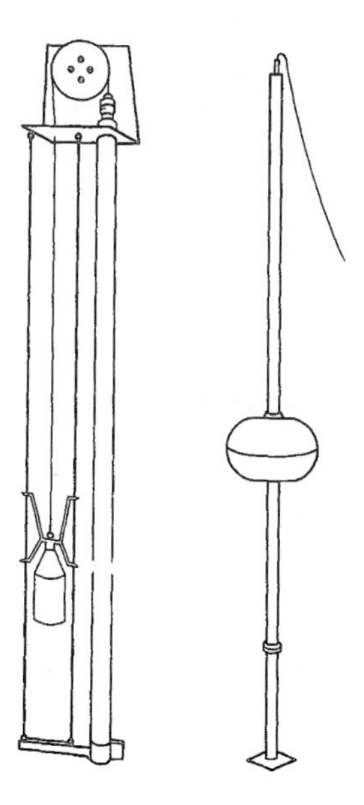

Fig. 8.6

Measuring the level in tanks: float system.

Two sensors are required on floating drilling rigs. They are laid out diagonally on each tank, in order to compensate for variations in level due to roll and pitch.

B. Ultrasonic wave system

The sensor transmits an ultrasonic wave which is reflected on the surface of the mud and bounces back to the sensor. The variations in level correspond to variations in wave traveltime.

8.1.8 Flow metering

8.1.8.1 Aim of the measurement

Flow rate is a parameter that is set in order to keep the borehole and bit clean and cool the bit down.

As far as the driller is concerned, flow influences the rate of penetration and it is also important for the geologist because:

- the time it takes information to come back up, or lag time, can be calculated,
- since the differential flow rate is known, i.e. the difference between the inlet and outlet flow rates, any mud losses or gains can be detected rapidly without resorting to measurements on the surface.

8.1.8.2 Principle of the measurement

A. Inlet flow rate (Fig. 8.7)

The simplest method is to count the number of pump strokes. With the volume pumped per stroke and the efficiency of the pump, the flow rate can be calculated. It is easy to measure the number of pump strokes by proximity detectors or electric contactors.

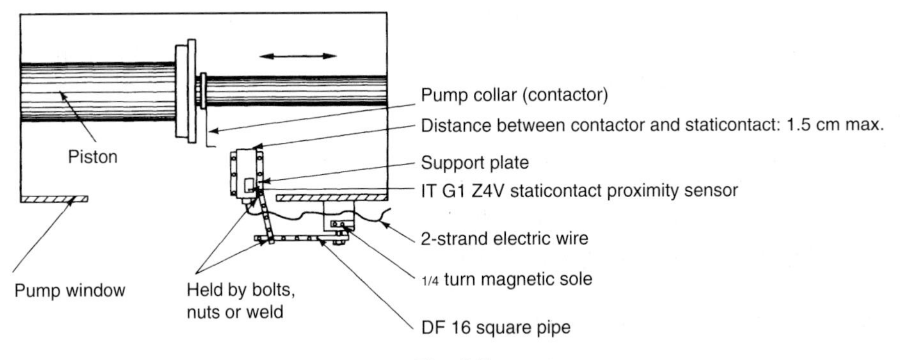

Fig. 8.7

Measuring flow rates with a pump stroke counter
(Source: Les mesures en cours de forage, Editions Technip, Paris, 1982).

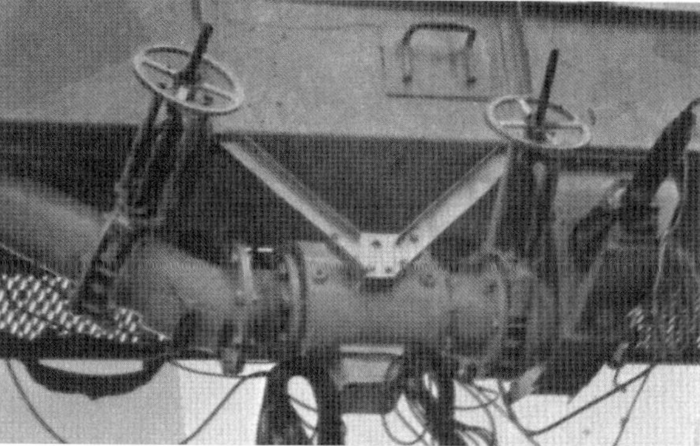

Fig. 8.7a

Magnetic flow meter *(Source: Geoservices).*

B. Outlet flow rate

Here measurements are made by a blade-type system with far from satisfactory accuracy. Flow rate can not be measured with the system, but variations in it can be assessed.

C. Inlet and outlet flow rate (Fig. 8.7a)

A magnetic field perpendicular to the axis of flow produces a difference in potential proportional to the fluid flow rate between two diametrically opposed electrodes. The system can work only with drilling fluid that conducts electricity. This means it can not be used with mud that has a continuous oil phase, for example.

8.1.9 Measuring mud temperature

8.1.9.1 Aim of the measurement

It is not really possible to measure the temperature of drilled formations by using conventional methods. The only possibility is to monitor the changing temperature of the mud as it comes out of the well and hope that the variations recorded have a geological origin.

Drillers and geologists are mainly interested in this measurement to:

- detect zones with abnormal pressure,
- see when an evaporite series is near,
- possibly help choose the type of fluid best suited to temperature conditions.

8.1.9.2 Principle of the measurement

Temperature can be measured either by means of a platinum thermocouple (pyrometric probe) immersed directly in the mud or a thermosensitive insert installed on piping (stand pipe, flowline).

8.1.10 Measuring mud conductivity

8.1.10.1 Aim of the measurement

By comparing inlet and outlet conductivity measurements, the geological phenomena that modify the mud's ion content can be detected.

Drillers and geologists keep track of the changes in the measurements to identify salt-bearing formations, water influxes or sour gas kicks.

8.1.10.2 Principle of the measurement

Conductivity is measured from an induced resistivity sensor immersed in the mud (flowline and suction tank).

8.1.11 Detecting gas

Among the many parameters that are recorded during drilling, gas must be the one that has, and always has had, the greatest impact on both drillers and geologists alike.

Gas detection is a constant concern because of the safety hazards for both personnel and well. Additionally, the amount and nature of the gas that is detected and analyzed are important data for the exploration geologist.

8.1.11.1 Principles of degassing (Figs. 8.8 and 8.8a)

The operation is performed by means of a degasser that is installed as close as possible to the point where the mud exits from the well. The mud coming from the well is mixed briskly in the degasser to separate the gas from the liquid phase.

Air is sucked through the degasser and conveys the gas show to the detector from the mud logging facilities.

8.1.11.2 Gas detection

The amount of gas (total gas) in the gas/air mixture coming from the degasser is measured:

- by a thermal conductivity system, or
- by a catalytic combustion system, or
- by a flame ionization detector.

8.1.11.3 Gas analysis

The gas is analyzed using the principle of gas chromatography. The different components of the gas (generally from methane, C_1, to pentane, C_5) are separated when the gas goes through chromatograph columns. Then they can be identified by the same detection systems as mentioned before.

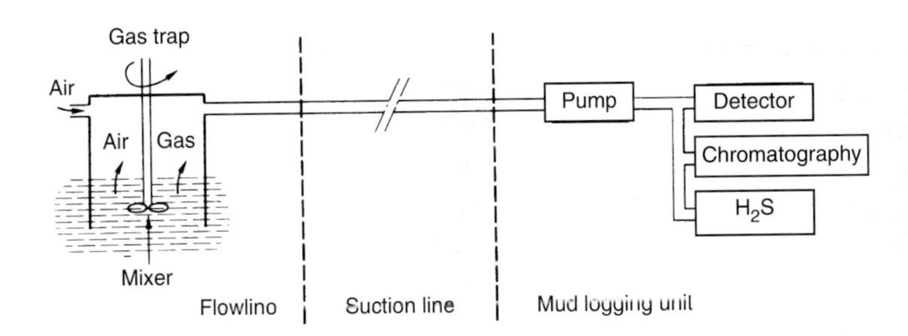

Fig. 8.8

Principle of gas analysis *(Source: Les mesures en cours de forage, Editions Technip, Paris, 1982)*.

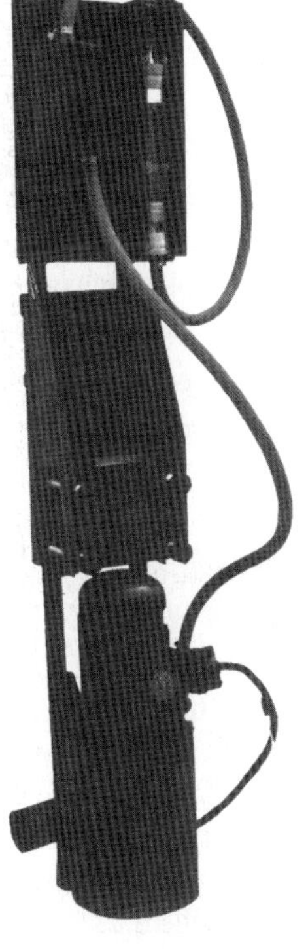

Fig. 8.8a

Degasser *(Geoservices* type).

8.1.11.4 Detecting H_2S

Hydrogen sulfide (H_2S) is often present in the formations that are penetrated by an oil well. However, this highly toxic gas can also be found associated with volcanism in geothermal exploration drilling.

Most well sites are provided with hydrogen sulfide protection equipment.

So that the detectors can work properly in their preventive function, each sensor on the rig must be checked periodically.

8.1.11.5 Detecting carbon dioxide

Specific detection of carbon dioxide is important during drilling operations, since the presence of CO_2 can signal a nearby reservoir. It can also be at the root of a blowout, however.

Carbon dioxide is sometimes produced by the breakdown of mud lignosulfonates with a resulting modification in mud properties.

8.2 WIRELINE LOGGING

8.2.1 Definition

Any graphic representation of the variations in one parameter versus another, generally depth, sometimes time, is termed a log.

Wireline logs are measurements of physical parameters in the formations penetrated by borehole. They are run when drilling has been stopped (after the drill string has been pulled out). They are therefore different from mud logging operations.

The logs are usually in the form of curves that show the measurement performed with a reduced sampling interval generally of one-tenth, sometimes one one-hundredth or even one one-thousandth of a meter (**Fig. 8.9**).

The measurements are made by sondes that are lowered into the wellbore on a wireline wound around the winch of a logging truck. There are a number of specialized service companies that do this type of work: the most well known worldwide are *Halliburton Wireline Services, Schlumberger, Western Wireline Services.*

There are three major types of logs:
- the ones used by geologists and reservoir engineers to evaluate the characteristics of the formations and fluids and quantify them,
- the ones for drillers that provide technical information (cement bond log, stuck-point indicator, etc.),
- the ones used by production staff to study fluid and fluid-flow phenomena.

8.2.2 Objectives

Drilled formations can be studied quantitatively and qualitatively by using the graphic transcription of the physical parameters that are measured by logging sensors. The aim is to:
- identify potential reservoir rocks and cap rocks and analyze sediment deposition conditions,
- determine the nature and amount of fluids contained in the rocks.

In contrast with the study of cuttings (too small in size) brought up by the drilling fluid and of core samples (usually few because of their cost), wireline logs allow formations to be studied continuously and in situ.

They are a sort of multiple X-ray of the formations the borehole has crossed through and therefore link up surface geophysics and subsurface geology. The two techniques both help define a reservoir's lateral extension and assess its available reserves.

By quantifying geological data, wireline logs also aid in describing the beds and representing sedimentary phenomena. They give an objective, permanent view of the formations.

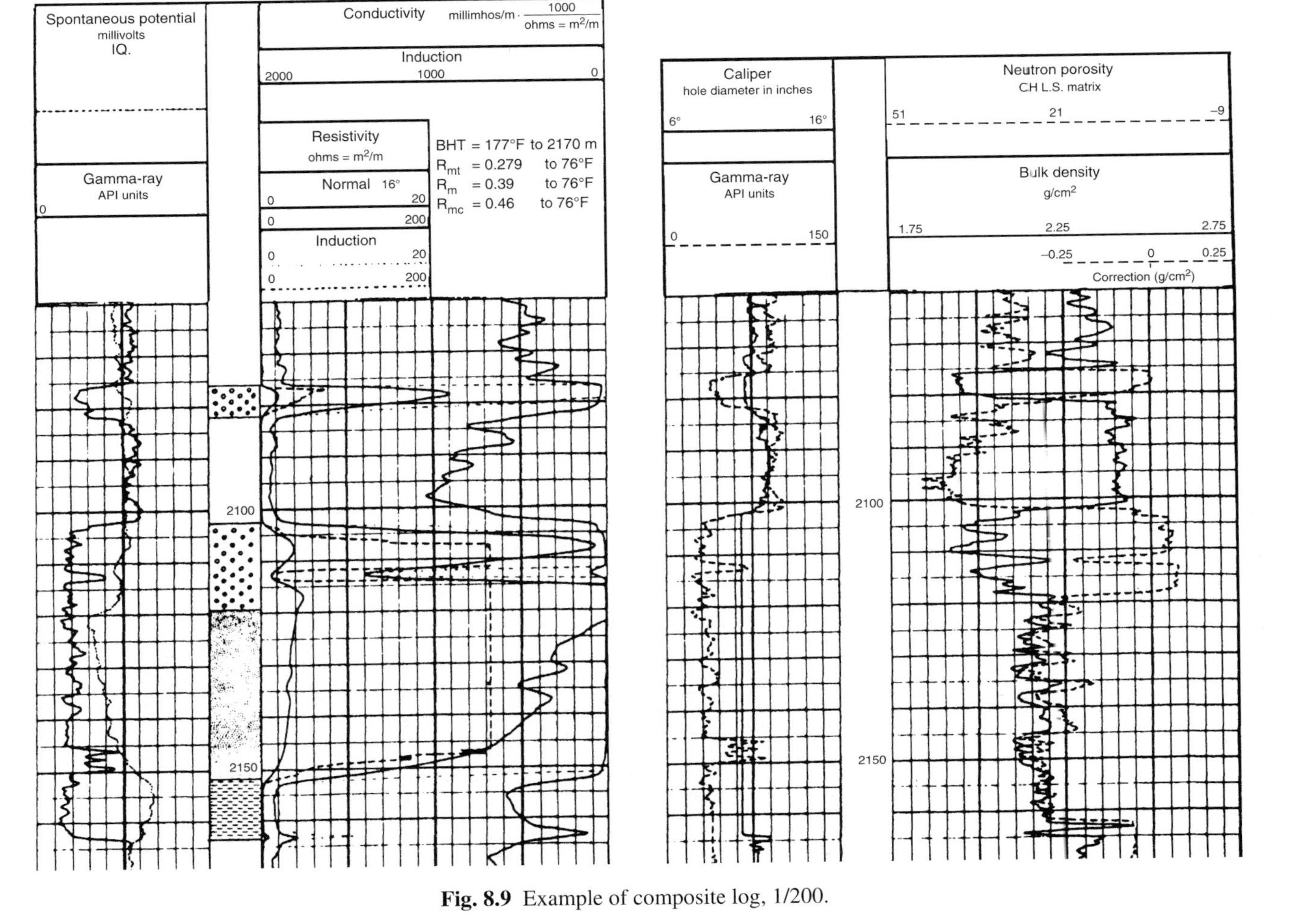

Fig. 8.9 Example of composite log, 1/200.

Wireline logs are recorded at several different times in the life of an oil field:

- During the exploration stage, the well program specifies a certain number of logs which will yield as much information as possible on the drilled formations. Further logs can be recorded at the request of the rig geologist to help solve problems encountered during drilling.
- Wireline logs during exploration serve as a decision-making tool because of the local data they provide on formations. They are used to determine whether exploration should be pursued, where wells will be located or on the contrary whether operations should cease locally or regionally.
- During the development phase, wireline logs can help refine the assumed reservoir model from the standpoint of its geological structure and the fluids it contains.
- During the production phase, certain logs provide data on how production is developing and how fluids are moving in the well and in the reservoir.

8.2.3 Making the measurements

The sonde, or measuring instrument package, is run into the borehole at the end of a wireline that connects it to recorders on the surface located in the logging unit. The unit is mounted on a truck parked near the well for operations on shore (**Fig. 8.10**) and has its own stationary position on platforms off shore.

The sonde is made up of a series of measurement instruments and an electronic cartridge together in a package. It may be about ten centimeters in diameter and from a couple of meters to thirty meters and more in length when tools are combined.

It has to withstand the most commonly found pressure and temperature conditions in oil wells. Each instrument includes one or more transmitters and/or receivers located either on the body of the sonde or on a pad that slides along the wall of the hole.

The wireline, which is the mechanical support for the tools, must be able to hold up to a number of stresses (pressure, temperature, corrosion, tension, etc.). It supplies the sonde with electricity and transmits several data to the surface at the same time. It is the only way to measure depth, as it has a marking system by passing on a calibrated pulley.

The standard wireline has a diameter of 11.8 mm and seven conductors plus double-spiral armor. It must meet severe mechanical and electrical standards. Several thousand meters long, it weights several tons. It is one of the more expensive pieces of equipment.

Panels on the surface serve to monitor, calibrate and process data, while recorders store the measurements on magnetic tape and photographic film. The data can be processed by a computer and quick interpretation programs in on-site computerized units or sent to large computer centers.

Recordings are usually made when the sonde is being brought back up out of the hole and different sondes are run at different speeds. The sonde may hang freely in the hole, be kept in place by metal centralizers or it may be pressed against the wall of the hole by an arm remote controlled from the surface.

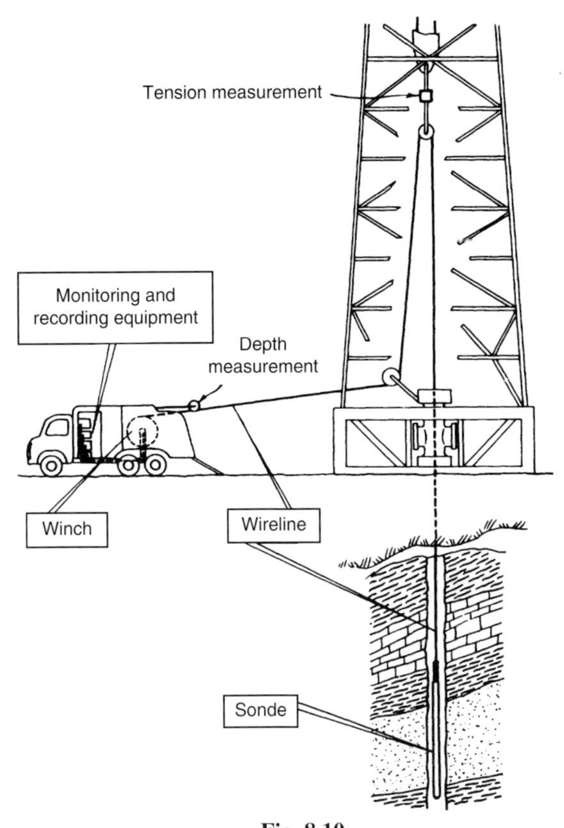

Fig. 8.10

Running well logs *(Source: Well Logging, R. Desbrandes, Editions Technip).*

Depending on the configuration of the tool, measurement may involve all of the formations around the sonde (sphere, sheet shapes) or it may be more directional (slice, cone or even part of a straight line) (**Fig. 8.11**).

Investigation depth varies from a few centimeters (induced radioactivity logs) to one or more meters (resistivity logs). The vertical definition, i.e. the smallest thickness of bed that can be measured, is about one meter for most tools and one centimeter for electric micrologs.

Wireline logging operations used to mean that all drilling had to shut down for several days. However, they are now much faster since the introduction of simultaneous recording of several parameters using a combination of measurement instruments. Even so, it is often necessary to make several repeat runs since the amount of data that can be transmitted by the wireline is limited by the number of available conductors. All data are stored and depth-adjusted.

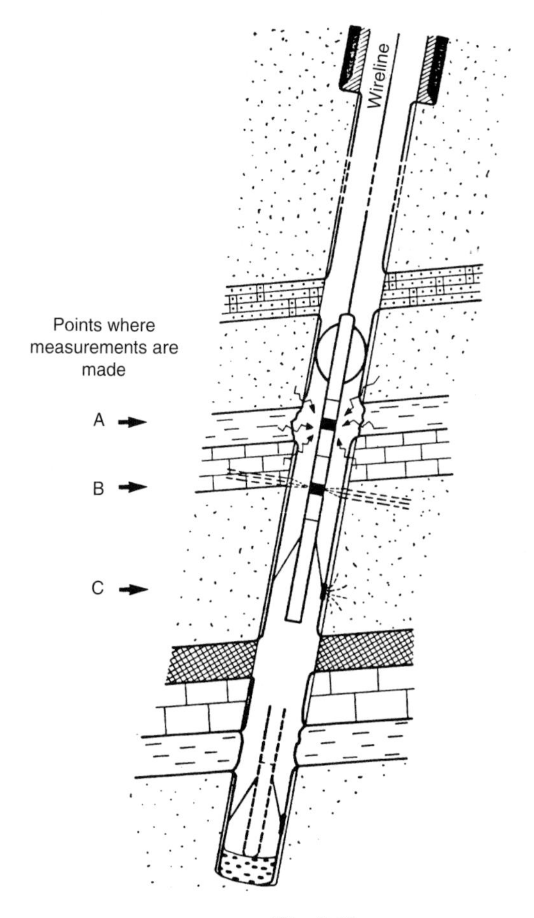

Fig. 8.11
Simplified diagram showing measurement.

8.2.4 Parameters that are measured

There are two types of phenomena that can be studied by wireline logging:

- natural ones (temperature, spontaneous radioactivity, etc.) that are measured by a sensor or a receiver with no signal transmitted beforehand,
- induced ones (radioactivity, electric logs, wave traveltime, etc.), generated by a transmitter or a source and measured by one or more receivers.

Table 8.1 lists the major phenomena, the tools that measure them and the geological parameters associated with them. Formation characteristics are determined only by combining the measurements made by the different tools.

TABLE 8.1

Major phenomena and parameters

Logs	Phenomena		Tools			Parameters measured	Parameters deduced
Conventional	Electric		Spontaneous polarization			ddp	R_w
			Resistivities	Macrologs	Inductions	C	S_w
				Macrologs	Electric logs	R	S_w
				Macrologs	Laterologs	R	S_w
				Micrologs	Electric logs	R	S_{xo}
				Micrologs	Laterologs	R	S_{xo}
	Radioactivity	natural	Gamma type		Gamma ray		
		induced	Neutron-neutron		Neutron	ϕ	ϕ
			Gamma-gamma		Density	b	ϕ
	Wave propagation				Sonic	t	ϕ
Non-conventional	Induced radioactivity		Neutron lifetime			Σ or	ϕ, S_w
			Spectral study of neutron/matter reaction			A_i/A_j	
	Electromagnetic		Nuclear magnetic resonance				ϕ, S_{xo}
			Electromagnetic propagation time				k, ϕ, S_{xo}
Auxiliary	Geometry	of the hole			Caliper	d	
	Structuring	of formations			Dipmeter		
	Sampling	rock/fluids			Sidewall coring		
					Formation-interval tester		k, R_w, hc
Drilling					Inclinometer		
					Cement bond log		
					Stuck-point indicator		
Production			Flow rate measurement		Flowmeter		
			Density measurement		Gradiometer-pressure gage-densimeter		
			Temperature measurement		Thermometer		

8.2.5 Background information and basic geology

8.2.5.1 Definitions

A rock is made up of solid components that constitute the matrix (simple or complex) and are arranged in such a way that there are "voids" (the pores). Porosity is the ratio of the volume of the pores to the total volume of the rock:

$$\text{porosity} = \frac{V_p}{V_t} \text{ (generally expressed in per cent);}$$

where

V_p = volume of the pores,

V_t = total volume.

Beginning at a certain depth (a few hundred meters at the most), the pore volume is considered to be totally occupied by fluids other than air. Commonly the fluids are water (fresh or salt), sometimes liquid or gaseous hydrocarbons, or other gases (CO_2, H_2S, etc.).

Water saturation is the ratio between the pore volume occupied by water and the total pore volume:

$$S_w = \frac{V_w}{V_p} \text{ (generally expressed in per cent);}$$

where

V_w = volume of water contained in the pores.

The oil or gas, i.e. hydrocarbon, saturation is defined in the same way:

$$S_{hc} = \frac{V_{hc}}{V_p}$$

$$S_{hc} = 1 - S_w$$

Permeability of a rock is its capacity to allow the fluids in its pore spaces to circulate. This involves the concept of communication between pores.

Reservoir rock is the term given to all porous, permeable rocks (e.g. sandstone, limestone). All compact (without pore spaces), nonfractured rocks (e.g. salt) and all rocks that are impermeable enough to keep fluids from migrating vertically (e.g. shale) are termed permeability barrier, or cap rock.

The amount of oil or gas in place in a reservoir rock with a total volume V_t and porosity Φ is equal to:

$$\Phi \times S_{hc} \times V_t$$

Wireline logging enables rock porosity and water saturation to be assessed. Permeability is a parameter that is more difficult to evaluate and logging can only give an order of magnitude.

8.2.5.2 The concept of invasion

During drilling, mud density is kept high enough so that the hydrostatic pressure of the mud column is greater than the pressure of the fluids contained in the reservoirs. The aim is to keep the walls of the borehole from caving in and prevent fluid influxes from intruding into the wellbore. The difference between the two pressures allows the liquid phase of the drilling mud (filtrate) to penetrate into porous, permeable formations. Meanwhile, the solid suspended particles in the mud are deposited on the borehole walls corresponding to these formations. Once the deposit of solid particles (mud cake) has reached a certain thickness, invasion slows down. It slows down even more suddenly when the mud contains fluid-loss additives. In a radius around the wellbore, all the formation water and part of the oil and/or gas are flushed out of the invaded zone and replaced by the filtrate. There is a transition zone containing formation water, filtrate and oil and/or gas which separates the flushed zone from the zone that has not been invaded by the filtrate.

Measurements of physical parameters are influenced by the invaded zone and therefore do not give exactly the same values as in the noninvaded zone.

Some tools, termed porosity tools, and resistivity micrologs have an investigation depth limited practically to the flushed zone. They measure its characteristics. Deep investigation resistivity tools measure parameters between the ones in the flushed zone and the ones in the noninvaded zone.

Well log interpreters use the measurements made at different investigation depths to get the data required for quantitative reservoir evaluation by means of relevant corrections and calculations. Only a wide and carefully selected range of logs can yield this result. One piece of information lacking can make it difficult or even impossible to quantify the reservoir properly and determine its lithology and the nature of its fluids.

8.2.6 The tools

(See **Table 8.2**: Trade names of the major measurement tools).

8.2.6.1 Conventional tools

A. *Resistivity logs*

a. Spontaneous polarization

Spontaneous polarization, or spontaneous potential (**Fig. 8.12**), is related to alternating shale layers and porous, permeable beds.

Ion exchanges between the formation water and the filtrate through the shale and the invaded zone generate a difference in potential in the mud column between shale beds and reservoirs. The difference in potential depends on the difference in salinity between the two fluids (**Fig. 8.12a**). It serves to evaluate the resistivity R_w of the formation water, a fundamental value required to determine the water saturation.

TABLE 8.2
Trade names of the major measurement tools

	Gearhart*		**Schlumberger**		**Welex***		**Western Wireline services**	
Laterolog	Dual Laterolog	DLL	Dual Laterolog	DLL	Dual Laterolog Logging Tool	DLLT	Dual Laterolog	DL
Induction	Induction Electric log Dual Induction Laterolog	IEL DIL	Induction Electrical Logging Induction Spherically Focused Dual Induction Laterolog Dual Induction SFL	IEL ISF DIL DIS	Dual Induction Guard Log	DIL	Induction Electrolog Dual Induction Focused Log	IEL DIFL
Microresistivity	Microlog Microlaterolog Microspherically Focused Log	MEL MLL MSF	Microlog Microlaterolog ou Proximity Microspherically Focused Log Derived Microlog (from MSFL)	ML MLL/PL MSFL MSFL-M	Microlog Caliper Tool Micro-Guard Logging Tool	MICRO MG	Minilog Microlaterolog Proximity Log	ML MLA PXL
Electromagnetic propagation	Dielectric Constant Log	DCL	Electromagnetic Propagation Tool Deep Propagation Tool	EPT DPT			Dielectric Log (47 MHz) Dielectric Log (200 MHz)	DCLL DCLH
Neutron	Sidewall Neutron Compensated Neutron Log	SNT CNS	Compensated Neutron Multidetector Neutron	CNL CNT-G	Sidewall Neutron Dual Spaced Neutron	SWN DSN	Sidewall Neutron Compensated Neutron	SWN CN
Density	Compensated Density Log	CDT	Formation Density Litho Density	FDC LDL	Spectral Density Logging Tool	SDLT	Compensated Densilog Z-Densilog	CDL ZDL
Gamma ray	Gamma Ray Spectral Gamma Ray	UGR SGR	Gamma Ray Gamma Ray Spectroscopy	GR NGS	Gamma Ray Log Compensated Spectral Natural Gamma Ray	GR CSNG	Gamma Ray Spectralog	GR SPL
Acoustic	Borehole Compensated Sonic Long Space Sonic	BCT BCT-EA	Sonic Log Sonic Long Spacing Digital Sonic	BHC SLS SDT	Compensated Acoustic Velocity Log Full Wave Acoustic Tool	AVL FWAT	Normal Space Acoustilog Long Space Acoustilog	AC ACL
Dipmetering	Four Electrode Dipmeter Omnigraphic Dipmeter Tool	FED	High Resolution Dipmeter Stratigraphic High Resolution Dipmeter	HDT SHDT	Resistivity Dip Log	DIP	4-Arm Diplog	DIP

* Halliburton Logging Services.
NB : 1) This list is not exhaustive. 2) Other contractors provide comparable services under other trade names.
3) Most of the tools can be combined together under names and rules specific to each contractor (see catalogs).

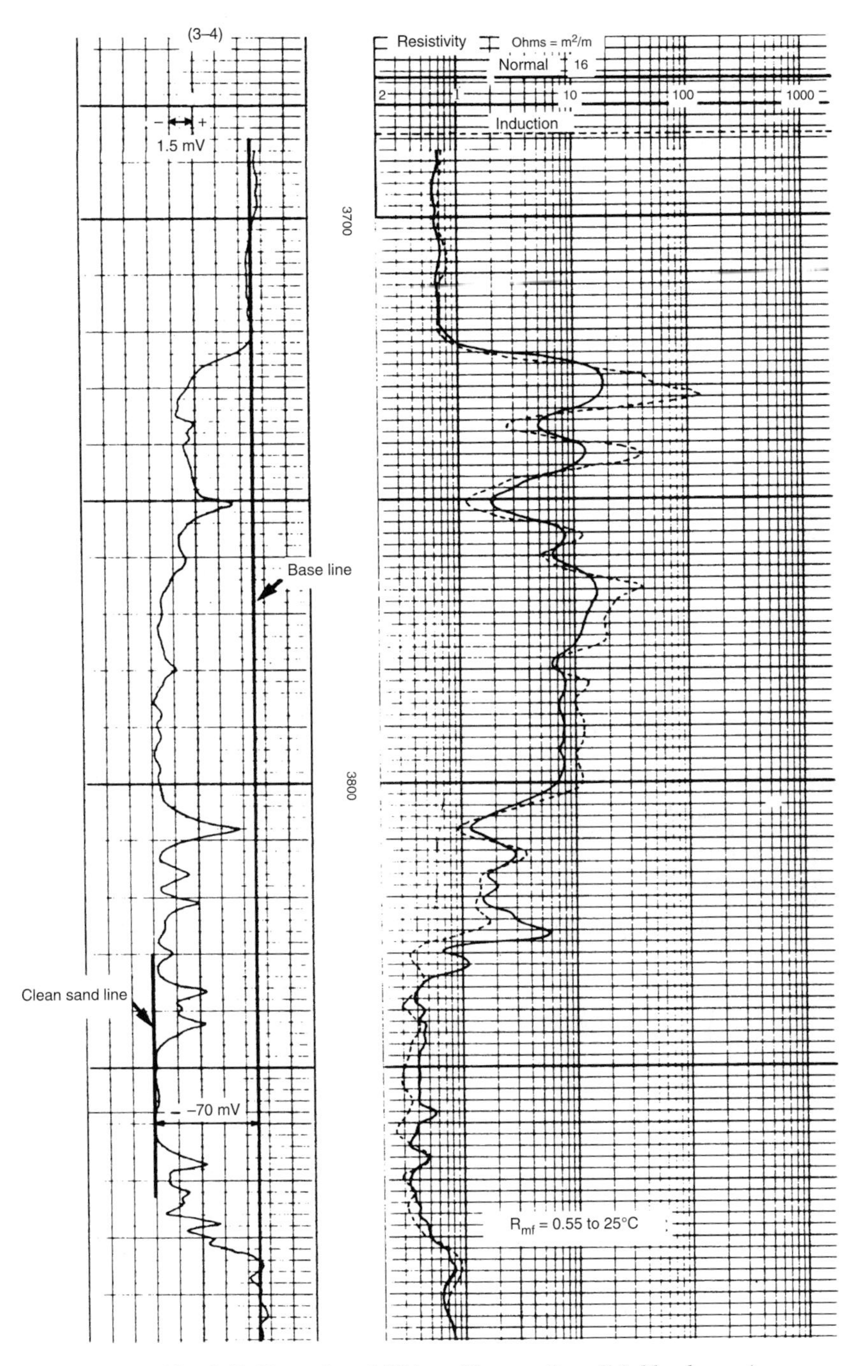

Fig. 8.12 Examples of SP logs *(Source: Dowell Schlumberger)*.

b. Resistivity logs

Tools that measure formation resistivity send current into the formation or generate induced currents (for inductions) by means of electrodes or transmitting spools.

Depending on the tools' depth of investigation, the following types can be distinguished:

- Macrologs that give information on the noninvaded zone:
  - induction (**Fig. 8.13**),
  - resistivity tools:
    - nonfocused,
    - focused, of the Laterolog type (**Fig. 8.14**),
- Micrologs that measure in the flushed zone; there is a nonfocused microlog (**Fig. 8.15**).

Modern micrologs are focused (**Figs. 8.16a and b**).

B. Radioactivity logs

A distinction is made between natural and induced radioactivity.

a. Gamma ray (GR)

The natural radioactivity that is measured is of the γ type. It is generally related to shales that contain the most radioactive matter (uranium, thorium, potassium). Recording the gamma radioactivity therefore makes it possible to identify shale beds.

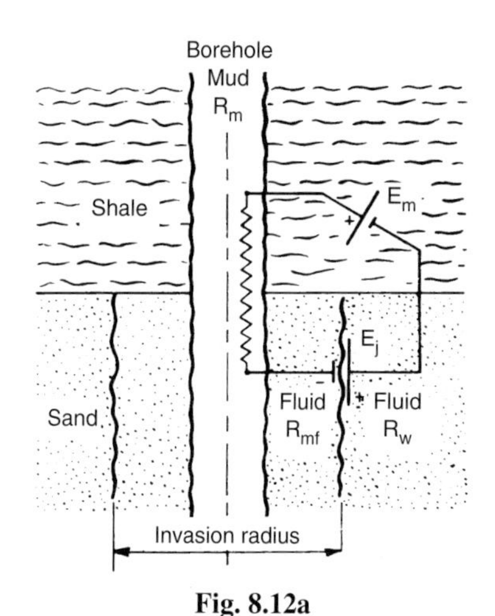

Fig. 8.12a

Origin of SP potentials
(Source: Dowell Schlumberger).

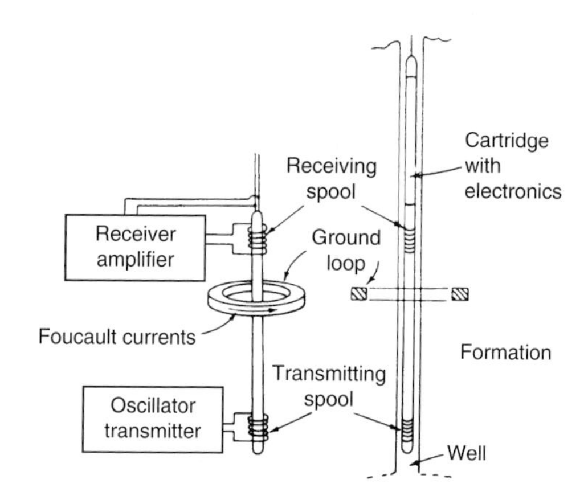

Fig. 8.13

Principle of the two-spool induction sonde
(Source: Well logging, R. Desbrandes, Editions Technip, Paris).

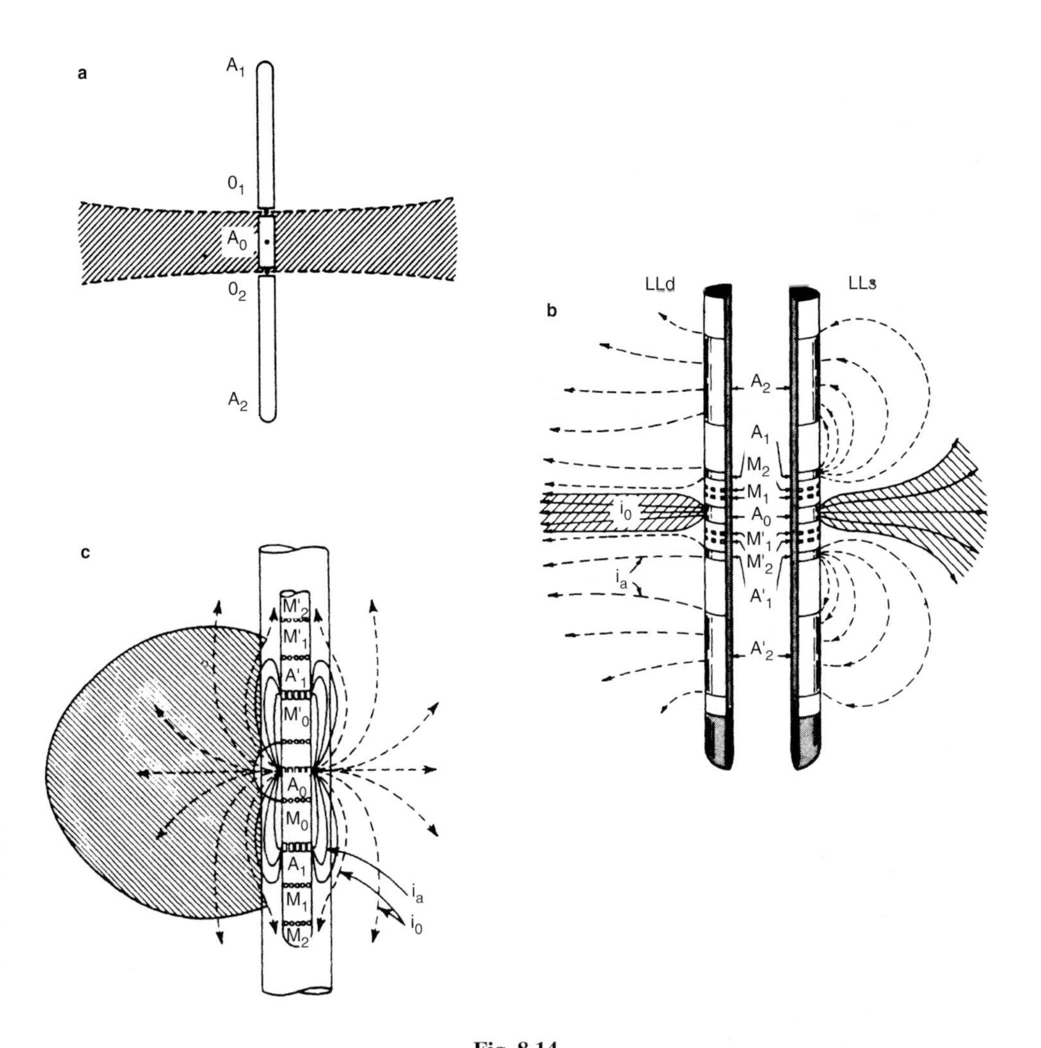

Fig. 8.14

a. Initial principle of focusing. **b.** Dual Laterolog. **c.** Spherically-focused log, SFL.
(Source: Dowell Schlumberger).

However, in certain formations other than shales, there can be concentrations of radioactive elements (potassium salts, radioactive ores). Differentiating between them can be done by correlation with other logs.

The natural radioactivity measurement is a good correlation tool in the casing. Radioactivity is recorded on a standard API scale (**Fig. 8.17**).

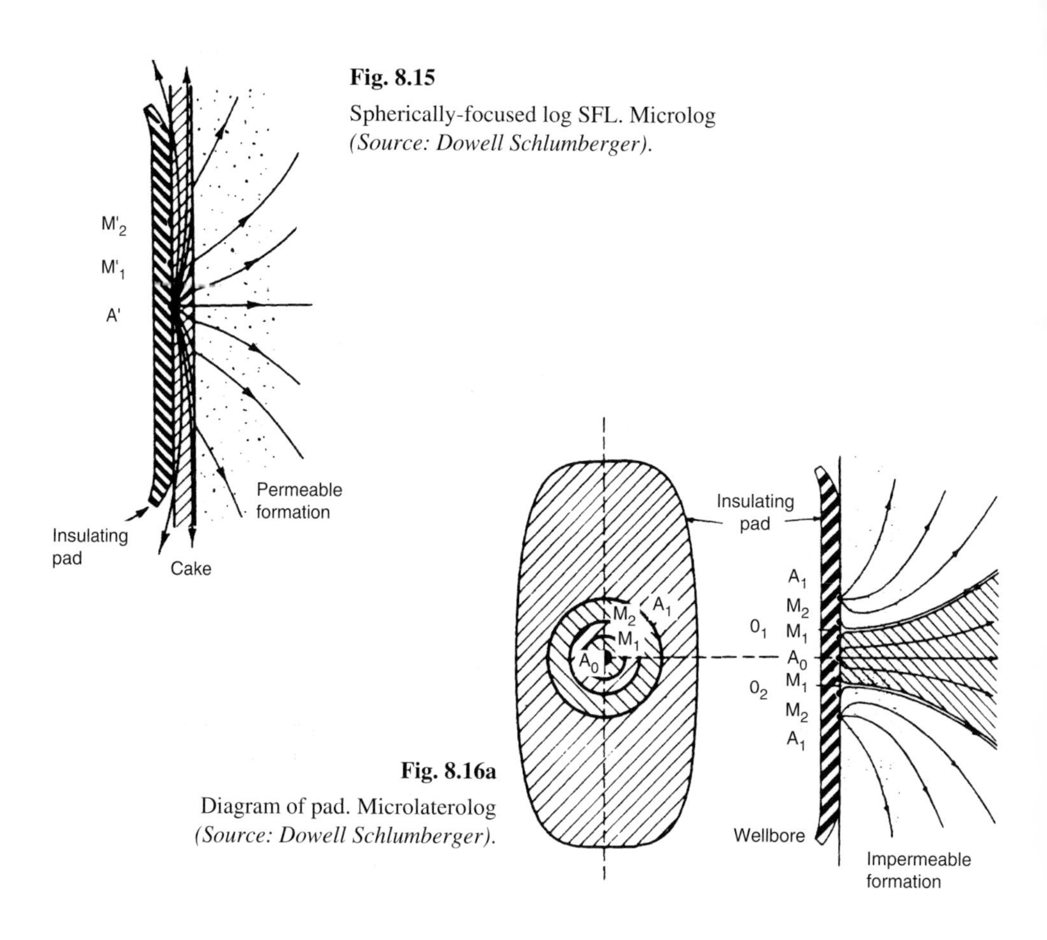

Fig. 8.15

Spherically-focused log SFL. Micrulog *(Source: Dowell Schlumberger).*

Fig. 8.16a

Diagram of pad. Microlaterolog *(Source: Dowell Schlumberger).*

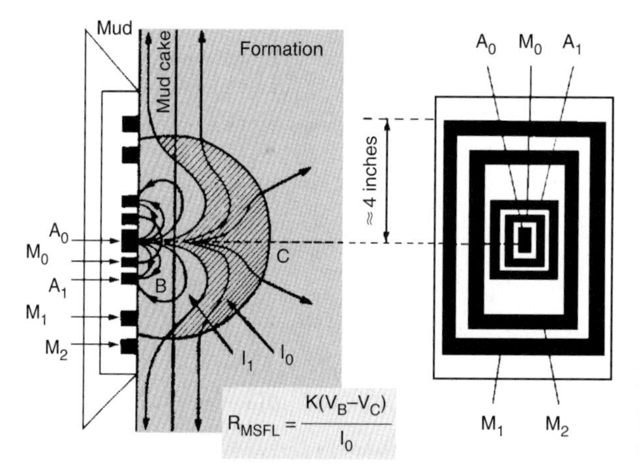

Fig. 8.16b

Micro SFL *(Source: Dowell Schlumberger).*

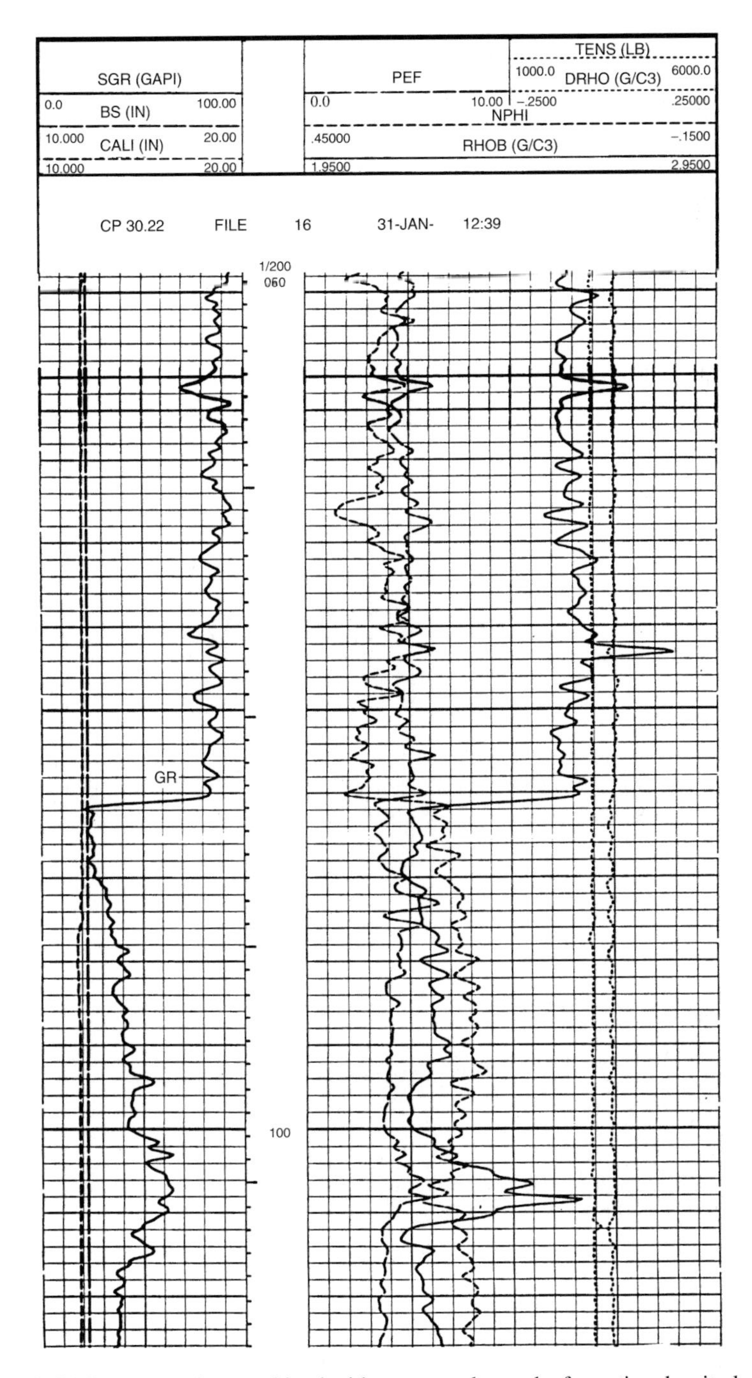

Fig. 8.17 Gamma ray log combined with a neutron log and a formation density log.

b. Radioactivity

Radioactivity is induced by bombarding formations with neutrons or gamma rays emitted by relevant radioactive sources.

- The neutron log (neutron-neutron type) (**Fig. 8.18**)

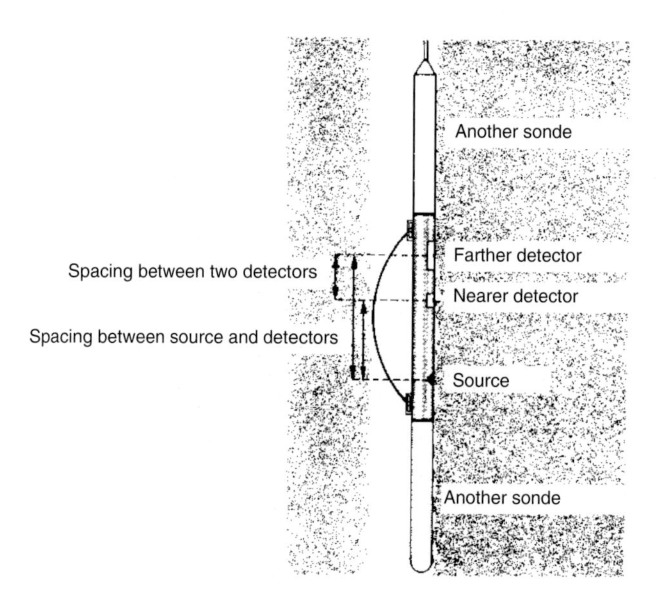

Fig. 8.18
Diagram of CNL sonde *(Schlumberger)*.

The tool measures the density of neutrons reaching one or two detectors. Neutron density chiefly depends on the number of hydrogen atoms in formation fluids. Accordingly, the log allows apparent porosity to be evaluated (**Fig. 8.19**).

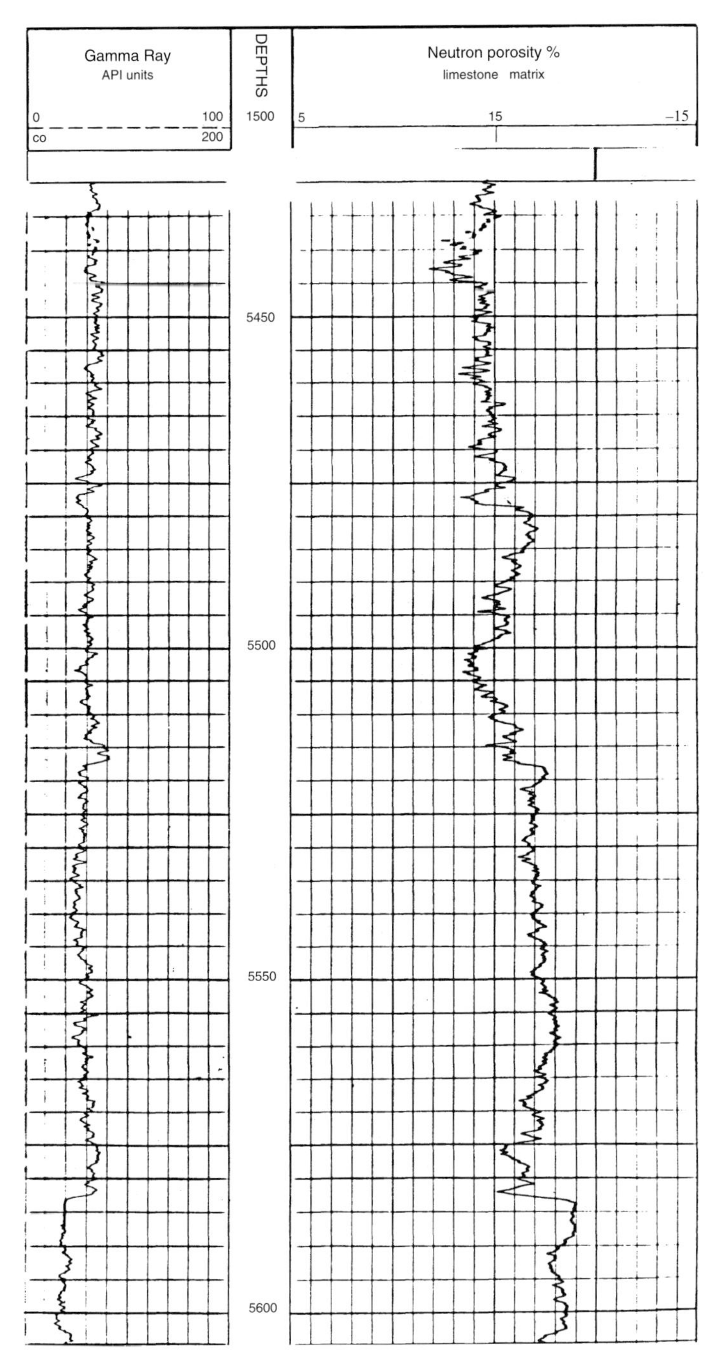

Fig. 8.19 Neutron log.

- The density log ($\gamma - \gamma$)

The tool measures the attenuation of incident gamma radiation, which is dependent on the electron density of the medium (**Fig. 8.20**).

A simple relationship gives the formation density, which is in turn related to the porosity. The curve is represented on a density scale (**Fig. 8.21**).

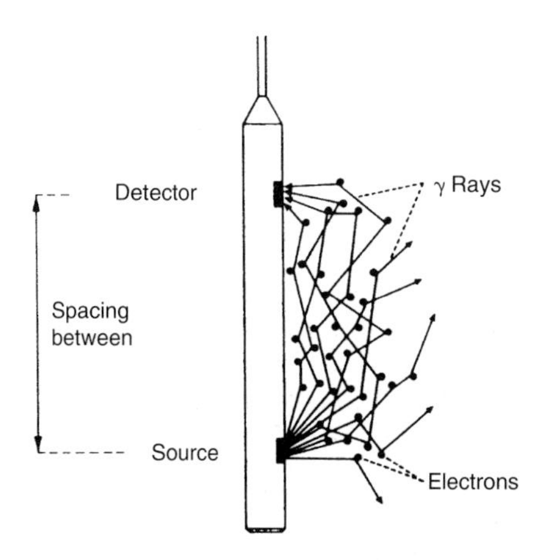

Principle of the gamma-gamma sonde

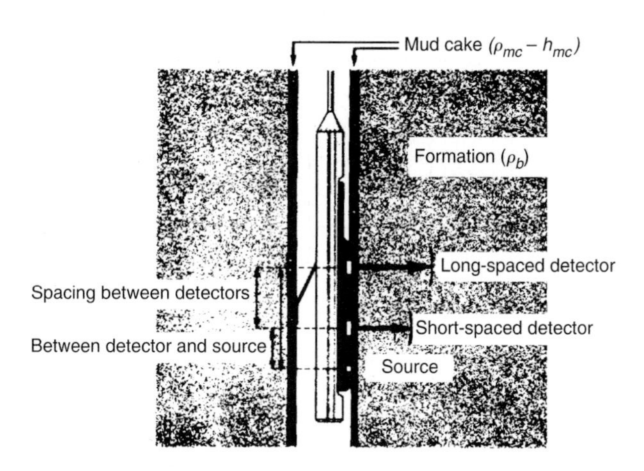

Principle of compensated logging tool

Fig. 8.20

(Source: Diagraphies différées. Bases de l'interprétation, O. Serra, SNEAP).

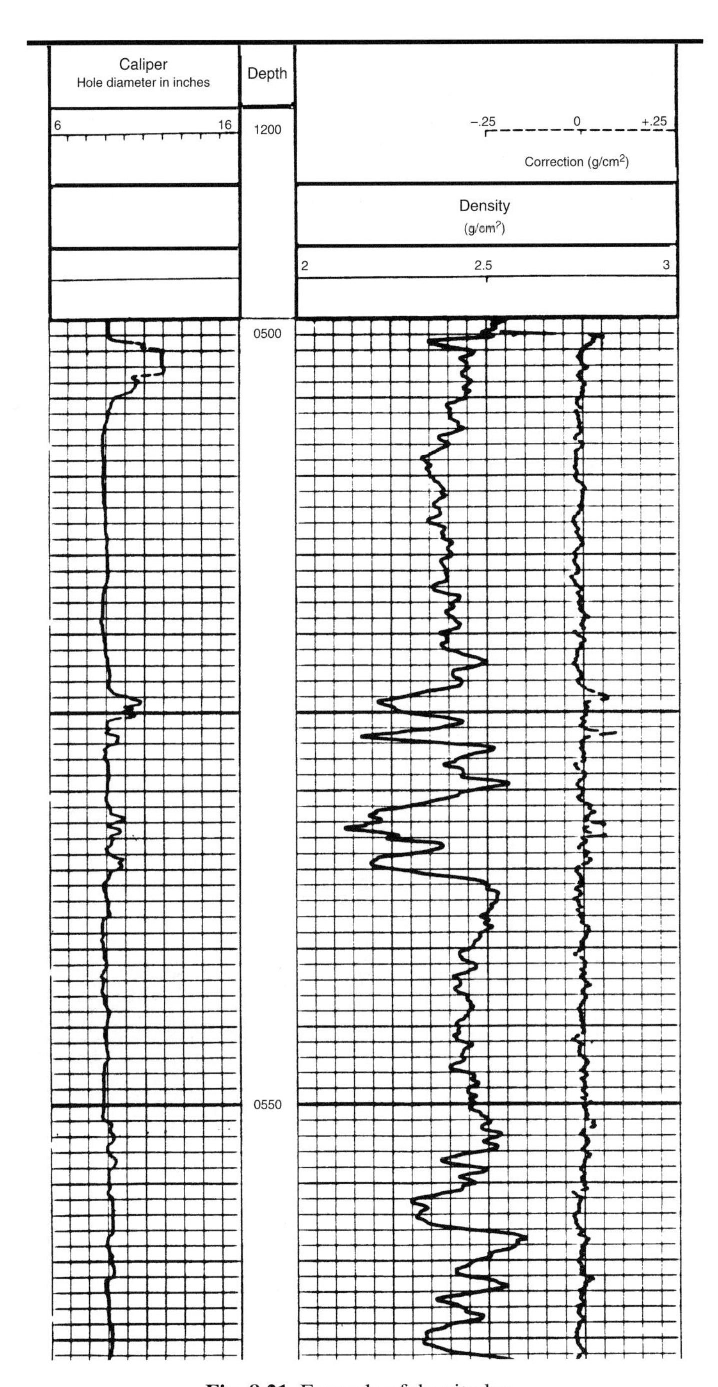

Fig. 8.21 Example of density log.

C. *The acoustic log*

The basic principle is to record the traveltime (inverse of the velocity) of a wavetrain that is propagated in the formation along the wall of the borehole between a transmitter and a receiver. The arrival of the fastest wave is detected (**Fig. 8.22**).

The complete recording of the wavetrain helps differentiate between different wave arrivals (compression, shear, mud, etc.).

There is a relationship between the traveltime and the porosity. The curve is recorded on a scale in microseconds per foot (**Fig. 8.23**).

Other applications of the acoustic log are possible:

- to study fracturing,
- to add further information to geophysical data,
- to check cement jobs,
- to evaluate formation permeability.

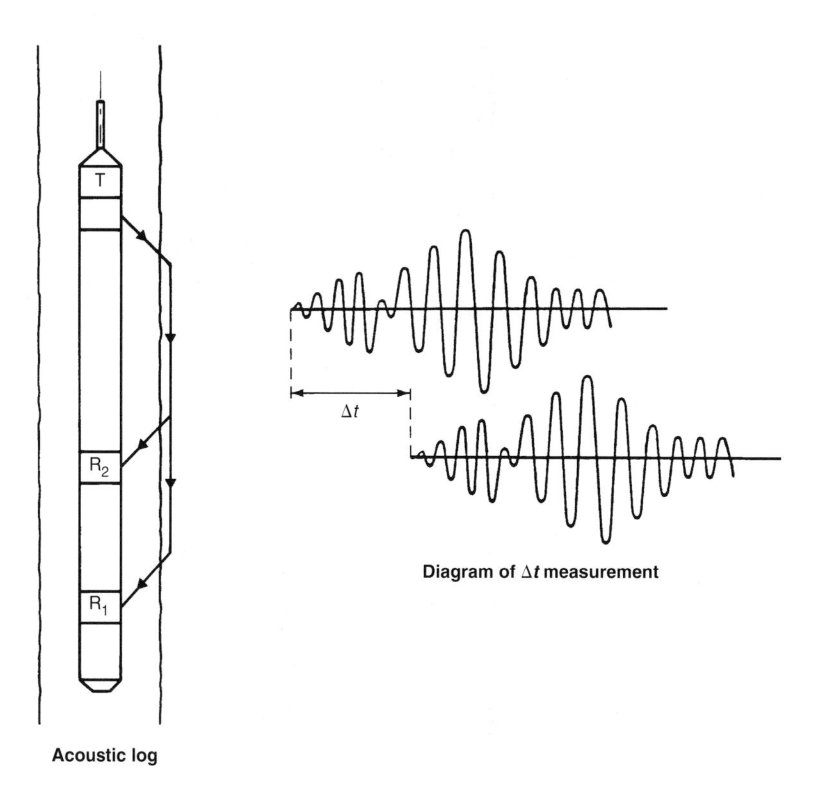

Fig. 8.22

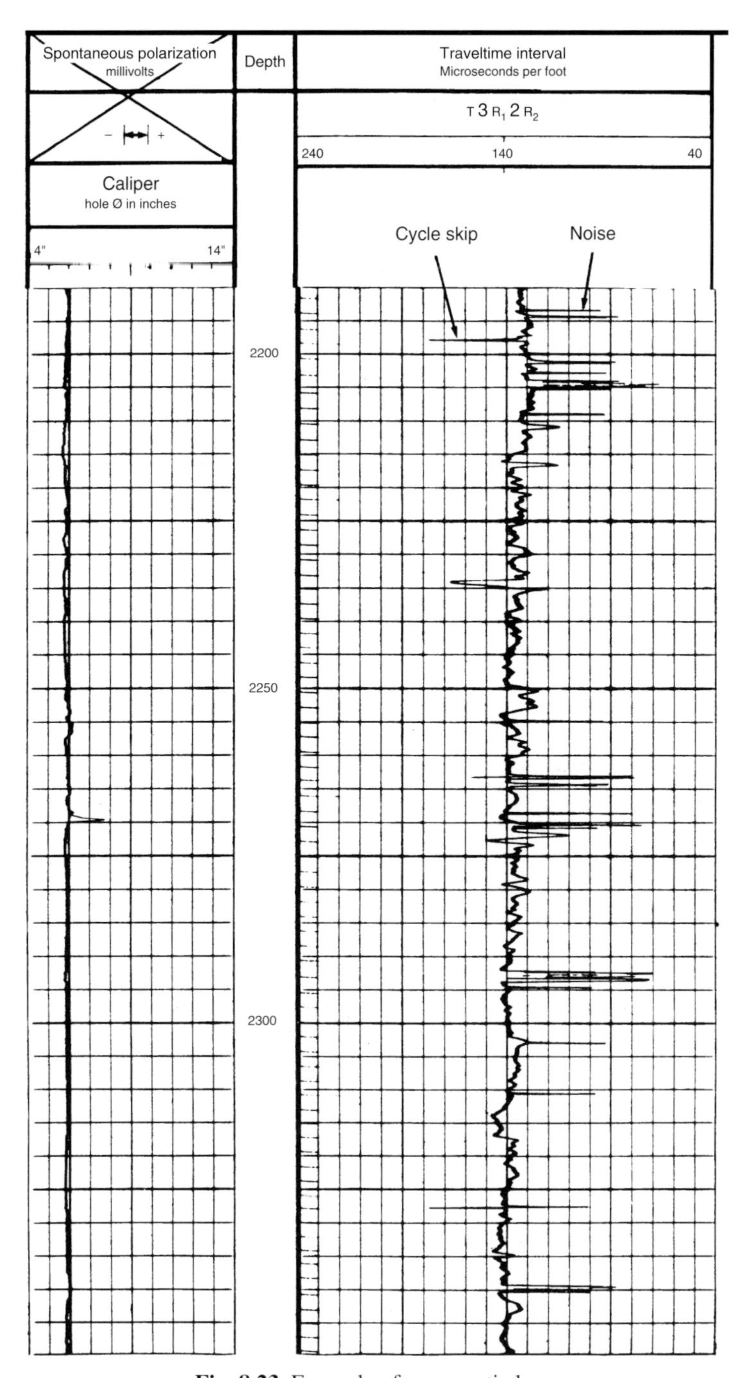

Fig. 8.23 Example of an acoustic log.

8.2.6.2 Nonconventional logs

A. *The neutron-lifetime log*

This is a particular measurement of the radioactivity induced by bombarding the rock with neutrons. The amount of chlorine contained in the rocks can be studied and yields data mainly on the formation water.

B. *The spectral study of induced radioactivity*

This method is already used by *Dresser Atlas* (Neutron Lifetime) and seems to hold a lot of promise for log interpretation. It gives direct access to determining the atoms and fluids present in formations (*Schlumberger*'s GST).

C. *The nuclear magnetic resonance log*

Here investigation involves the time it takes hydrogen atoms to get reoriented after they have been forced by a magnetic field generated by the tool.The measurement gives an indication of the free water and the nonviscous hydrocarbons contained in the formations.

D. *The electromagnetic propagation log*

The tool measures the propagation time of an electromagnetic wave and its attenuation. Porosity and water saturation values in the flushed zone can be deduced.

E. *Well shooting*

A geophone lowered into the well records the arrival of a wave emitted on the surface at a given distance from the well. The wave goes through all of the formations (**Fig. 8.24**). Comparison of this seismic time and the traveltime of the acoustic wave serves to correlate between well logs and seismic profiles.

8.2.6.3 Auxiliary logs

A. *Sidewall sampling*

a. Sidewall coring

Core samples 25 mm in diameter and 20 to 50 mm long can be taken by firing hollow bullets into the wall of the borehole (**Fig. 8.25**).

b. Measuring formation pressures, sampling fluids

Production geologists and drillers are both interested in finding out formation pressure gradients. With this data, drillers can improve drilling conditions and safety when approaching an undercompacted zone.

The measurements can be made by repeat formation testers that are run in on a wireline. The testers give the pressure at different depths and may take one or two fluid samples.

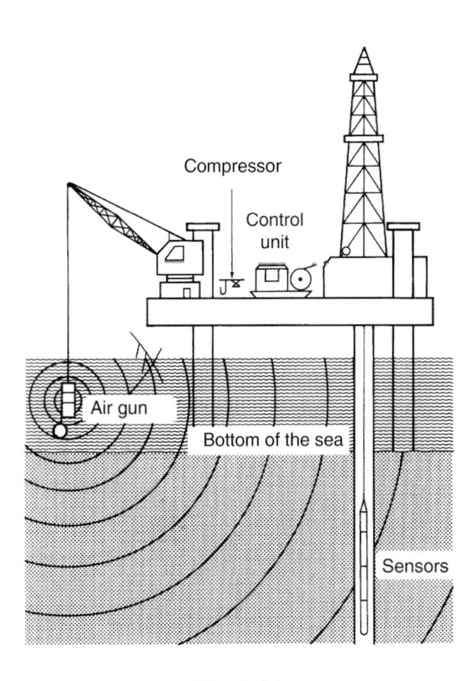

Fig. 8.24
Well shooting.

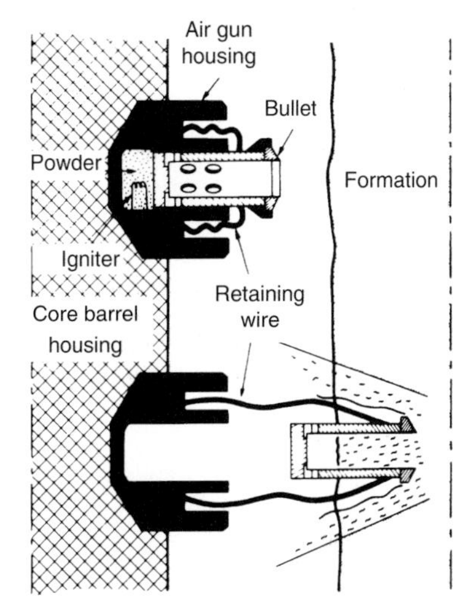

Fig. 8.25
Sidewall coring.

The measurements are made in the open hole. However, some testers can be run into a cased hole but allow only one or two pressure measurements to be made.

A gamma ray log can usually be combined with these tools to make correlations with other wireline logs easier and to place the sonde in the right position.

The tool

- Repeat Formation Tester (RFT) by *Schlumberger* (**Fig. 8.26**),
- Formation Multi-Tester (FMT) by *Dresser Atlas*.

The tool is held against the wall by a hydraulically-controlled sealing pad and counter-pad. Two pretest chambers open one after the other and take samples of the fluid coming from the formation. A strain gage measures the flow pressures, then the pore pressure.

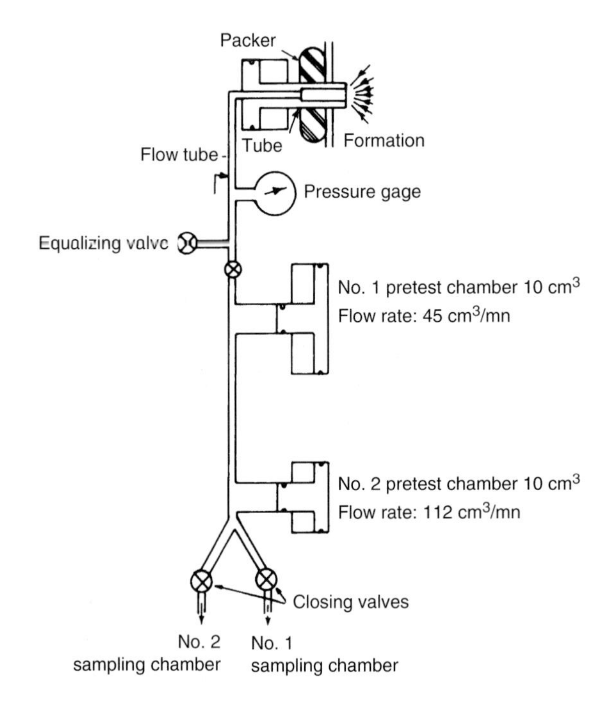

Fig. 8.26

Sidewall sampling with RFT *(Source: Dowell Schlumberger).*

Then the measurements can be made again at another depth where a sample of formation fluid can be taken. Two chambers are available and they can be filled at two different depths or at the same depth. An initial draw-off into the larger chamber gives a more representative fluid sample collected in the second chamber.

B. Stuck-point indicator

When pipe gets stuck during drilling, the lowest free point needs to be known so that as much of the drill string can be retrieved as is possible.

The tool

- Stuck-Point Indicator Tool (SIT) by *Schlumberger* (**Fig. 8.27**),
- Free-Point Indicator Back-Off Combined Tool (FPIT) by *Schlumberger*,
- Free-Pipe Indicator Tool by *Dresser Atlas*.

Detecting where the pipe is stuck is done by tools that are run step by step into the drillpipe with two anchoring systems (arms or guides) located on either side of a measurement point. A strain gage records the axial and angular strain that the portion of

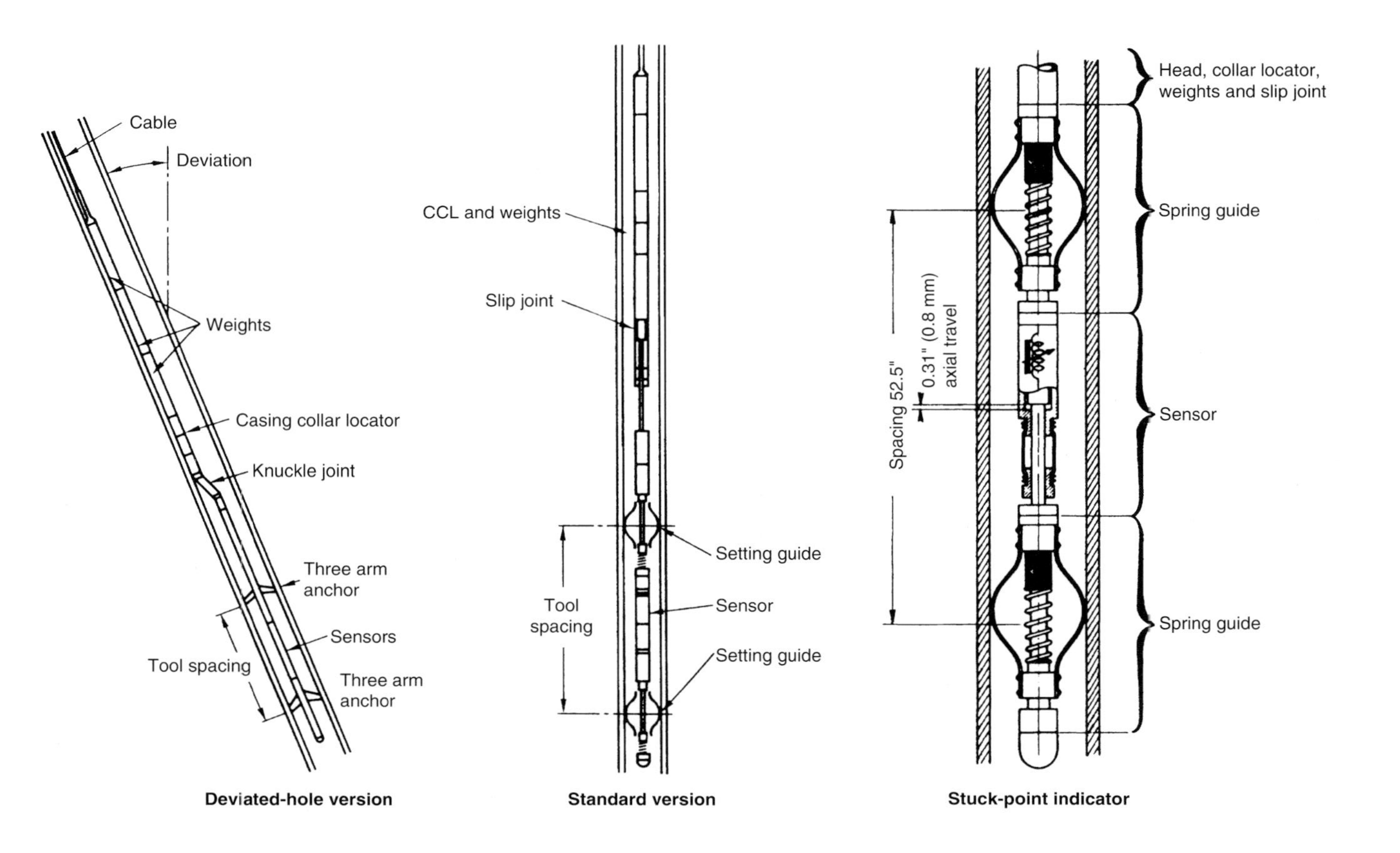

Fig. 8.27
The stuck-point indicator tool (SIT) *(Source: Schlumberger)*.

pipe between the anchoring systems undergoes when tension and torque are exerted under constant conditions. The tension and torque are transmitted via the hook and the rotary table along the drillpipe to the stuck point. Below this point, the stresses are not transmitted to the stuck portion.

Torque transmission is measured by applying right-hand torque in the pipe-screwing direction. The pipe should be "worked" to get good transmission of the stresses to the stuck point.

Measurements and interpretation

The gage gives readings from 0 to 100 that are proportional to the torsion and stretch of the pipe up to an extreme value which should preferably not be exceded. The measurement is converted into percentage of free pipe: 0% when no pulling force is exerted and 100% in free pipe subjected to a stress condition defined on the basis of the pipe's characteristics.

The chances of unscrewing after backoff are usually considered good when the strain gage reading gives a degree of freedom greater than 85% on torque and tension. However, depending on the type of sticking, lower figures can still allow the pipe to be unscrewed. Torque is generally more readily transmitted than pulling force:

- In a straight well, absolute sticking (of stabilizers on drill collars) can be recognized by a sudden drop in the two transmitted stresses.
- When differential pressure sticking occurs, the drop in stress transmission happens gradually, as pulling force is generally not transmitted as well as torque.

A backoff can be attempted at the joint located just above the depth where 80% torque freedom is read.

- In deviated boreholes, sticking also makes itself felt by a gradual decrease in transmitted stresses. The backoff point can be chosen according to the pulling force applied to the pipe (even if the torque reading ranges between 70% and 85% freedom), since torque is transmitted better when pulling force is greater.

C. Cement logs

When cement sets, heat is released and there is an increase in cement bond stress, so a casing cement job and its quality can be assessed by the following:

- Temperature well logging,
- Wavetrain amplitude and recording logs (CBL, VDL, etc.),
- Cement evaluation tool (*Schlumberger*'s CET).

a. Temperature well logging

The heat released when cement sets causes an increase in temperature in the cemented zones. The recording is made as the tool (resistance thermometer) is lowered in and can be made during a six to 36-hour time frame depending on the type of cement. This log locates the top of the cement in the annulus.

b. *Measuring an acoustic wave attenuation: CBL and VDL*

Measuring the amplitude and the traveltime

The amplitude of an acoustic wave decreases as it moves in the medium it is going through. The attenuation depends on the elastic properties of the medium. It is measured and applied to determine the quality of a casing cement job.

Principle

A transmitter periodically generates a wavetrain with a frequency varying between 15,000 and 30,000 Hz, depending on the type of tool. The wave goes through the mud, the casing, the cement and the formation if these different media are acoustically coupled. It is then detected by a receiver which is located on the body of the tool (usually 3 ft away from the transmitter).

The velocities of the different waves that are transmitted and produced from one medium to another in succession depend on the physical characteristics of the medium and the type of wave. The acoustic energy traveling along the casing is propagated faster than the formation waves (P then S waves) and the formation waves are in turn faster than the waves in the mud. **Fig. 8.28a** shows a schematic recording of such a wavetrain arriving at a receiver.

The first arrival is detected according to the same principle as for measuring the propagation time of an acoustic wave in a formation (sonic), by a minimum threshold of detectable energy (bias). The amplitude of the first arrival (usually the casing wave) is measured by positioning a window.

This log is called a cement bond log or CBL and is used to study and quantify cement job quality.

Interpretation

- When a casing is not cemented, all the acoustic energy circulates along the steel. There is very little wave attenuation and the amplitude of the first arch of the signal is considerable (**Fig. 8.29a**).
- If the casing is properly cemented, the energy is propagated through the cement to the formation. The casing wave is then greatly weakened (**Fig. 8.29b**).
- If a casing is poorly cemented, the energy is divided up between the casing and the formation (**Figs. 8.29c and 8.29d**).

Recording the wavetrain

The assessment of a cement job can be thrown off by a number of phenomena that will be discussed again later. It is useful to record a complete wavetrain picked up by a receiver, which is usually located 5 ft away from the transmitter, to differentiate between the series of arrivals.

The recording is presented:

- In the form of a complete wavetrain or solely its positive part (wave form, signature curve) (**Fig. 8.28a**).

- As variable density (VDL), with only the positive arches reproduced on a gray scale (the greater the amplitude, the darker the gray) (**Fig. 8.28b**).

This presentation is used to check for the presence of formation waves and monitor their variations versus depth, as well as the amplitude of the casing waves.

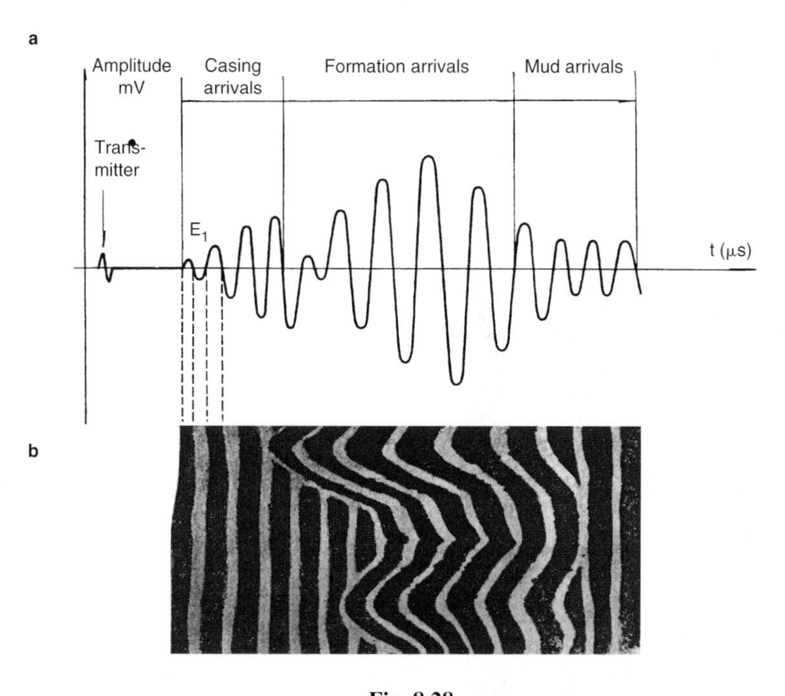

Fig. 8.28
Wavetrain recording *(Source: Dowell Schlumberger).*

Interpretation

- In uncemented casing, the casing waves appear very clearly. They are parallel and straight on all of the cement-free portion. No formation waves can be seen.
- In properly cemented casing, they are greatly weakened and can even practically disappear. Formation waves appear very clearly.
- In a portion with a mediocre cement job, casing waves can be seen (they can be relatively dark) along with formation waves.

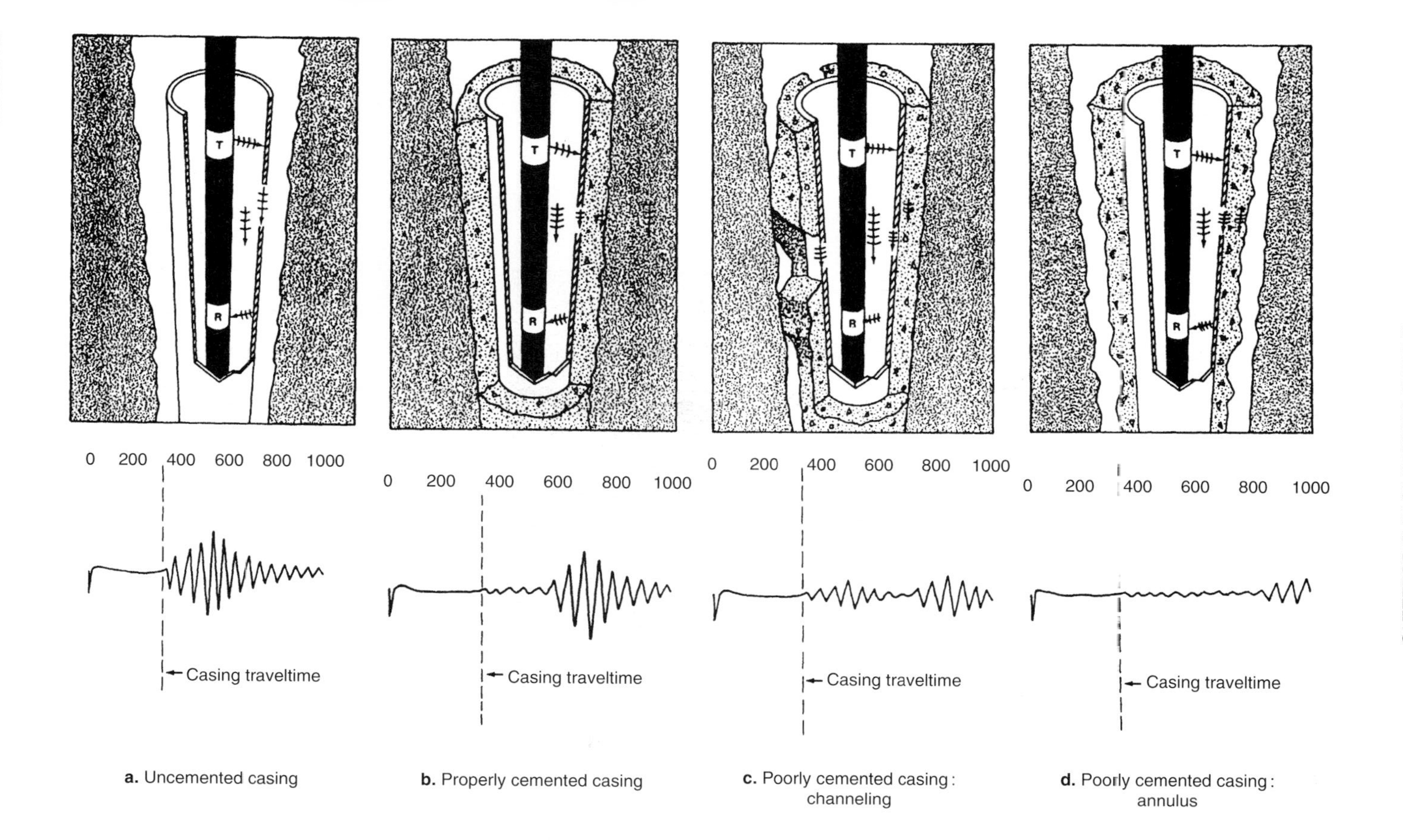

Fig. 8.29
Cement logs *(Source: Dresser Atlas)*.

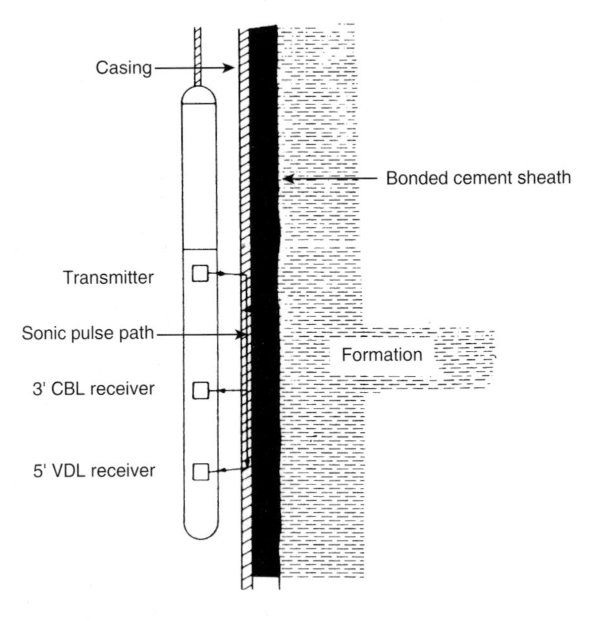

Fig. 8.30
Example of a CBL-VDL tool *(Source: Dowell Schlumberger)*.

Tool and presentation of the recording

The traveltime, the CBL and the VDL are now all recorded by the same tool. An example of the tool is shown in **Fig. 8.30**. It can be combined with a casing collar locator (CCL) and with a natural radioactivity tool (GR). The three measurements can be presented on the same film (**Fig. 8.31**).

Prerequisites for a good recording

- Laboratory studies show that from a cement thickness of 3/4" and up, attenuation no longer depends on thickness. Interpreting in a zone where the cement thickness is less than this value would give a pessimistic indication.
- Cement must be completely set, i.e. at least 24 to 36 hours must have elapsed before running a log, sometimes even more.
- Variations in pressure in the casing between the cementing operation and the logging operation may cause a reduction in casing diameter compared to the cement. A microannulus can be formed or there may be a microseparation in the cement itself.
- The tool must be well centered, otherwise the amplitude of the signal may be greatly attenuated.

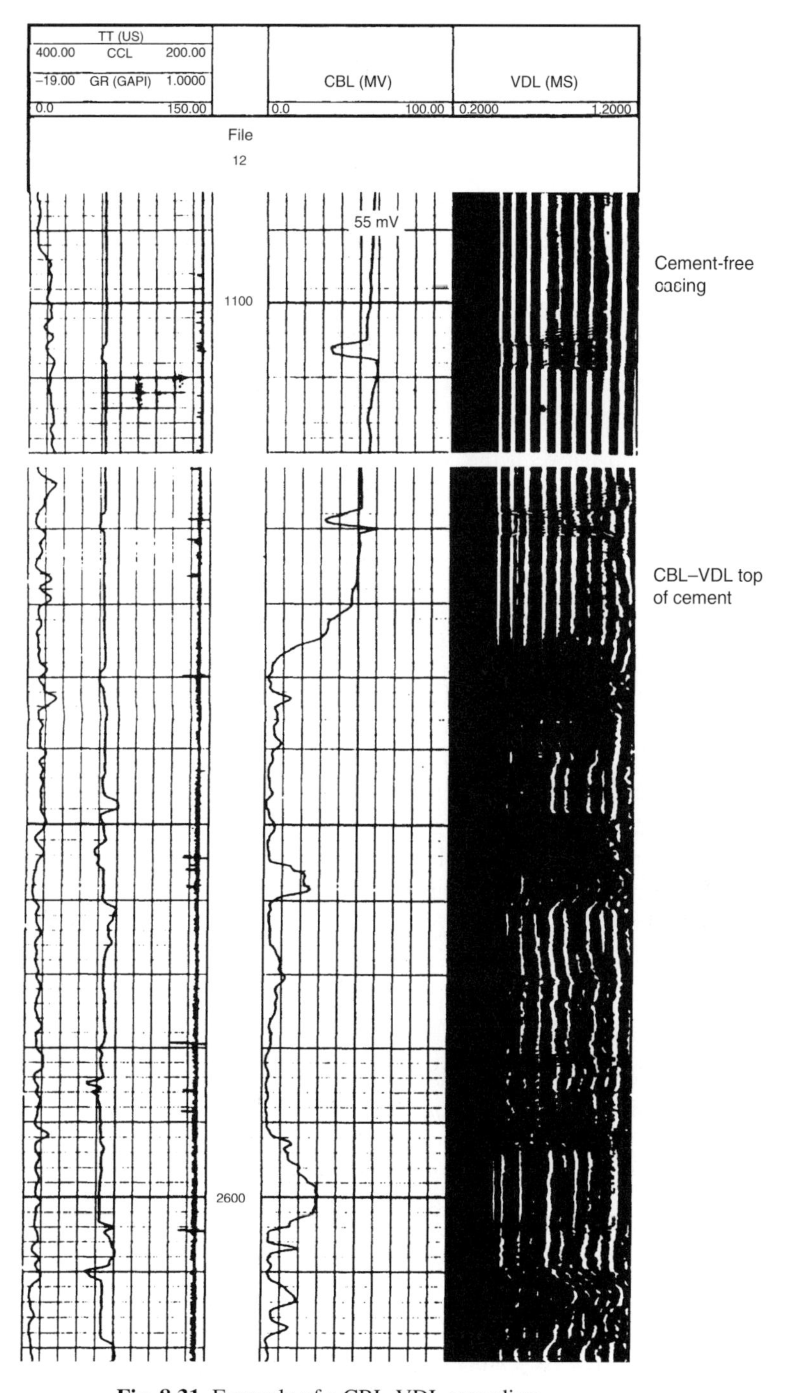

Fig. 8.31 Example of a CBL-VDL recording.

Precautions in interpreting

Before the cement log is interpreted, it is a good idea to gather all the information on recording conditions and on the cement job itself:

- type of cement, additives and fluid in the casing,
- type of casing, characteristics of the borehole and type of formation,
- characteristics of the sonde (type of centralizers, spacing between transmitters and receivers, frequencies used, etc.),
- recording mode (fixed or floating window),
- calibrations.

These data are used to find out the maximum compressional strength of the cement, the traveltime of the wave in the mud and the arrival time of the first casing wave.

A comparison between the VDL and an acoustic log (or another log if necessary) recorded in an open-hole situation can help assess the quality of the recording when there are any doubts.

c. *Other evaluation tools*

Other tools give a more sophisticated picture of the quality of the cement job by focusing the measurement on portions of casing.

The cement evaluation tool (CET) (**Fig. 8.32**) is equipped with eight high-frequency acoustic transmitter-receivers that are used to study the vibration through the casing. The amplitude of the signal recorded on each transmitter-receiver is related to the acoustic impedance of the cement, and therefore to the quality of the cement job in the portions of casing under study. The frequency used makes this tool less sensitive to the microannulus phenomenon. Its principle and utilization also make it less prone to centering problems, so that it can record in highly deviated wells (up to 70°).

The wave's return-trip time between the transmitter-receiver and the casing gives a good indication of the different radii (with an accuracy of 0.1 mm). This in turn gives measurements of four casing diameters as well as how far the casing is out of round and the degree of sonde eccentering.

The measurement technique also serves to get information on the quality of the cement job, even in high-velocity formations.

The results are presented in the form of a log (**Fig. 8.33**): the average diameter and how far the casing is out of round, sonde eccentering and quality of the cement job. Cement job quality is assessed by calculating the cement's compressional strength at each transmitter-receiver depth. Two curves represent the minimum and maximum compressional strength values. The eight values are transcribed on a gray scale, with black standing for a good cement job and white for a poor one.

The tool is oriented by means of an borehole-profile log cartridge that gives the position in space of poorly cemented zones.

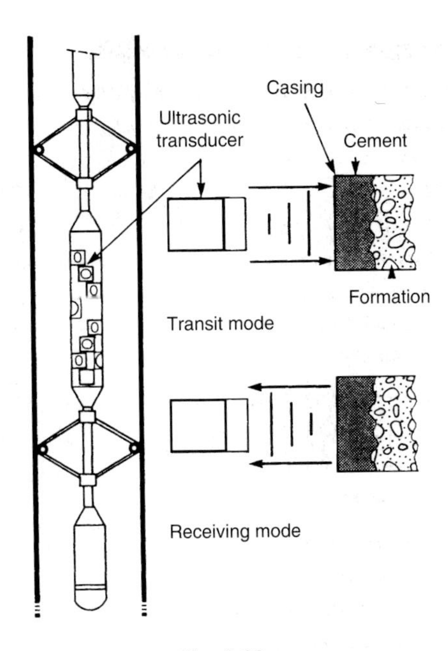

Fig. 8.32

CET cement evaluation tool *(Source: Dowell Schlumberger)*.

Other tools use the principle of the CBL–VDL, with focused transmitters and receivers to give indications of cement quality by sectors (eight for *Halliburton*'s Pulse Echo Log tool).

The advantage of these tools is that they afford a better understanding of the causes of poor-quality cementing and how to remedy them: e.g. landing casing and emplacing cement. They also give insight into the relationships between cementing and the formation.

D. Logs to measure borehole geometry and deviation

Borehole geometry and deviation measurements are parameters of interest for drillers, geologists and production engineers.

Diameter is usually recorded with most well logs (**Fig. 8.34**) and serves to calculate the corrections that need to be made in the measured parameters. The diameters can be more or less accurate according to the tool that is used. Microresistivity logs generally give more sensitive diameters. The measurement depends on the shape of the sonde and its position in the borehole.

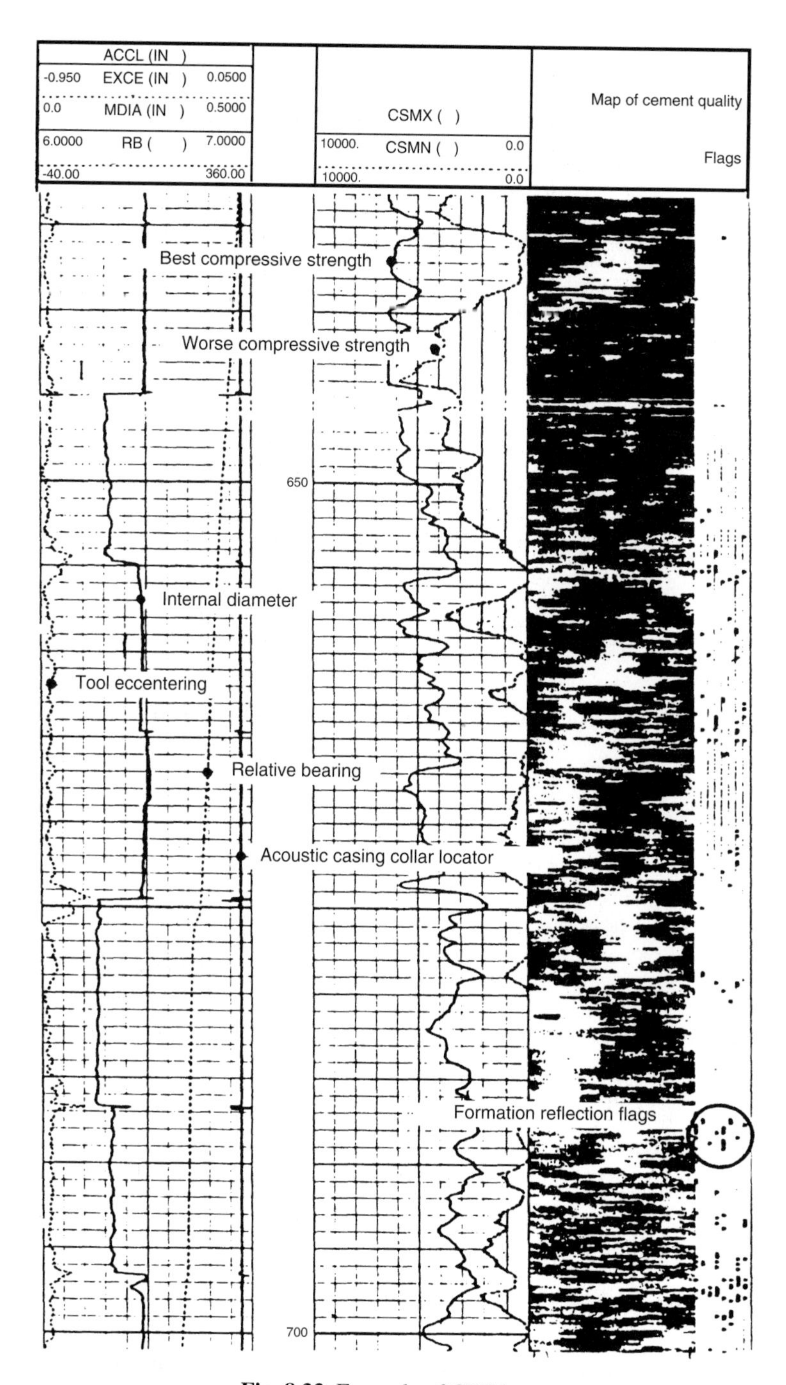

Fig. 8.33 Example of CET log.

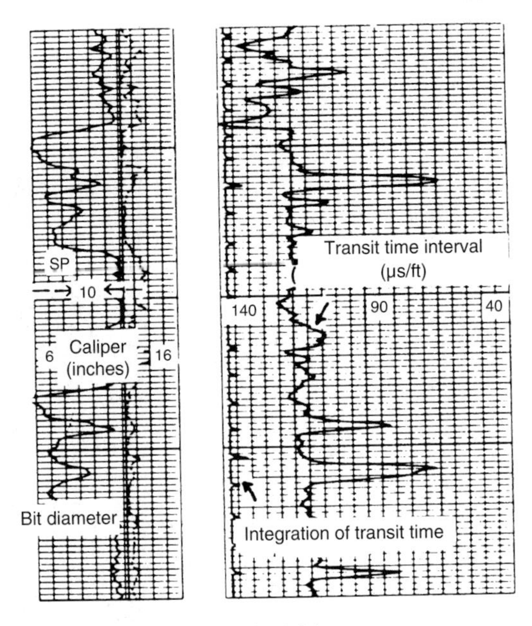

Fig. 8.34

BHC tool to measure diameter *(Source: Schlumberger)*.

Other tools are used to measure deviation (**Fig. 8.35**):

- Borehole Geometry Tool (BGT).
- Dipmetering (Diplog, HDT, etc.).
- Guidance Continuous Tool (GCT).

Tools that measure borehole geometry use the same principle as the dipmeter. The dipmeter is a tool used by geologists to determine formation and fault dips in an open hole. The principle for calculating dips and their accuracy require measurements of borehole diameter and deviation.

Four arms at right angles, remote controlled from the surface, measure two diameters and keep the sonde centered in the borehole. A borehole-profile log cartridge measures the borehole angle and deviation continuously by means of a reference pad's orientation.

A GR can be combined with borehole geometry tools (*Schlumberger*'s BGT, etc.) to allow correlations with other well logs.

The tools generally use magnetic orientation and for this reason can only be run in uncased boreholes.

Recording in computerized units on site allows results to be available quickly:

- the volume of the borehole (based on integrating the diameter measurements),

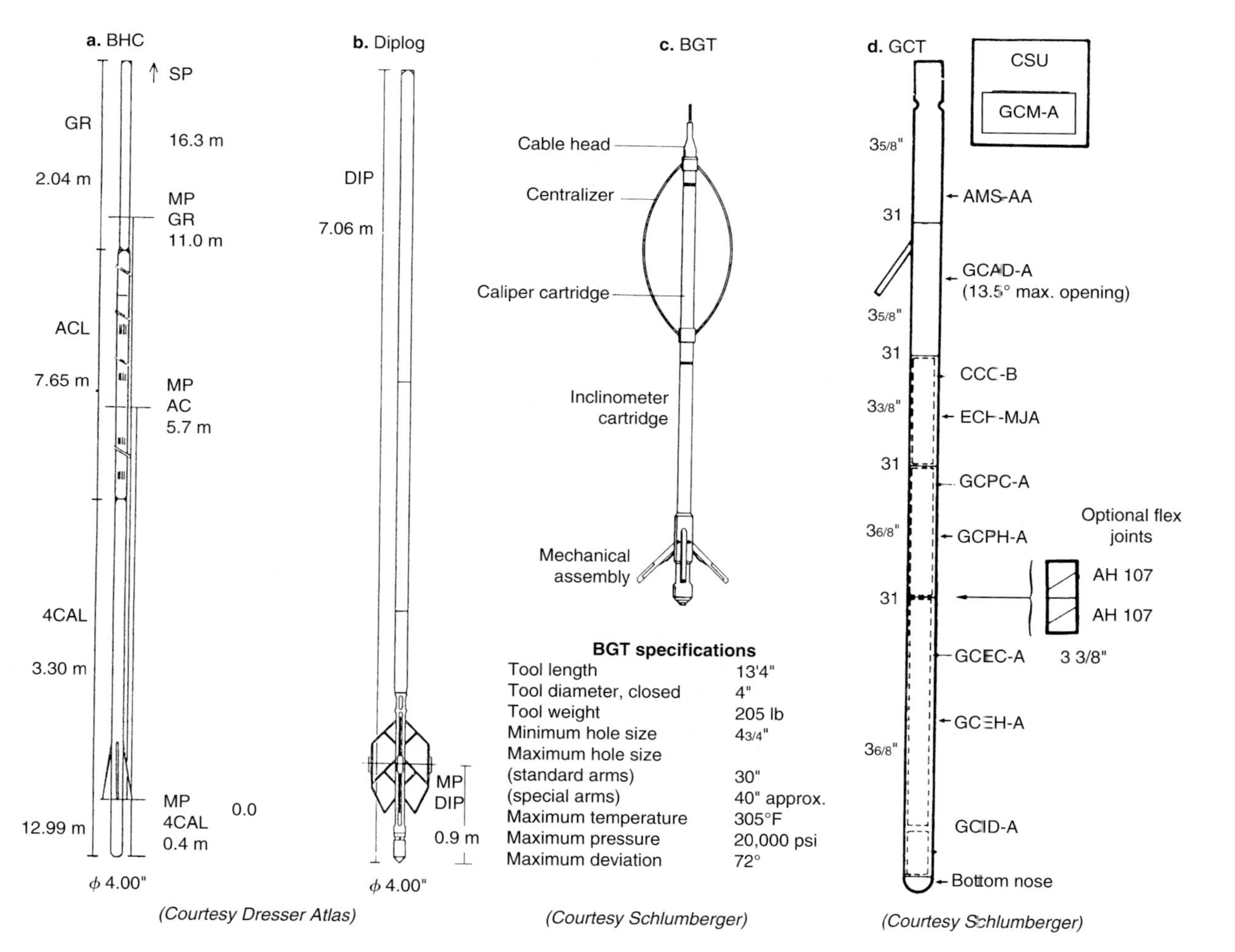

Fig. 8.35 Other measurement tools *(Source: Dresser Atlas, Schlumberger).*

- the volume required to cement a casing,
- the degree to which the hole is out of round.

Based on deviation measurements, it is also possible to:

- calculate vertical depths (TVD) from log depths (TMD),
- give an image of the borehole on a horizontal or vertical plane.

Diameter orientation can give useful information on the direction of the borehole and of the fractures.

A new generation of deviation measurement sondes appeared with *Schlumberger*'s GCT, with the advantage that it is operational in a cased borehole. The tool has a gyroscope that defines a stable vertical plane in space and accelerometers that give the sonde's movement, i.e. position, in the borehole. Error is said to reach a maximum of about ten feet in relation to the vertical depth of the bottom of the hole and about twenty to forty feet in relation to its horizontal position in a very inclined well 10,000 ft long.

Drift phenomena and measurement require lengthy tool calibration and short logging time. The sonde gives its position every ten seconds, so it can be raised at a speed of 8000 ft per hour. This shortens logging time provided there are no sudden variations in the speed. The quality of the measurement is monitored by recording when the tool is being lowered and then when it is being raised back up to the surface.

This tool's characteristics should be improved with the development of gyroscopes and when they can withstand temperature better. GCT use is restricted to wells with a deviation of less than 70° and latitude zones of less than 70°. It does not measure diameters.

8.2.7 Use of well logs

Rapid interpretation, called quick look, based on the examination of raw logs recorded on the proper scale, gives a rough idea of the formations and fluids.

The method consists in:

- identifying reservoirs by eliminating shales and compact beds,
- comparing reservoir resistivity and porosity logs:
  - comparing the resistivity logs (a macrolog and a microlog giving approximate values of R_t and R_{xo} respectively) shows the contact between the water and hydrocarbons and gives an approximate value of the water saturation.
  - comparing the neutron and density logs serves to determine the lithology in the water zone, identify the type of fluid in the hydrocarbon zone and assess formation porosity.

The widespread use of computerized processing units enables rapid interpretation on well sites. They calculate and supply a number of important results. The conclusions that can be drawn from these initial studies are usually sufficient to answer the questions that may arise before drilling is resumed or casing is run in (sampling, testing, setting a packer, well shooting and well completion, etc.)

8.2.8 Billing

Pricing and billing systems depend on:

- the logging company,
- the type of contract (services needed only for logging time or full time as is the case offshore),
- the location of the site (country, onshore or offshore).

As a result, it is fairly difficult to give a cost for logging operations, not least because of the impact of the number of times tools are run in and the depths they reach.

Generally speaking, the cost includes:

- a mobilization fee,
- the daily rental (or monthly rental for equipment that remains on the well site for offshore operations) of the processing unit, the winch and the sondes,
- the personnel, whether operational or on standby (transportation, accommodation and catering are also billed to the client),
- transportation,
- the price of running in to the lowest depth,
- the price of the measurement (minimum charges are applied here and for the depth item),
- any loss of equipment (fishing for tools left in the borehole or loss of tools is billed to the client, except when the contractor is patently to blame),
- interpretation on the well-site computer (interpretation in the computer center is billed separately from operations).

Depending on the type of well (exploration or development) and the location of the site (onshore or offshore), the cost of well logging alone can be estimated to be about 5% to 10% of the total cost of the well (sometimes 20% in very special cases). Since the objective of an exploration well is to get the most geological and technical information possible, the cost of well logging is very low compared to the conclusions that can be drawn.

A good understanding of logging techniques and of the billing system can not help but improve the quality of the information that is gathered. It would be pointless to record a set of logs and leave out a crucial curve to "save money", when this would prevent valid interpretation. The minimum mobilization fee also means that it is not advisable to restrict the section of well bore measured, a basic rate is in any case applied. This is true except if the aim is to shorten the time the drilling rig is occupied, for safety reasons for example. Money saved in the wrong places can mean faulty information and jeopardize the whole prospecting operation.

Chapter 9

PRINCIPLES OF KICK CONTROL

9.1 BACKGROUND INFORMATION

9.1.1 Causes of kicks

A kick is an unintended influx into the well of a fluid contained in the formation. The causes are listed below:

- Drilling mud density is too low. The hydrostatic pressure on the bottom of the hole can become lower than the formation pore pressure.
- Swabbing occurs when the drill string is being pulled upward in the well. This can happen when a joint of pipe is added or when a bit is changed.
- The drilling mud level in the annulus gets too low, with the resulting lower pressure on the bottom of the hole. The low mud level may be because the hole is insufficiently filled during a trip or because of mud losses into the formation.
- The bit grinds up porous rocks. The risk is obviously not the same since as soon as drilling has ceased, there is no more influx.

9.1.2 Warning signs

Several signs can indicate that there is an impending risk of a kick or that a kick is occurring. These warning signs are listed below and are extremely important to drillers, since the methods of control and the degree of risk are directly influenced by how quickly control procedures are started.

A. Increased rate of penetration (drilling break)

A considerable increase in this parameter may indicate:

- a change in formation drillability because the bit penetrates a porous or fractured formation,
- a decrease in differential pressure between the pressure exerted by the column of mud in the well and the formation pore pressure.

Drilling technique dictates an overbalanced well, i.e. a slight overpressure on the working face. The influence of overpressure on rate of penetration is shown by the curve in **Fig. 9.1**.

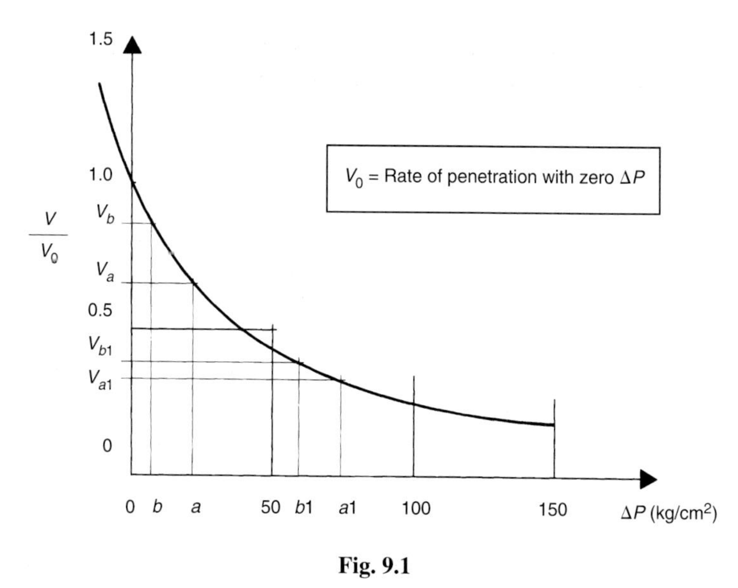

Fig. 9.1

Influence of differential pressure on rate of penetration *(After W.C. Maurer).*

When drilling at point a and the formation pore pressure rises because the bit drills into a transition zone, the differential pressure decreases and becomes b. The rate of penetration increases considerably from V_a to V_b. If the overpressure was a_1, the difference between V_{b1} and V_{a1} would possibly not be significant.This means that wells should be drilled with minimum differential pressure.

Also important is the fact that annular pressure losses, though often small, exist and are added onto hydrostatic pressure. As a result, an actual kick or an upcoming kick may not be detected during drilling (i.e. with mud circulation) but make itself felt when a joint of pipe is added (with circulation stopped).

B. *Anomalies when the well is being filled*

When the drill string is being pulled out of the hole, the driller replaces the volume of steel removed by drilling mud, thereby keeping the mud level at a maximum. He does this by means of a special mud tank called a trip tank or possum belly tank that makes it easier for him to monitor the volume of mud used to fill up the borehole. The driller compares this volume to the volume of steel that he has pulled out of the hole. If the volume of mud is smaller, it means there is a fluid kick at the bottom of the well.

A kick that does not take place at the beginning of the trip (the driller checks the well under static conditions before starting to pull the drill string out of the hole) is caused by the swabbing effect due to the speed of the drill string being pulled out. The faster the tripping speed and the higher the mud viscosity and gel strength are, the greater the swab effect, especially if the bit is balled up and the drill collars are oversized.

The kick may stop if the swab effect is curbed. However, the kick may unbalance the well when the gas has displaced part of the mud by expanding. Likewise with a fluid such as water, the fluid influx may be high enough in the annulus to "prime" the well, i.e. make it flow.

C. *Lost circulation*

Mud losses can be caused by substantial amounts of mud filtering into a highly porous and permeable or fractured formation, whether the porosity and permeability characteristics are natural or induced by hydraulic overpressures in the well. Lost mud causes a lower hydrostatic level to exist in the well which may in turn cause a kick.

In a very thick gas-bearing reservoir, the mud density required to control the pore pressure in the upper part of the reservoir may be too high for the lower layers. This is where the pore pressure gradient is lower than the hydrostatic gradient. As a result, the reservoir could be fractured when further drilling is done to deepen the well.

D. *Gas-cut mud*

Gas mixed in with the mud should also be considered as the sign of a kick, but it is necessary to determine how this phenomenon has appeared.

The causes may be:

- Drilling into a permeable gas-bearing formation, with adequate drilling mud density. This is not really a kick, but gas associated with the cuttings gives gas-cut mud. The concentration of gas in the mud is directly related to the bit diameter, the rate of penetration, the mud flow rate, the porosity of the rock and the pore pressure. The problem can become serious if the rate of penetration is too fast and the percentage of gas in the annulus becomes very high. This would decrease the hydrostatic pressure considerably and could trigger off a kick.
  This phenomenon is not usually dangerous. Since gas expands dramatically only near the surface, the decrease in hydrostatic pressure is slight under normal drilling conditions. For example, when drilling at 3000 m with a mud density of 1.2 and gas-cut mud density of 0.8 at the surface, the decrease in bottomhole pressure is estimated at 3 bar. If the density of the mud returns decreases to 0.6, the drop in bottomhole pressure becomes 6 bar.
- Drilling into shales containing high-pressure gas but having no permeability. When the pumps are switched off, the well does not flow. Other identical but permeable layers may be encountered and they will give rise to gas kicks. These signs should be taken into account when the minimum drilling mud density is being evaluated and the future casing shoe depth is being determined.
- Kick when circulation is stopped.
- A slug of gas-cut mud often appears when the annular volume has been circulated. The gas comes from swabbing when a joint of pipe is added or from gas diffusing through the mud cake. Gas diffusion occurs independently of differential pressure and increases when there is oil in the mud, i.e. at a maximum for oil-base mud.

The slug of gas-cut mud should be considered as something normal, but should not be disregarded as it is a warning sign of something more serious and also provides information on bottomhole pressure.

- Air in the drill string coming from an added joint of pipe, H_2S or CO_2 coming from the breakdown of mud products.

E. Increased flow rate at the flowline. Increase in mud tank level

These signs are the result of a volume gain in the well and unmistakably indicate that a kick is occurring. The increase in mud tank level is the measurable parameter on all well sites, but has some drawbacks: system inertia due to large-size flowlines, large-size tanks, additions to mud volumes, unstable tanks on floating drilling supports.

Technically speaking, it is preferable to monitor the increased flow rate on the mud return line, but this requires two flowmeters (inlet and outlet) and a means of comparing them.

9.1.3 Observation with the well closed

Once the driller has realized a kick is taking place, he closes the well according to the driller's procedures for well control (**Fig. 9.2**) which are posted on all rig floors.

The first step is to get into a position that will allow all operations that might be necessary to control the well. The operations vary according to the situation at the time when the kick is detected:

- bit on the bottom (drilling, coring, circulation), i.e. with the kelly screwed to the drill string,
- tripping (bit far off bottom), i.e. with the drill string open.

However, in any case an effort is made to get the first tool joint into a position about one meter above the rotary table so that it is accessible to the crew. Before shutting in the well with the annular BOP, care is taken to open the valve on the choke line to prevent erosion damage to the BOP packing element. In this way the well flow will bypass it to the choke manifold through the open drilling choke. To measure the pressures when the well is shut in, the driller closes the choke line by the drilling choke itself while monitoring the pressure. This method is called a soft shut-in. What is termed a hard shut-in consists in closing the drilling choke as well as the upstream valve. The well is shut in directly by a BOP.

If the drill string is not equipped with a means of closing it, a safety valve is immediately screwed onto the top of the string. Then an attempt is made to run the bit down to the bottom if conditions, i.e. allowable gain, acceleration in flow rate and presence of gas, etc., allow. The well is shut in with the same precautions as those discussed above.

The formation is decompressed in the vicinity of the well and recompresses until the bottomhole pressure offsets formation pressure. The shut-in drillpipe pressure will then increase. A shut-in well with the bit on the bottom is schematically represented in **Fig. 9.3**.

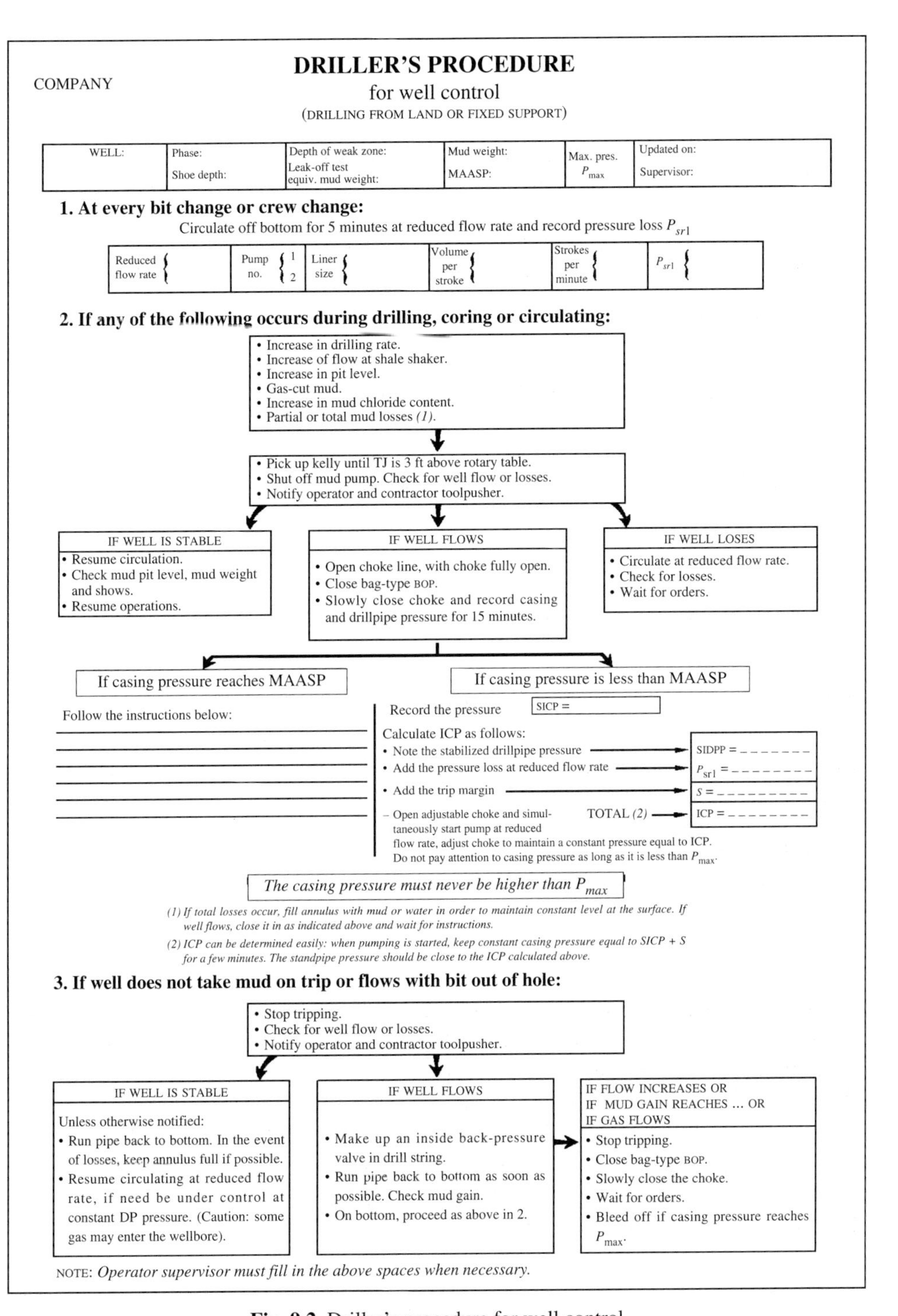

COMPANY

DRILLER'S PROCEDURE

for well control

(DRILLING FROM LAND OR FIXED SUPPORT)

WELL:	Phase: Shoe depth:	Depth of weak zone: Leak-off test equiv. mud weight:	Mud weight: MAASP:	Max. pres. P_{max}	Updated on: Supervisor:

1. At every bit change or crew change:

Circulate off bottom for 5 minutes at reduced flow rate and record pressure loss P_{sr1}

Reduced flow rate	Pump no. 1 / 2	Liner size	Volume per stroke	Strokes per minute	P_{sr1}

2. If any of the following occurs during drilling, coring or circulating:

- Increase in drilling rate.
- Increase of flow at shale shaker.
- Increase in pit level.
- Gas-cut mud.
- Increase in mud chloride content.
- Partial or total mud losses *(1)*.

- Pick up kelly until TJ is 3 ft above rotary table.
- Shut off mud pump. Check for well flow or losses.
- Notify operator and contractor toolpusher.

IF WELL IS STABLE
- Resume circulation.
- Check mud pit level, mud weight and shows.
- Resume operations.

IF WELL FLOWS
- Open choke line, with choke fully open.
- Close bag-type BOP.
- Slowly close choke and record casing and drillpipe pressure for 15 minutes.

IF WELL LOSES
- Circulate at reduced flow rate.
- Check for losses.
- Wait for orders.

If casing pressure reaches MAASP

Follow the instructions below:

If casing pressure is less than MAASP

Record the pressure SICP =

Calculate ICP as follows:
- Note the stabilized drillpipe pressure → SIDPP =
- Add the pressure loss at reduced flow rate → P_{sr1} =
- Add the trip margin → S =
- TOTAL *(2)* → ICP =

– Open adjustable choke and simultaneously start pump at reduced flow rate, adjust choke to maintain a constant pressure equal to ICP. Do not pay attention to casing pressure as long as it is less than P_{max}.

The casing pressure must never be higher than P_{max}

(1) If total losses occur, fill annulus with mud or water in order to maintain constant level at the surface. If well flows, close it in as indicated above and wait for instructions.

(2) ICP can be determined easily: when pumping is started, keep constant casing pressure equal to SICP + S for a few minutes. The standpipe pressure should be close to the ICP calculated above.

3. If well does not take mud on trip or flows with bit out of hole:

- Stop tripping.
- Check for well flow or losses.
- Notify operator and contractor toolpusher.

IF WELL IS STABLE

Unless otherwise notified:
- Run pipe back to bottom. In the event of losses, keep annulus full if possible.
- Resume circulating at reduced flow rate, if need be under control at constant DP pressure. (Caution: some gas may enter the wellbore).

IF WELL FLOWS
- Make up an inside back-pressure valve in drill string.
- Run pipe back to bottom as soon as possible. Check mud gain.
- On bottom, proceed as above in 2.

IF FLOW INCREASES OR IF MUD GAIN REACHES ... OR IF GAS FLOWS
- Stop tripping.
- Close bag-type BOP.
- Slowly close the choke.
- Wait for orders.
- Bleed off if casing pressure reaches P_{max}.

NOTE: *Operator supervisor must fill in the above spaces when necessary.*

Fig. 9.2 Driller's procedure for well control
(Source: Blowout Prevention and Well Control, Editions Technip, Paris, 1981).

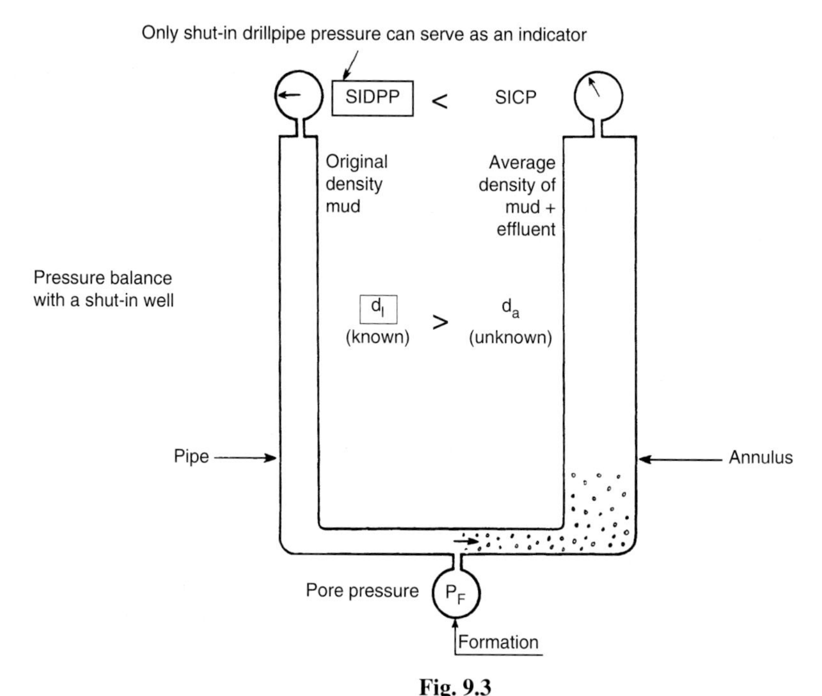

Fig. 9.3
Diagram of a well as a U-shaped tube.

The driller can calculate the pore pressure by using the following equation:

$$P_F = 9.81\ Zd_1 + \text{SIDPP}.$$

All procedures provide for keeping bottomhole pressure slightly higher than formation pressure:

$$\text{BHP} = 9.81\ Zd_1 + \text{SIDPP} + S\ (\text{kPa}).$$

S = safety margin with a common value of 500 kPa per 1000 m depth.

9.1.4 Well safety

When a gas kick has occurred, the well can not be considered as perfectly safe since the gas is going to migrate in the annulus by difference in density. As the gas bubble rises in the annulus, the shut-in pressure will increase and approach the formation pressure. At the same time (**Fig. 9.4**), the bottomhole pressure will increase by the same amount and approach twice the formation pressure.

The uncased portion of the hole will therefore be subjected to the formation pressure. With all other conditions remaining the same, the uncased section will break down if its fracture pressure is not high enough compared to the formation pressure.

The maximum allowable casing pressure involves two boundary values:

- one (P_{max}) that must in no way be exceeded: this is the working pressure of the wellhead assembly or the bursting pressure of the last casing string,
- an allowable value (MAASP) related to the fracturing pressure of the formations located under the shoe of the last string of casing.

In order to calculate the allowable pressure, the fracture gradient of the formations located in the uncased portion of the hole needs to be determined by:

- local statistical data,
- leak-off test,
- evaluation.

The deeper the shoe of the last casing string and the lower the density of the mud, the higher the allowable casing pressure. The risk of breaking down a homogeneous formation is highest at the point of greatest stress, i.e. often at the top of the uncased part of the hole. In this case, the allowable pressure is:

$$MAASP = 9.81\ Z_{shoe}\ (G_{frac} - d_1)\ (kPa).$$

If a weak layer has been penetrated, the fracturing risk may become higher there than at the last casing shoe. MAASP must therefore be recalculated when anything special happens during drilling (e.g. drilling into a porous layer, modification in mud density).

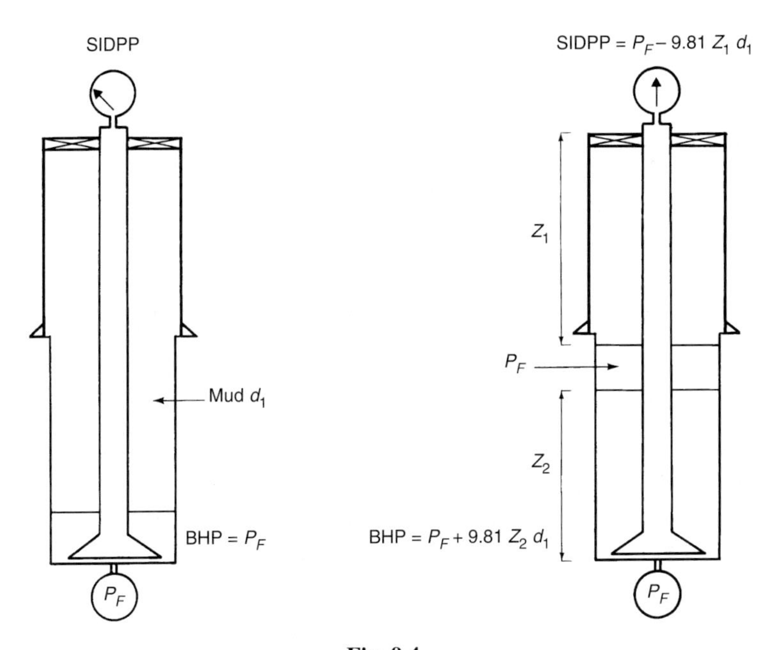

Fig. 9.4

Pressures while a volume of gas is migrating.

While a gaseous effluent is being circulated out with the original mud and the bottomhole pressure is being held constant, the pressure at the top of the uncased part of the borehole or at the weak point:

- increases proportionally to the increase in height occupied by the gas in the annulus, until the gas reaches the weak point,
- decreases as the gas passes by the weak point,
- remains constant while the gas is rising above this point.

If weighted-up mud of density d_r has reached the annulus before the gas bubble gets to the weak point, the pressure at this point is reduced by a value P:

$$P = 9.81\ h\ (d_r - d_1)\ (\text{kPa}),$$

with h the height of heavy mud d_r present in the annulus when the gas reaches the weak point.

It should be noted that:

- If the casing pressure reaches or exceeds MAASP while the well is shut in or the gas bubble is rising up to the weak point, then the control method discussed below may fail. The reason is that the architecture of the well may be destroyed at the depth of the uncased part of the hole. Here there is a choice between two situations:
  - keep the casing pressure equal to MAASP and accept the kick flowing into the well,
  - accept the risk of breaking down the formation and continue applying the method; this risk can not be run if there is any danger of cratering on the surface.
- When the gas bubble has passed above the weak point, the casing pressure can be allowed to rise as long as it does not exceed P_{max}.

In conclusion, an effort must be made to circulate weighted-up mud as soon as possible so as to decrease casing pressure. The lowest gain will help keep casing pressures low.

9.2 PRINCIPLES OF CHOKE LINE CIRCULATION, BIT ON BOTTOM

9.2.1 Control principle

When the kick is caused by a pressure imbalance at the bottom of the well, control means replacing the original mud by a fluid with a density of d_r that is higher than d_1. It is calculated such that:

$$\begin{aligned} 9.81\ Zd_r &= P_F + S \\ &= 9.81\ Zd_1 + \text{SIDPP} + S, \end{aligned}$$

$$d_{r2} = d_1 + \frac{(\text{SIDPP} + S)}{9.81\, Z}$$

There are therefore two problems to be solved:

- make up the volume of mud with a density of d_r;
- flush out the lighter mud by circulating the weighted-up mud while balancing the reservoir with a bottomhole pressure higher than P_F. The mud must therefore be circulated so that a dynamic pressure is created which, added to the static pressure, will be equal to $P_F + S$. The choke on the choke manifold will be used to produce the required counterpressure which is applied on the bottom of the hole.

9.2.2 Shut-in circulation with the choke

This type of circulation is broken down into two phases:

1. Circulating the volume that serves to get rid of the effluent (the annular volume of the well),
2. Replacing the lighter mud by mud of the required density when the annulus is clean, i.e. when the effluent has been circulated out (at least the inside volume of the drillpipe).

Two procedures are used:

- The driller's method where the effluent is circulated out with the original mud with its density d_1, by adjusting the choke so that the drillpipe pressure is:

$$\text{ICP} = P_{sr1} + \text{SIDPP} + S$$

  where P_{sr1} is the pressure losses in the drill string with the well control pumping rate (usually half the drilling flow rate). The pressure losses are measured and noted down on the procedure sheet (**Fig. 9.2**) regularly as the well is drilled. Using less than half the normal drilling flow rate is warranted because:

  - It is easier to keep the mud pumps at a constant flow rate. It is absolutely indispensable for head losses to remain constant if BHP is to be kept constant by using the pressure gage reading of the discharge pressure. If P_{sr1} inadvertently increases and ICP is still kept constant, the bottomhole pressure then decreases.
  - Annular head losses can be disregarded. This holds true except on a floating support (subsea wellhead) where head losses in the choke line must be taken into account.

  The effluent will be circulated out (after circulating the annular volume) at constant ICP. The final operation will be to flush out the light mud with the heavy mud when it has been made up on the surface.

  During circulation, the driller adjusts the choke so that discharge pressure decreases linearly from the initial value ICP to a value FCP. FCP represents the pressure losses of the d_r density mud inside the drill string (with the same control flow rate) until the inside of the drill string has been filled. The heavy mud will rise in the annulus at FCP (**Fig. 9.5**).

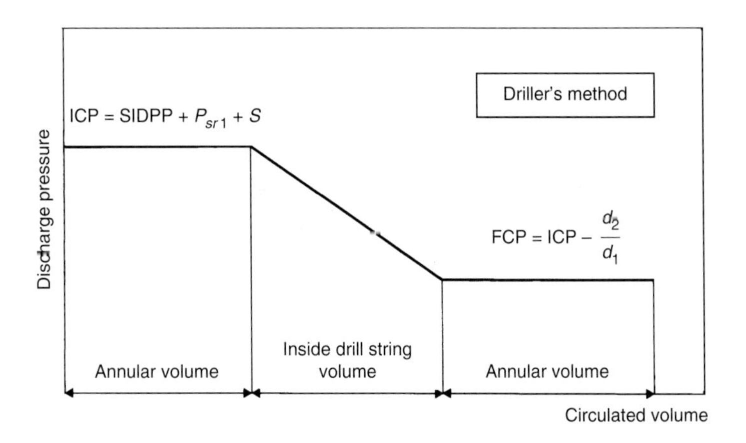

Fig. 9.5
Discharge pressure diagram during the "driller's method".

- If the rig set up allows, the fastest and most reassuring method is to pump the d_r density mud down directly: the wait and weight method. The two phases of circulating the effluent out and flushing the light mud out are done at the same time. While the inside volume of the drill string is being circulated out, the driller adjusts the choke according to the linear decrease in discharge pressure from ICP to FCP. He completes the cycle with discharge pressure constant at the FCP value (**Fig. 9.6**).

9.2.3 Changes in annular pressure during well control (bit on bottom)

When the driller has decided to get rid of the effluent by circulating it out with the original mud, he can at any time during the operation stop circulation to measure the shut-in pressures:

- the shut-in drillpipe pressure is constant and equal to SIDPP,
- the casing pressure varies depending on the pressure exerted by the column of annular fluid.

9.2.3.1 Influence of the type of effluent

If the effluent is a liquid, i.e. practically incompressible, the annular pressure remains roughly constant until the effluent gets up under the BOPs. Some variations can be seen if a substantial modification in the cross-sectional area of the annulus causes significant variation in the height of the effluent. While the liquid effluent is being circulated out through the choke, the casing pressure falls gradually. If the well is shut in after the

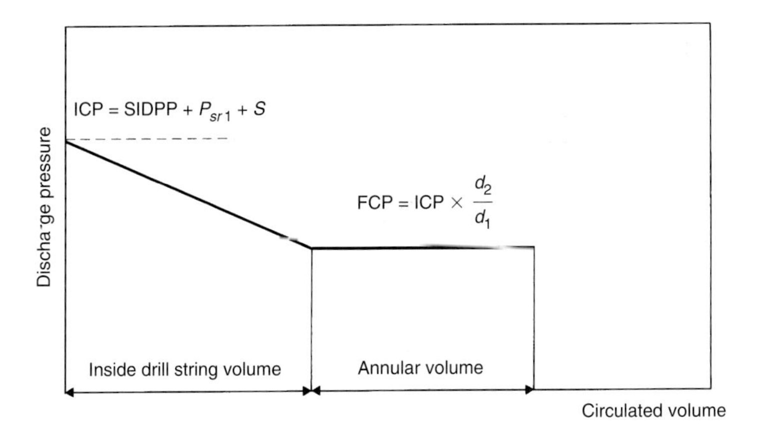

Fig. 9.6

Discharge pressure diagram during the "wait and weight method".

effluent has been completely circulated out, the shut-in drillpipe and casing pressures are equal to the shut-in drillpipe pressure originally measured (SIDPP).

A gaseous effluent behaves in a much more complex way. The volume occupied by a given amount of gas depends on its pressure, its temperature and its composition. The gas expands as it approaches the surface and the return flow rate is therefore higher than the input flow rate. The pressure increases to compensate for the expansion of the gas which has decreased the static pressure. An increase in this pressure can be noted when the gas approaches the surface and a maximum is usually reached when the effluent gets up under the BOPs. While the gas is being released through the choke, the annular pressure drops quickly. If the well is shut in after the gas has been released completely, the shut-in drillpipe and casing pressures are equal to the originally measured shut-in drillpipe pressure (SIDPP).

9.2.3.2 Influence of the volume of effluent

It has been established that detecting a kick too late is one of the reasons a driller loses control of it. The larger the volume of effluent that has penetrated inside the annulus, the lower the pressure exerted by the column of annular fluid and the higher the shut-in casing pressure.

If the effluent is a gas, the larger the initial kick volume is, the larger the volume occupied by the gas when it reaches the BOPs and the higher the casing pressure. The amount of gas that has entered the well is the major factor that determines the maximum casing pressure while the gas bubble is being circulated out.

9.3 CONTROLLING A KICK WHILE TRIPPING

In this case, control is much more complicated since the procedures depend first of all on the well conditions. The following methods can be mentioned:

- run the drill string down through the closed annular BOP (stripping) or snubbing with the aim of getting into the position described in section 9.2,
- squeeze the kick back into the formation (pump into the shut-in well via the kill line or the drillpipe),
- control the kick (well shut in) by adjusting shut-in pressure without circulation,
- control the kick by circulating weighted-up mud.

Chapter 10

DIRECTIONAL DRILLING

Directional drilling is now an integral part of conventional drilling techniques, it has become very common, or rather just about systematic in field development. Directional drilling does not only include choosing and designing the course of the borehole. It also means defining the measurements, measurement instruments, deflection tools, appropriate drilling parameters, the well architecture compatible with the course of the well, the most effective cementing techniques and the completion methods that are acceptable in a slanted or even horizontal drain hole.

10.1 APPLICATIONS OF DIRECTIONAL DRILLING

10.1.1 Inaccessible sites (Fig. 10.1)

The objective may be underneath a spot on the surface that is hard to get to (sea, lake, river, mountain, etc.) or in a populated area. In these two instances, it is physically or economically impossible to set up a rig vertically above the target in the subsoil. A directional well will solve the problem.

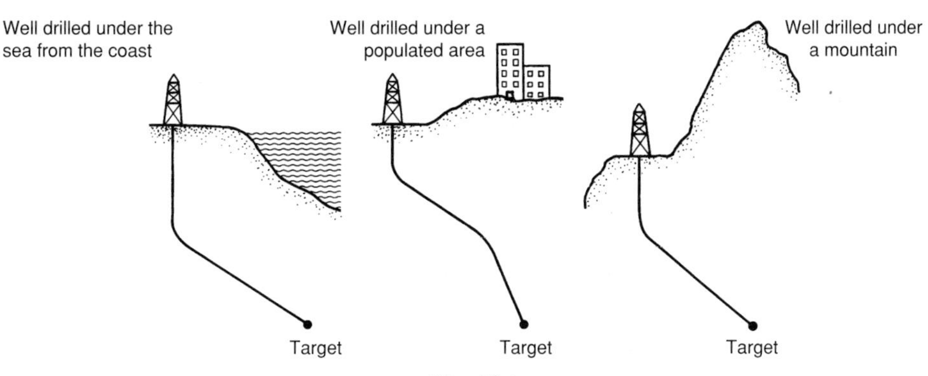

Fig. 10.1

Directional drilling toward objectives located under inaccessible sites.

10.1.2 Many wells on the same site

This is what is termed a well cluster and is only valid in development drilling. The most common examples are offshore where building a fixed platform can be cost-effective only with a group of production wells (**Fig. 10.2**). Onshore the technique is also used in regions where it is difficult or expensive to have one drilling site per well (**Fig. 10.3**). This type of development well also simplifies the infrastructure required to gather the produced effluent.

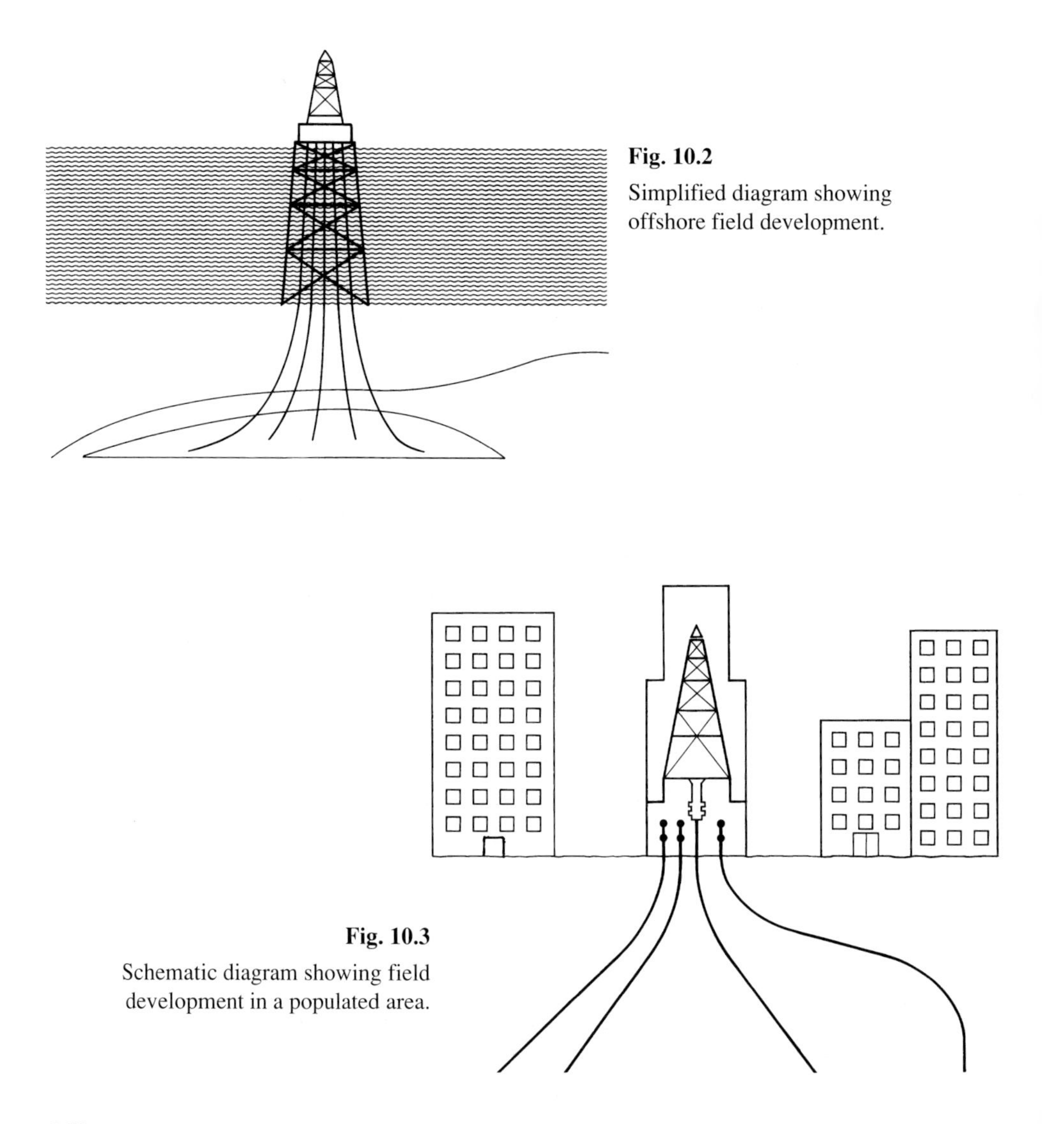

Fig. 10.2
Simplified diagram showing offshore field development.

Fig. 10.3
Schematic diagram showing field development in a populated area.

10.1.3 Sidetracking

Sidetracking is discussed in the chapter on fishing jobs because it is frequently used when the lower portion of a well has to be abandoned due to drill string failure (**Fig. 10.4a**). However, it may also be used for geological reasons (drilling rig located in the wrong place or further exploration drilling planned (**Fig. 10.4b**)).

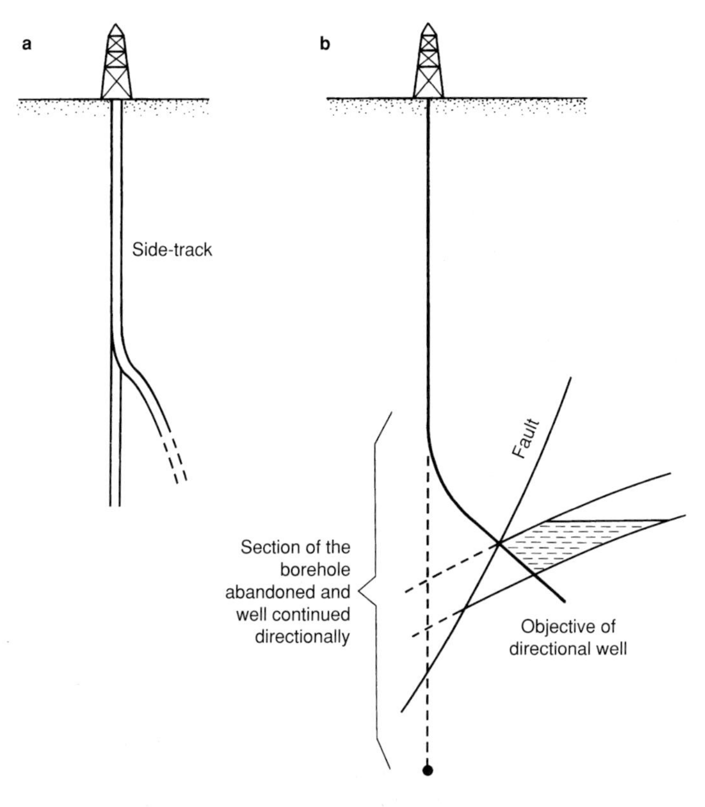

Fig. 10.4

Well deepened by directional drilling for technical or geological reasons.

10.1.4 Relief wells (Fig. 10.5)

This is the most spectacular and sophisticated application, since it requires a high level of accuracy to intercept or get close to a well where a blowout is occurring.

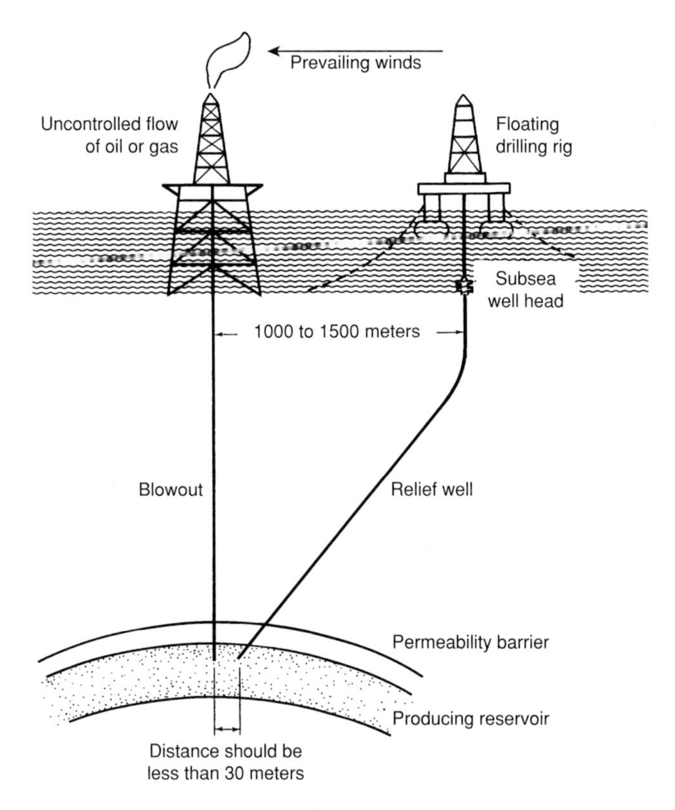

Fig. 10.5
Relief well drilled to control a blowout.

10.1.5 Miscellaneous applications

10.1.5.1 Paired geothermal wells (Fig. 10.6)

Two wells are usually required to produce geothermal resources: a producing well that draws hot water from the reservoir and an injection well to send the water back into the layer. The second well should reach total depth at an optimum distance from the producing well to prevent the produced water from undergoing too sudden a drop in temperature.

10.1.5.2 Coring after nuclear testing (Fig. 10.7)

10.1.5.3 Horizontal drilling (Fig. 10.8)

This application uses directional drilling techniques at an angle of up to 90° and is designed to increase productivity per well drilled.

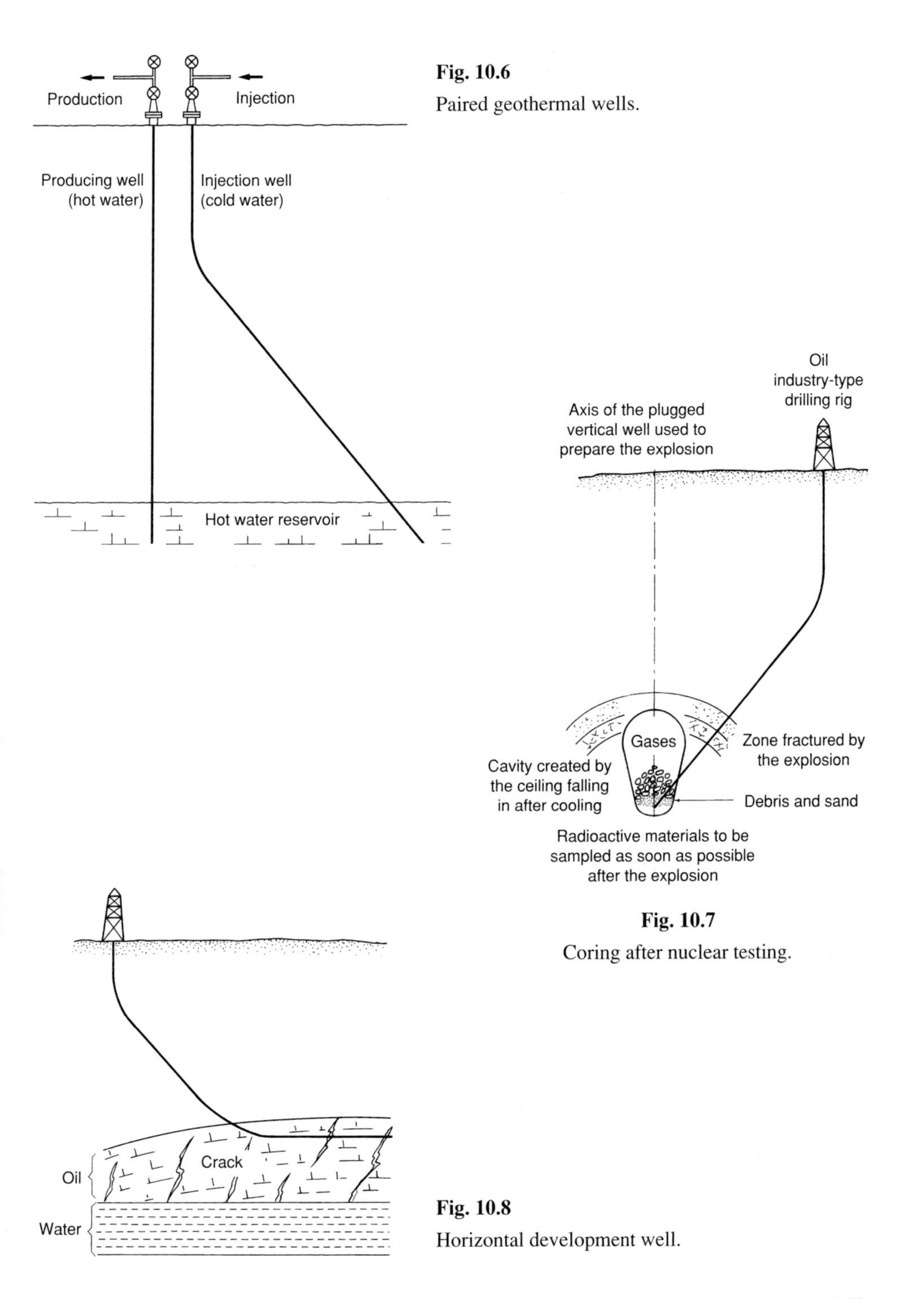

Fig. 10.6
Paired geothermal wells.

Fig. 10.7
Coring after nuclear testing.

Fig. 10.8
Horizontal development well.

10.2 PARAMETERS IN PLANNING A DIRECTIONAL WELL

Designing and preparing for directional drilling start with rig location and identifying and defining the position of the target. Its position is often defined only as a sphere centered on the point of impact. It is generally considered economically feasible to choose one twelfth of the drift angle (+ or –5° of azimuth angle in relation to the theoretical position of the target) as the radius of the target. The target is usually fairly wide and must often be narrowed down due to development constraints.

10.2.1 Well course

Once the departure and arrival points have been defined, the well course must be plotted (**Fig. 10.9**). J-shaped courses are the most common and economical. They comprise an initial vertical part down to a depth called kickoff point (KOP). This is where directional drilling begins. The drilling phase with an increase in angle of inclination is called the buildup and requires specialized drill strings.

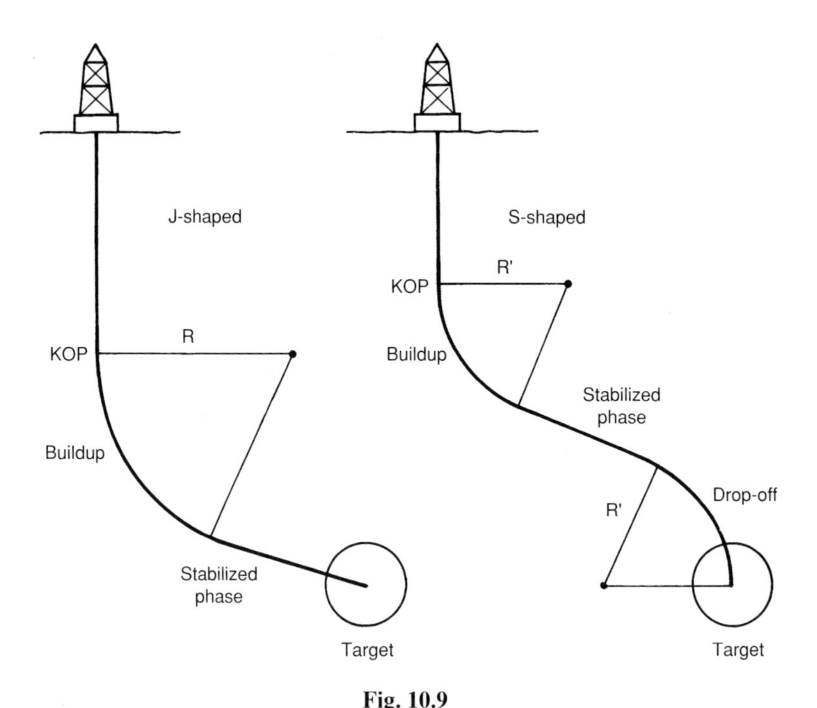

Fig. 10.9
Common well courses in directional drilling.

When the planned curve has been completed and if the direction is right, the driller starts using a stabilized rotary drill string to drill in a straight line toward the target.

Production engineers sometimes need the most vertical drain hole possible which means drilling an S-shaped well course. Here the stabilized phase is not the final one, it is followed by a curve, the drop-off, that decreases the angle of inclination.

10.2.2 Kickoff point

The KOP is determined according to the geology of the formations drilled during the buildup phase. The zone where deviation begins must especially be taken into consideration, since this phase is always critical.

10.2.3 Buildup gradient

A normal value for the buildup or deviation gradient is between 0.75 and 1° per 10 m (for a drop-off between 0.3 and 0.4° per 10 m maximum).

Buildup gradient (°/10 m)	0.5	1	1.5	2
Curvature radius (m)	1146	573	382	286

These common buildup gradients are compatible with conventional drilling and casing programs.

The maximum angle of deviation must take account of the obvious fact that the greater the angle of inclination, the more technical problems will arise and the lower the chances for success will be. For angles less than 15°, it is often very difficult to control the azimuth, and therefore the direction of the borehole. An angle of around 30° is considered to be an optimum deviation range. Beyond this figure, there will be trouble due to excessive drag, inadequate cuttings removal, and problems with wall stability in the uncased part of the hole and in logging and completion (cementing, gravel pack, packer landing, etc.).

10.2.4 Other well courses

When it is geometrically impossible to plot a J- or S-shaped course to reach the target, either the KOP is too deep or the buildup gradient is too small. The borehole can be spudded in with curved or slanted conductor pipe (**Fig. 10.10**). This requires a special drilling rig called a tilt rig or a slant rig that has the mast at an angle along the axis of the conductor pipe.

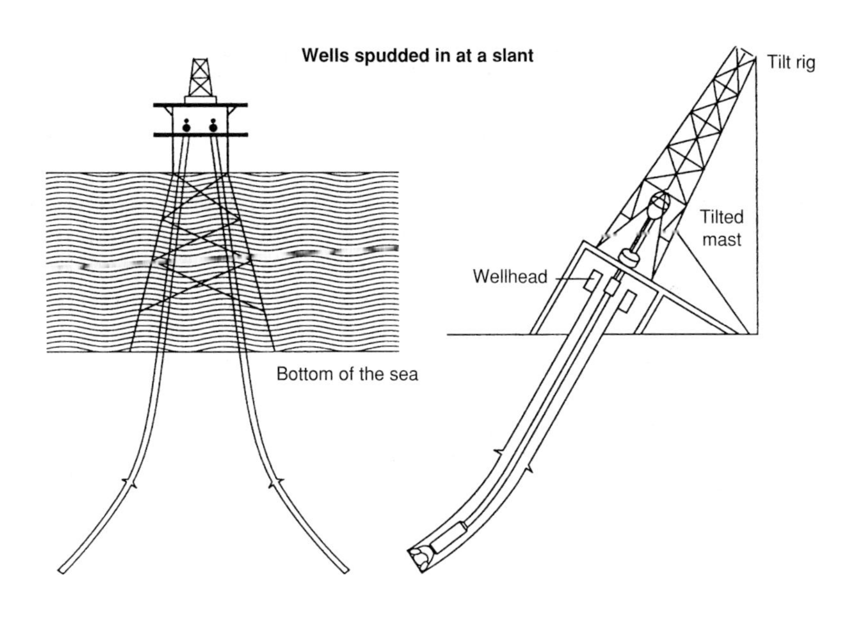

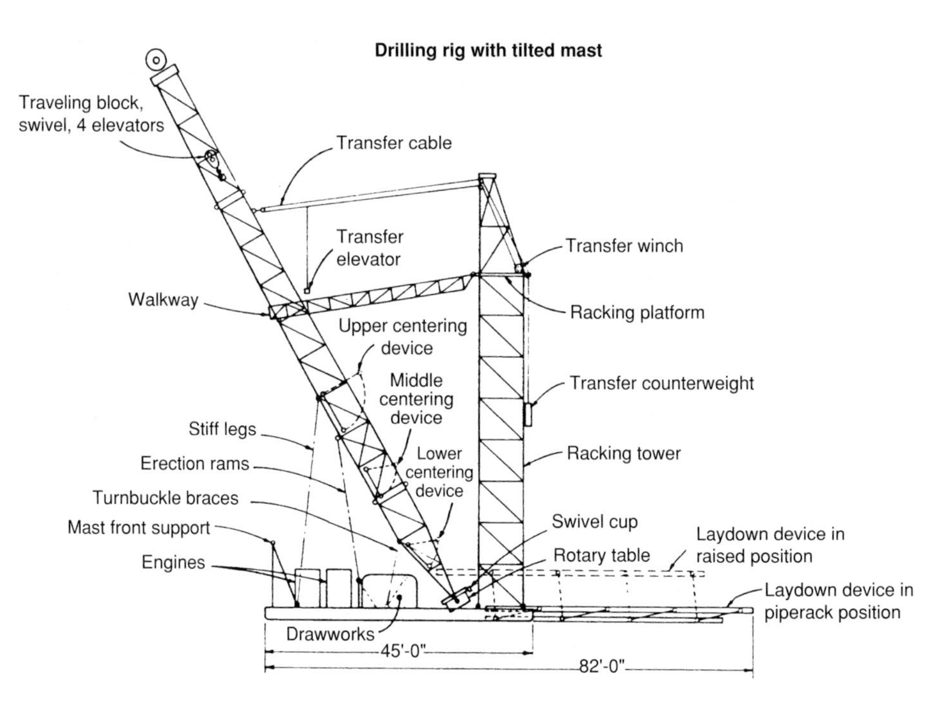

Fig. 10.10

(Source: Skytop Brewster).

10.3 MONITORING THE WELL COURSE

The course of the drilling bit must be monitored with a sufficient degree of accuracy for the following reasons:

- To locate the doglegs or elbows that are much too crooked in relation to the allowable deviation gradient. Doglegs generate considerable drag and can result in stuck drill string or prevent a string of casing from being lowered. Based on the size of the angles involved, the driller decides whether the hole needs to be reamed out with a special drill string.
- To check on the theoretical well course and reach the planned target.
- To plot the borehole's position as accurately as possible so that the well can be killed by a relief well in the event of a blowout. The minimum distance required between the two wells is generally 100 ft and this shows the degree of accuracy needed in plotting the well at total depth and the precision demanded of directional drilling techniques.

10.3.1 Measurements

The course followed by the drilling bit can be reconstructed based on measurements of inclination, azimuth and drilled length. However, these measurements are limited in time and spaced about every 10 m. The calculated well course is in fact only the most probable one in the statistical sense, given the systematic and random error involved.

10.3.1.1 Measuring the length

During drilling, joints of drillpipe are added and the total is known within a few millimeters' accuracy. When the measurement is made after drilling has ceased, the length of the borehole is considered to be the length of the wireline used to run the logging sonde.

10.3.1.2 Inclination

This is the angle between the vertical and the axis of the wellbore. Measurement is based on detecting the angle of deviation in relation to the gravitational field by a pendulum device or by an accelerometer.

10.3.1.3 Magnetic azimuth

This is the angle between the vertical plane going through the axis of the wellbore and the vertical plane going through the magnetic north. The measurement is made by a magnetized needle or by a floating compass pointing in the direction of the earth's magnetic flow. It can also be made by a magnetometer that measures the three projections of the magnetic field vector. Whatever the type of magnetic measurement, instruments must be placed in nonmagnetic tubulars so that measurement will not be thrown off by the mass of the drill string. They can not operate inside casing.

10.3.1.4 Measuring azimuth with a gyroscope

This type of azimuth measurement is relative with respect to a reference position on the earth's surface. The principle of the gyroscope is shown in **Fig. 10.11**. A spinning wheel stays in the same position as long as there is perfect balance in two mobile frames with double universal joints. If the balance is upset by drag or by variation in temperature, the inner frame A turns causing the outer frame B to turn at an angle. This angle is the gyroscope drift that will have to be taken into account at the end of the operation.

10.3.1.5 Tool face

This is a measurement that is needed only to orient the deflection tool in the wellbore in order to correct the well course. It detects the angle between the drill string axis and the azimuth direction.

10.3.2 Measurement tools

10.3.2.1 Inclinometer (Totco) (Fig. 10.12)

This is the tool that is found on all well sites. It is not designed for use in directional drilling, but only to check that the borehole remains close to vertical. The compass card punched by the pendular needle is read directly when the tool is pulled out of the hole, either when it is fished out by wireline or at the end of the trip when the bit is changed.

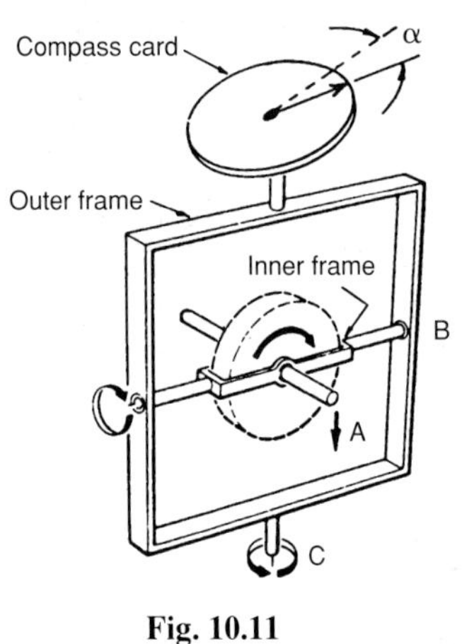

Fig. 10.11
Principle of the gyroscope.

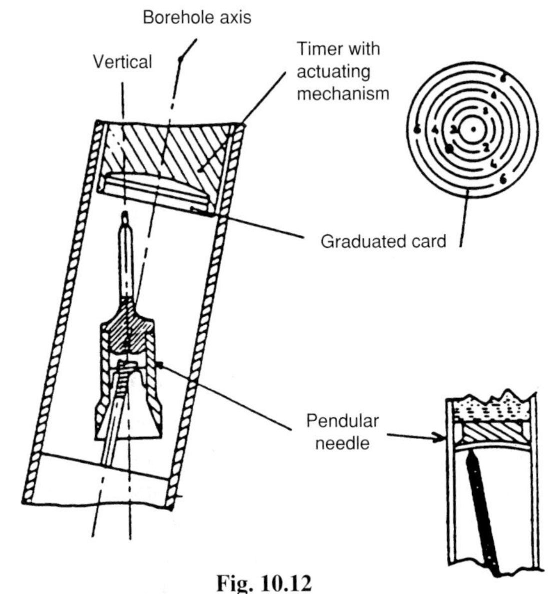

Fig. 10.12
Principle of the inclinometer.

10.3.2.2 Magnetic single-shot or multishot survey

Here the measurement is recorded on a photographic disk (**Fig. 10.13a**) (single-shot) or on a timer-activated film. The same recording gives inclination, azimuth and tool face if the survey tool is oriented in relation to the deflection tool.

The principle of the single shot survey is shown in **Fig. 10.13b.**

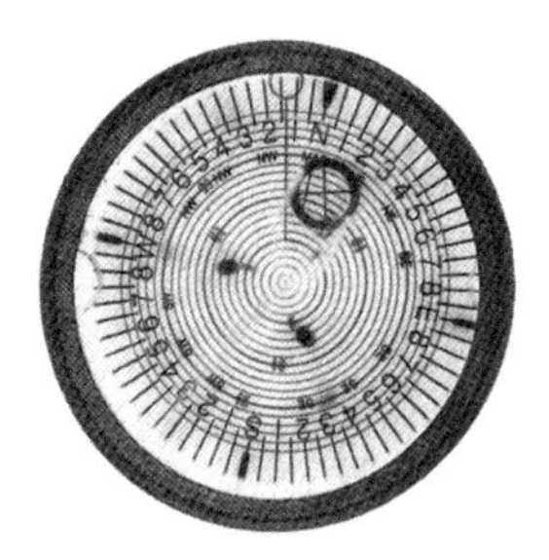

Fig. 10.13a

Photograph of inclination and direction *(Source: SII Servco).*

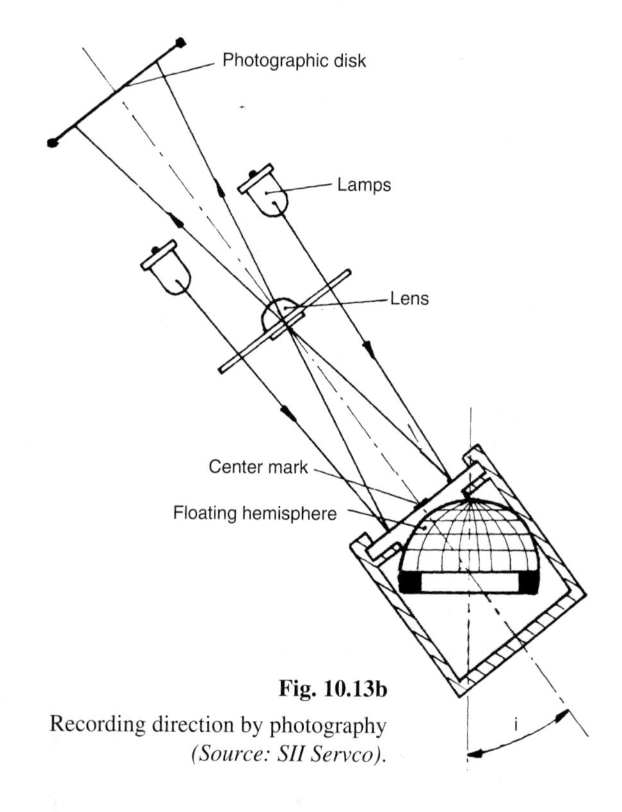

Fig. 10.13b

Recording direction by photography *(Source: SII Servco).*

10.3.2.3 Magnetic wireline steering tool *(Sperry Sun)*

This is an electronic instrument for instantaneous measurement of inclination, azimuth and tool face (**Fig. 10.14**). It includes a directional sonde made up of electronic sensors to measure inclination and azimuth, with an orientation ramp to maintain a position with a fixed orientation in the mule-shoe sub. A conductor line of the same type as the *Schlumberger* logging wireline lowers or raises the sonde and supplies the power required for the measurement. The line conductors are connected to a computer that processes the measurements before sending them to a display unit.

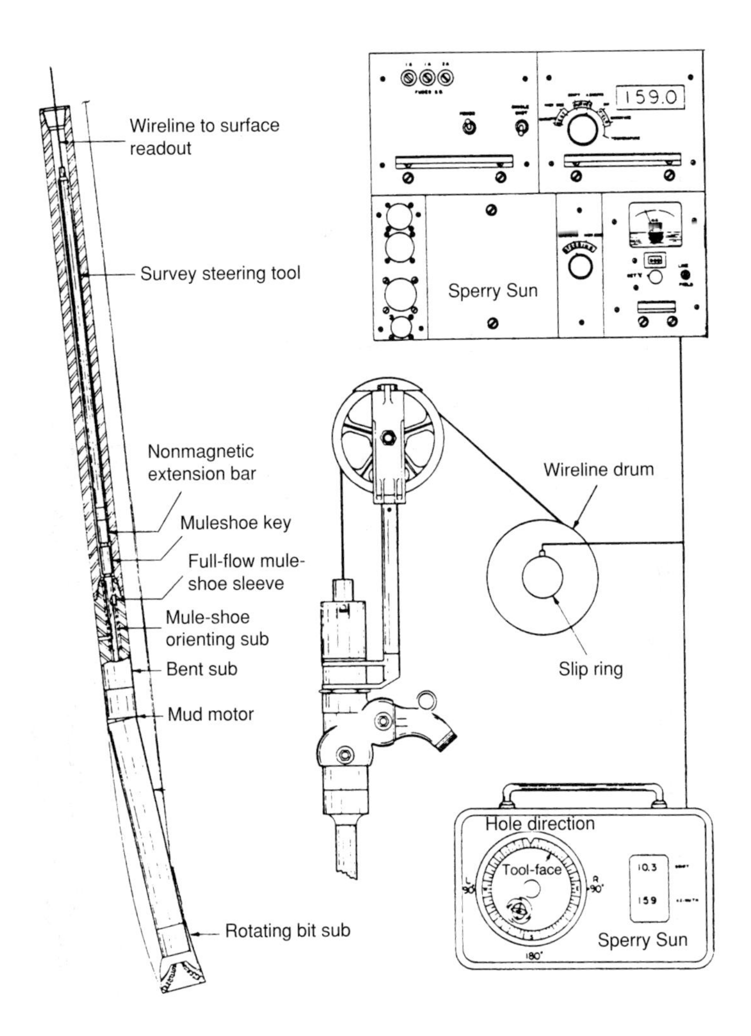

Fig. 10.14

Directional measurement tool with wireline transmission *(Source: Sperry Sun)*.

The rotary table does not turn of course when the sonde has been run in. The tool is usually restricted to when the buildup phase is being drilled or for azimuth corrections with a downhole motor. But each time a joint of pipe is added, the sonde must be pulled up to the surface by the wireline winch. An improvement in the technique was to use a side entry sub screwed into the drillpipe. It allows the conductor line to come out into the annulus between the casing and the drillpipe (**Figs. 10.15a and b**).

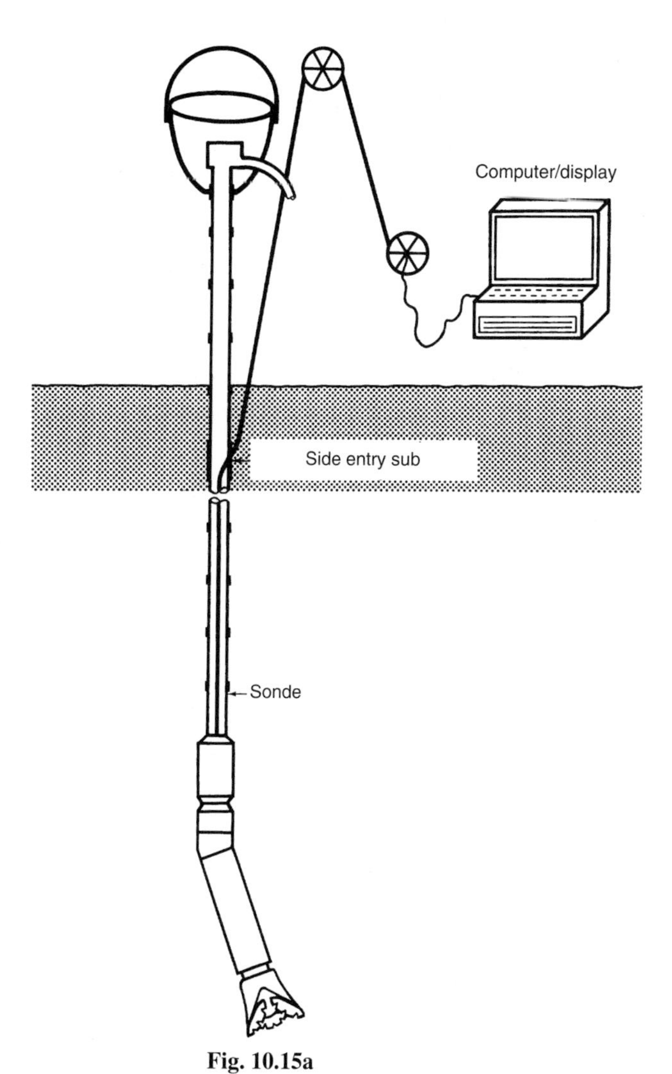

Fig. 10.15a

Principle of side entry sub operation
(Source: Side entry sub, SII Datadrill).

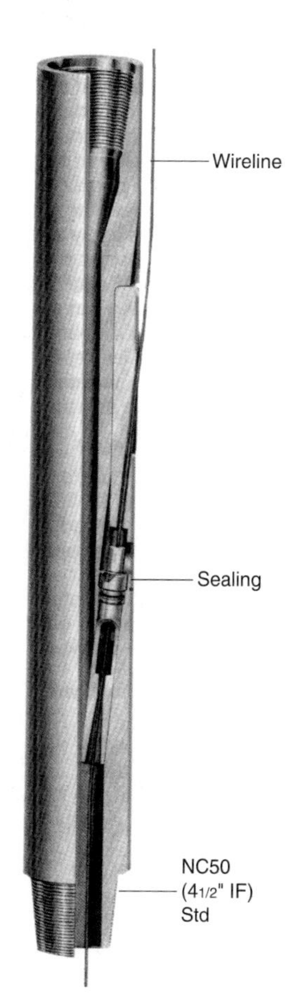

Fig. 10.15b

Side entry sub
(Source: Side entry sub, SII Datadrill).

10.3.2.4 Measurement while drilling (MWD) systems

The latest generation measurement tools are designed for much more than monitoring the well course. In particular, they can supplement mud logging information by instantaneous downhole measurements (gamma ray, resistivity, temperature, pressure, etc.). They look like nonmagnetic drill collars in that they have no specific connections or interference and are therefore no trouble to use for drillers. There are systems with two operating principles currently available on the market:

- transmission by pressure waves through the mud column that is inside the drillpipe (**Figs. 10.16a, b and c**),
- transmission by electromagnetic waves (**Fig. 10.16d**).

The tools that use transmission through the drilling mud come in three types: overpressure, continuous pressure and negative pressure waves. One or more surface sensors mounted on the standpipe read the coded words and transmit them to the computer that processes the measurements.

The electromagnetic tool is based on transmission of electromagnetic signals via the drill stem. The data measured downhole are transmitted by the phase modulation of a carrier wave. The major advantage of the technique is easier two-way transmission. Waves can be sent from the surface downhole by the same transmission channel.

10.3.2.5 Gyroscopes

The principle of these devices was touched on in paragraph 10.3.2.4. The advantage of these systems is that measurements can be made inside the casing.

There are two main categories:

- independent, with recording on film or cassette,
- systems with results read on the surface.

A. *The gyroscopic multishot*

This is the most common type (*Sperry Sun, Eastman, Humphrey*). It must be calibrated on the surface with a fixed reference, then measurements are made by stopping at specific depths with regular drift checking step by step. It requires a significant length of running time and fairly heavyweight surface processing.

B. *The gyroscope with surface readings*

This type of tool is lowered on a logging wireline connected to a computer on the surface. Measurements are almost instantaneous and triggered from the surface.

10.3.2.6 Inertial systems

These systems derive from the inertial guidance systems used in aeronautics. The aeronautical systems can detect the geographical north by their sensitivity to the earth's rotation, thereby avoiding problems of drift and calibration error.

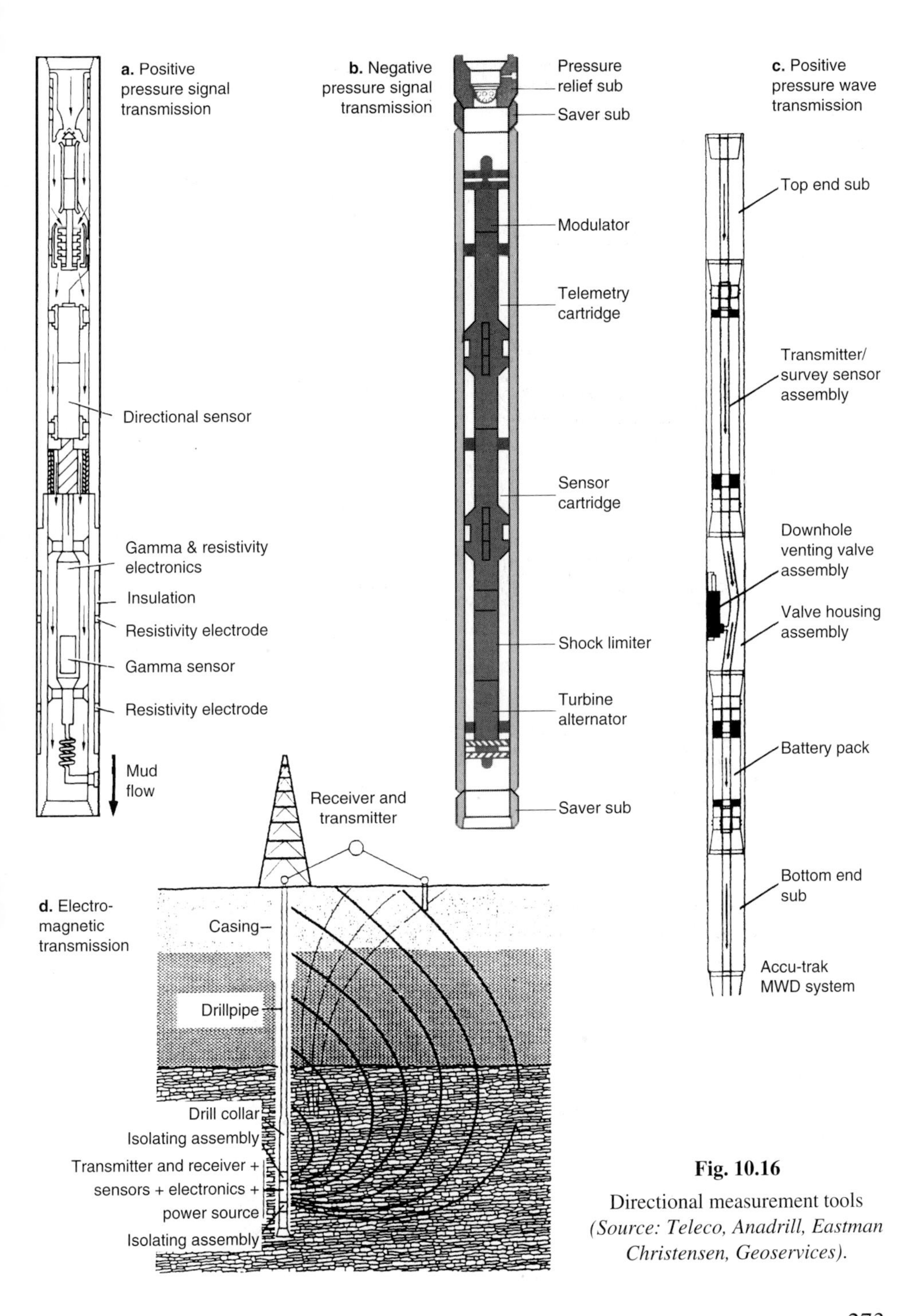

Fig. 10.16

Directional measurement tools
(Source: Teleco, Anadrill, Eastman Christensen, Geoservices).

10.3.3 Calculating the well course

There are several geometric models used in calculating well courses (**Fig. 10.16b**):

- the tangent method,
- the secant or average angle method,
- the balanced tangent method,
- the curvature radius method.

At any measurement point, the inclination and azimuth define a vector which is tangent to the well course.

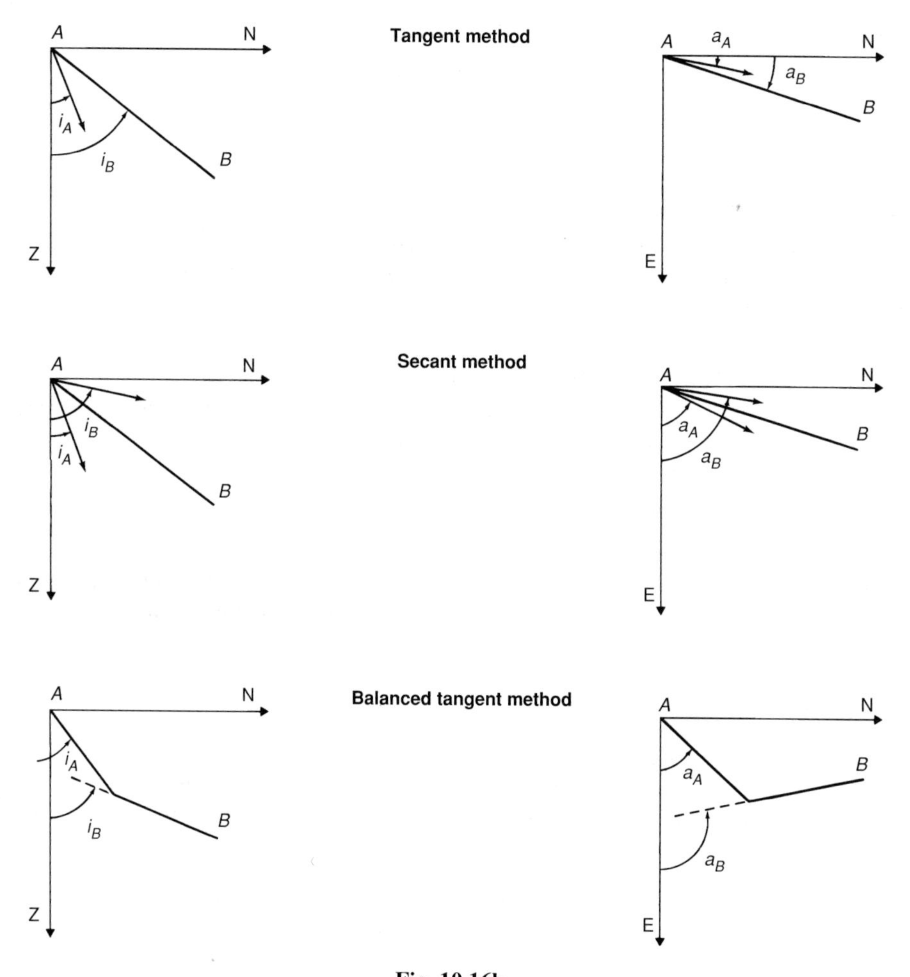

Fig. 10.16b

Principles used in calculating well courses.

The aim of a calculation method is to plot the shape of the well course between two measurement points:

- point *A*: inclination i_A, azimuth a_A,
- point *B*: inclination i_B, azimuth a_B,
- drilled length *AB*.

10.3.3.1 The tangent method

Here a segment with a length of *AB* is drawn from *A* parallel to the vector that is tangent to the well course at *B*.

10.3.3.2 The secant method

A segment with a length *AB* is drawn from *A*, but parallel to the vector with:

- inclination: $(i_A + i_B)/2$,
- azimuth: $(a_A + a_B)/2$.

10.3.3.3 The balanced tangent method

The section of the borehole that has been drilled is likened to two consecutive segments of a straight line with a length of *AB*/2. The first segment is parallel to vector (i_A, a_A) and the second is parallel to vector (i_B, a_B).

10.3.3.4 The curvature radius method

The section that has been drilled is likened to a curve inscribed on a cylindrical surface with a vertical axis. The vertical and horizontal projections of the curve are assumed to be arcs of circles whose radii are calculated according to:

- the gradient of inclination $(i_B - i_A)/AB$,
- the gradient of azimuth $(a_B - a_A)/AB$.

All the formulas corresponding to these methods can be found in the *Drilling Data Handbook*, 6th edition, Chapter J, pp. 433-435, Editions Technip, Paris, 1989.

10.3.4 Dogleg severity

This theoretical concept helps evaluate additional fatigue on the drill stem due to crooked wellbore geometry. The fatigue is generated by changes in azimuth and inclination direction which are expressed in degrees/100 ft. Maximum values go from 4 to 6°/100 ft. The API formula gives:

$$DLS = \frac{100}{\Delta L} \cos^{-1}(\cos i_A \cos i_B + \sin i_A \sin i_B \cos(a_B - a_A))$$

with

ΔL in feet,

DLS in degrees/100 ft.

10.4 DEVIATION TECHNIQUES

Figure 10.9, previously discussed, shows that directional drilling is made up of several phases: kickoff, buildup, stabilized phase and possibly drop-off. Each one requires different equipment and techniques.

10.4.1 Tools to initiate deviation

If the drill string is perfectly vertical, there is no lateral force on the bit except for heterogeneities and formation dip. The wellbore remains vertical due to the governing pendulum effect of the drill collars.

To deviate the wellbore, some way of creating a lateral force must be used. Pointing this force in the right direction for the well course will make it possible to drill toward the target. There are three types of commonly used tools to do this.

10.4.1.1 The whipstock (Fig. 10.17)

This is a trough-shaped implement that tapers to a wedge. The lower base is pointed so it can be inserted into the formation thereby keeping it from rotating at the same time as the drilling bit. The whipstock is oriented by a conventional measurement tool then anchored in the rock by exerting weight on it. Applying weight breaks off a shear pin and at the same time releases the small-diameter gage drilling bit. Drilling must proceed carefully. When five to six meters have been drilled, the bit and whipstock assembly is pulled out. A hole-opening assembly with the same diameter as the phase is then run in and tends to increase the inclination.

10.4.1.2 Jetting (Fig. 10.18)

This technique is very cost-effective and suited to soft, high-drillability formations (over 20 m/h). The drill string is of the angle-building type, i.e. made up of:

- a drilling bit with one nozzle, two nozzles (if there are two cones), or three nozzles with one main, larger-diameter nozzle,
- a near-bit stabilizer with a diameter identical to that of the drilling bit,
- nonmagnetic drill collars.

Jetting proceeds as follows: after the main nozzle has been oriented, the pumps are adjusted so that the fluid velocity at the main nozzle hole is around 120 m/s. The aim is to break down the formation in the shape of an off-center pocket with the stream of drilling fluid. The drilling bit rests on the working face with a little weight to avoid any variation in azimuth during pumping. The driller pulls up slightly on the bit and then drops it (jarring effect) exerting several tons on the bottom until there is no more penetration. Rotation is resumed with maximum weight and a length of pipe is drilled. After the deviation has been checked, the whole process can be repeated if necessary.

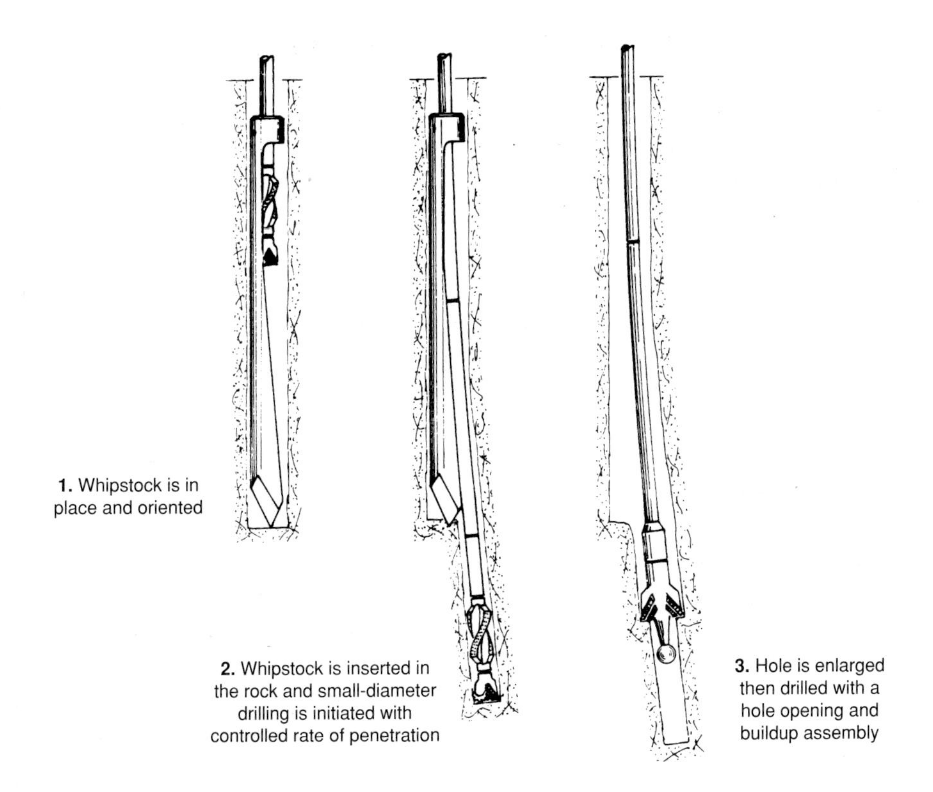

Fig. 10.17

Whipstock-type deflecting tool

(Source: Réalisation des forages dirigés et contrôle des trajectoires, Editions Technip, Paris, 1985).

10.4.1.3 The bent sub and downhole motor (Fig. 10.19)

The principle is simple and similar to the action of the whipstock: the drilling bit has to be pushed laterally in the direction required for the deviation. The bent sub decentralizes the weight on the bit produced by the drill collars and causes bending moment on the seat, i.e. the bit on the working face. The reaction of the formation tends to push the drilling bit sideways in the plane defined by the bent sub. The lateral force is therefore dependent on:

- the weight on the bit,
- the distance between the bit and the bent sub,
- the reaction of the formation,
- the bit diameter,
- the rigidity of the drill string at the corresponding depth.

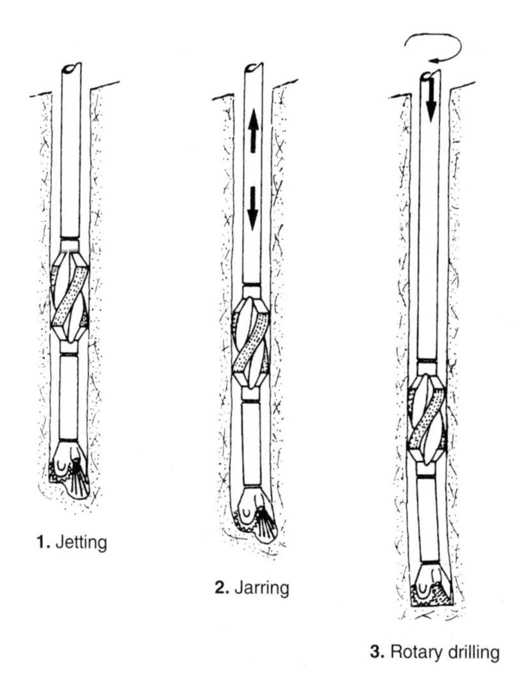

Fig. 10.18

Deviation by jetting
(Source: Réalisation des forages dirigés et contrôle des trajectoires, Editions Technip, Paris, 1985).

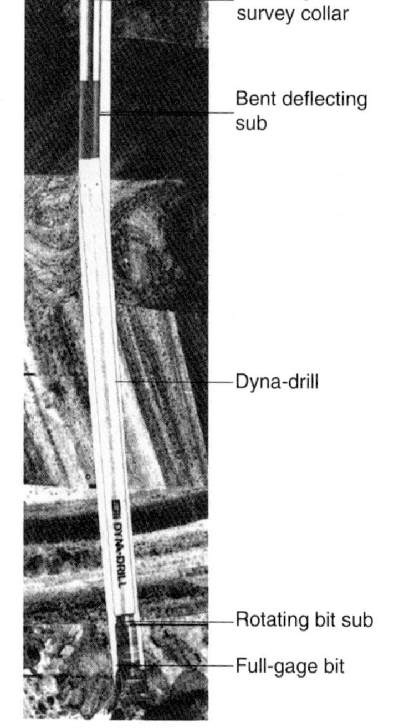

Fig. 10.19

Deviation drill string with a bent sub
(Source: Smith 1.1.).

The rotary drilling technique can in no way be used since the plane of the bent sub must remain fixed. A hydraulic downhole motor is therefore screwed on under the bent sub. Present-day industrial motors are of two types: the turbine and the positive-displacement motor based on the Moineau principle.

A. Drilling turbines (Fig. 10.20a)

Torque is supplied by the impact of the fluid stream deflected by a crown of stationary blades (stator) onto the blades of a crown connected to the drive shaft (rotor). The mechanical principle is hydrodynamic and to get enough power to rotate the bit, blade rotors and stators are piled up sandwich-style. Rotation speeds range between 500 and 1000 rpm. For good efficiency, considerable pumping power is often required, or at least pumping units that work much more intensively than usual. This has for some time been a sticking point for more systematic use of drilling turbines, since duplex pumps were limited. More recent triplex pumps provide better performance, but drillers are often reluctant to push them to their limits because it raises maintenance costs significantly. Bearings hold the rotor in the stator axially.

The stator is subjected to hydraulic thrust in the opposite direction to the reaction force of the weight applied onto the bit. The bearing system is made up of disks and wear pads whose lifetime can be between 50 and 200 hours depending on drilling conditions (temperature, type of fluid, difficulty in drilling ahead, etc.).

The motor is expensive and it must yield a gain in productivity. This type of motor allows the bit to rotate much faster than the conventional rotary table. It is often an indispensable extra if a diamond bit —when it is chosen mainly for drilling rate performance— is to be optimized. This is the prime use in fact that turbines are put to, but they are currently important in directional drilling due to more common use of diamond bits for KOP and azimuth corrections.

B. Positive displacement motors (Fig. 10.20b)

This motor is an application of the principle of the Moineau pump that is used industrially to displace viscous, heavy fluids, sand-bearing effluents, or even in oil industry applications as a downhole production pump. All it needs is to be made to work reversibly by injecting drilling mud to rotate the helicoid rotor. The motor works on a positive displacement basis, since the steel rotor isolates volumes of fixed cylinder capacity in the rubber stator. When the fluid moves downward under the pump discharge pressure, it applies torque onto the rotor. The torque provided is directly related to the pressure, and the rotation speed to the circulating flow rate, without disregarding efficiencies.

The Moineau principle technique has the particular feature of a rotor that turns with a lateral or circular movement, called nutation. This means a knuckle joint is required at the base of the rotor and limits the power of this type of downhole motor.

The PDM (positive displacement motor) is used in areas such as: directional drilling with a bent sub, downhole motor drilling with rock bits, and horizontal drilling. The main drawbacks are that they do not withstand high temperatures well and are very sensitive to gas in the mud. Generally speaking, however, they do not require higher than usual flow rates and can be incorporated into a drill string without being too hydraulically demanding.

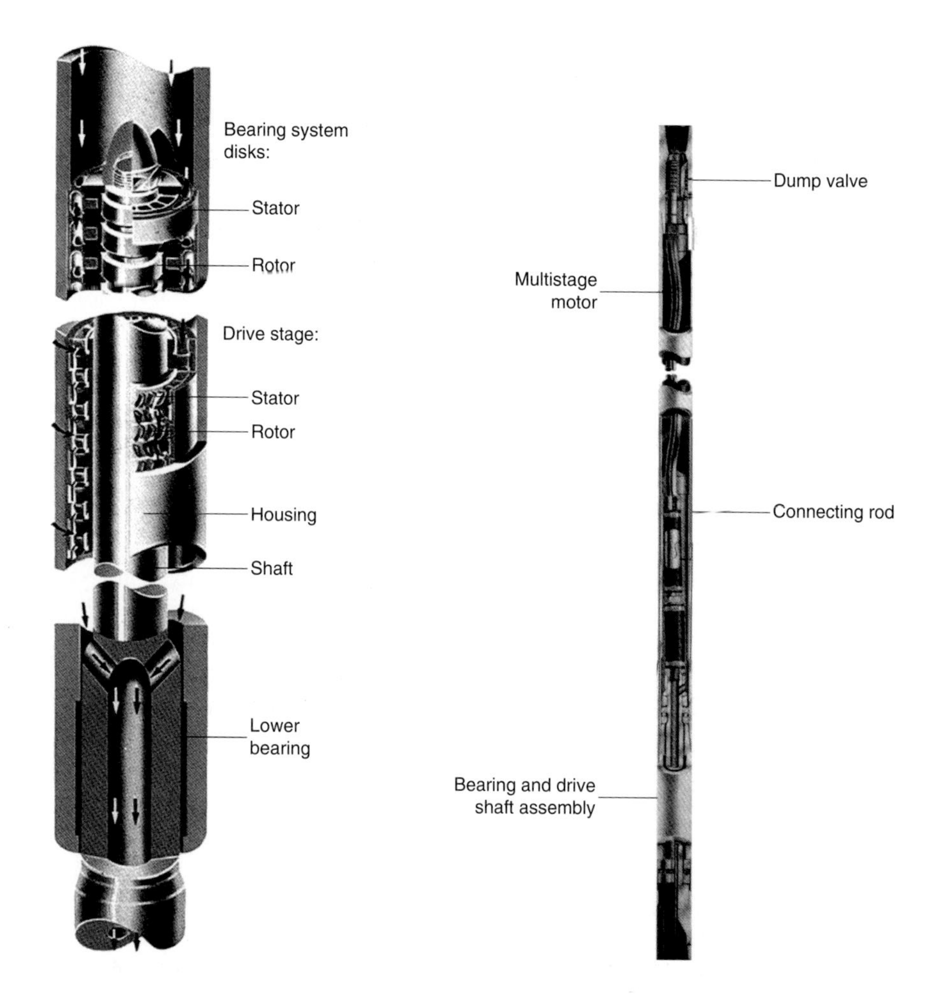

Fig. 10.20a
Turbine-type downhole motor *(Source: Neyrfor).*

Fig. 10.20b
Positive displacement-type downhole motor (Moineau).

10.4.2 Rotary assemblies in deviated wells (BHA) (Fig. 10.21)

As soon as possible, drillers stop using the original bits or deviation correction bits so that they can come back to a conventional bottomhole assembly. However, the azimuth can no longer be varied. The inclination can be accentuated, decreased or stabilized during drilling by using stabilizers placed at appropriate positions on the first 30 meters. Upper drill collars mainly serve only to add weight on the bit.

10.4.2.1 Buildup assemblies

Once the buildup has been initiated in one of the ways described above, the phase can be continued with an appropriate BHA, bottomhole assembly. Assembly A (**Fig. 10.21**) composed of a bit, a near-bit stabilizer and nonmagnetic and standard drill collars, provides a buildup gradient that depends on the weight on the bit, the rotation speed and the distance between the working face and the near-bit stabilizer blades (lever effect). The distance is generally about 1.5 m.

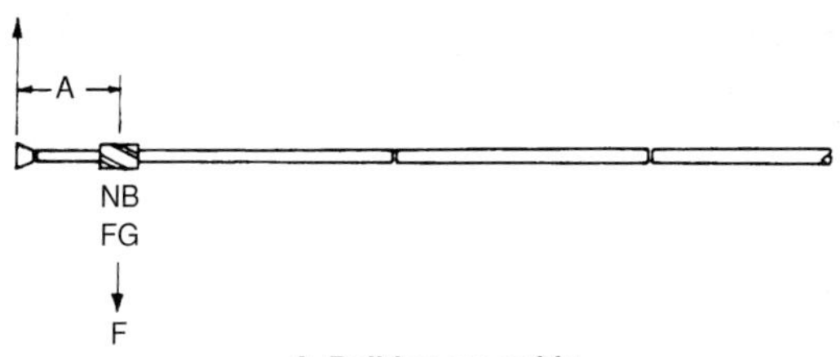

A. Buildup assembly

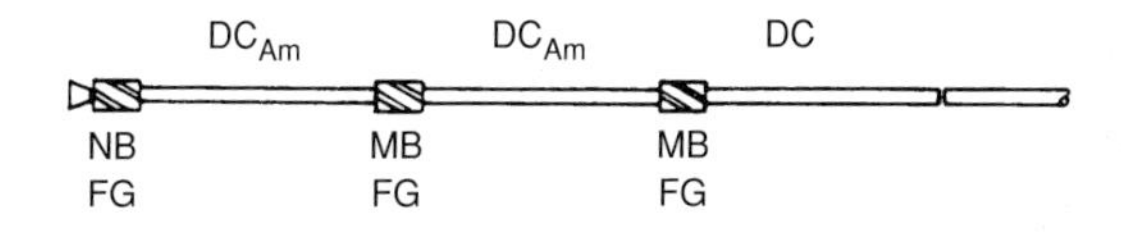

B. Stabilized assembly

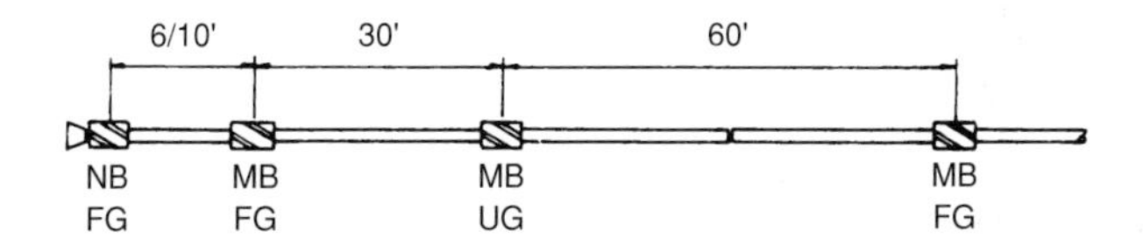

C. Stabilized assembly for soft formations

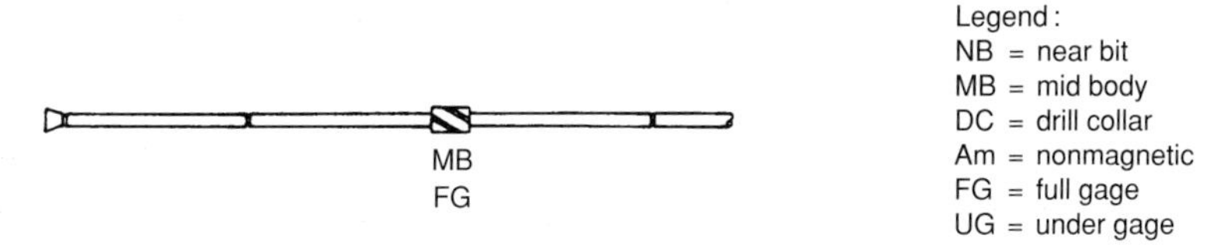

D. Drop-off assembly (pendular)

Fig. 10.21

Deviation assemblies used in rotary drilling.

10.4.2.2 Stabilized assemblies

The stabilized phase is usually the longest and also the most troublesome. Since the assembly must be rigid and as straight as possible, three full-gage stabilizers are placed at each end of the first two drill collars. If the formations are soft, a slight upward direction is given by undersizing the diameter of a stabilizer (**Figs. 10.21B** and **C**).

10.4.2.3 Drop-off assembly

This is a typically pendular assembly used also when the borehole must remain vertical in vertical drilling (**Fig. 10.21D**). It should be remembered that a rotary assembly has a natural azimuth tendency to turn to the right. This can be greatly modified, even reversed, according to dips, formation heterogeneities, parameters, bit types, diameters, etc. Drilling with a downhole motor usually reverses this tendency.

10.4.3 Horizontal drilling (After J.F. Giannesini, IFP, *Revue Forage*, n° 123)

10.4.3.1 What is a horizontal well?

Drilling horizontally is never an end in itself, the aim is to produce. The term horizontal well or horizontal drilling is therefore inaccurate. Practically speaking, it is the general category of a range of production systems which all have a horizontal or subhorizontal section in the producing reservoir.

This discussion can not be exhaustive, so it will be limited to wells drilled from the surface and having a portion with an inclination of at least 80° in the reservoir. If the drain hole has a continuous connection with the formation, it will be assumed to be long enough so that a parallel type of fluid flow is generated as illustrated in **Fig. 10.22.**

10.4.3.2 When are horizontal wells drilled? (Fig. 10.23)

A. *Fractured reservoirs*

When vertical fractures are dispersed and rare in these reservoirs, they are excellent objectives for horizontal drilling. The development of the Rospomare field in the Adriatic Sea by *Elf Aquitaine (SNEA(P))* was the original industrial precursor for this technology.

B. *Thin beds*

Beds less than 15 to 20 meters thick are considered to fall into this category. If there is a gas cap or an underlying aquifer, the situation is even more favorable for horizontal wells.

C. *Tight formations*

A horizontal well can be drilled in a tight formation and then four to eight vertical fracture planes can be opened up by a hydraulic fracturing job.

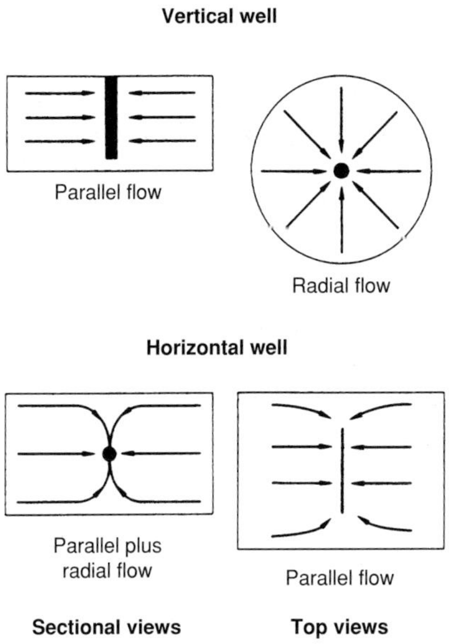

Fig. 10.22
Parallel flow production.

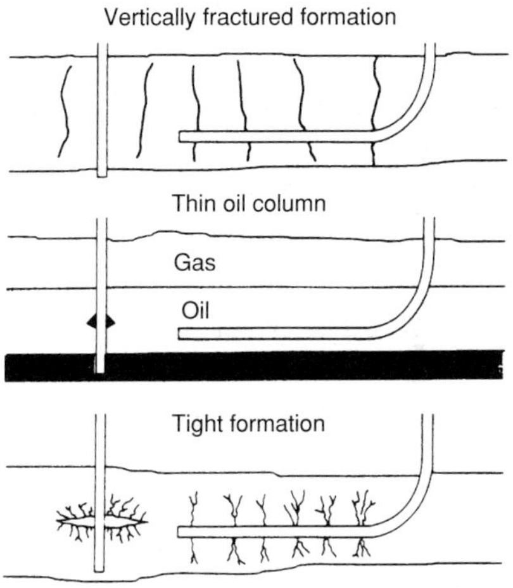

Fig. 10.23
Production by horizontal drain holes.

10.4.3.3 Geometry of horizontal wells

A. Horizontal section

The longer the horizontal part, the greater the productivity and the more oil can be recovered per well. Lengths of 300 to 600 meters are commonly drilled. The record held by Esso in Canada is one mile (about 1600 meters). Difficulties in drilling and completion of a well are clearly compounded, the longer the horizontal section is.

B. Altitude control

Altitude control is expressed by the radius of the cylinder that the horizontal drain hole is kept inside and is one of the most important parameters. Many horizontal wells require good control because they are drilled either in thin beds or vertically near an aquifer or a gas zone. If inadequate altitude control pushes the well course out of the reservoir (see **Fig. 10.24**), there will be a loss in productivity.

Controlling the altitude of the drain hole is therefore a must. Initially, it is a problem of cost. If there is no aquifer or gas zone, accuracy of about three meters is sufficient. Here, tools available on the market may be able to manage this type of accuracy. If it is not necessary to change the bottomhole assembly during horizontal drilling, the cost can be kept down within reasonable limits.

If on the contrary the nearness of water or gas demands accuracy of about one meter in altitude control, the only solution is frequent measurements and multiple bottomhole assembly changes.

The operation remains possible, but the costs will spiral. This problem of accuracy in positioning the drain hole should find a solution in the development of new guide sensors and new techniques to orient bottomhole assemblies.

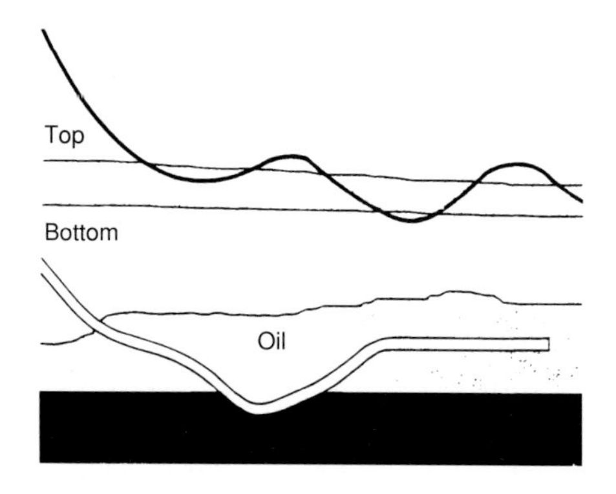

Fig. 10.24

Horizontal drain holes with poor altitude control.

C. Radius of curvature

The curvature radius of the part of the well course that connects the vertical section to the horizontal section depends on several factors:

- the horizontal distance between the surface well location and the point where the borehole enters the reservoir,
- the vertical position of the kickoff point,
- the length of horizontal drain hole to be drilled,
- completion constraints.

Generally speaking, a long curvature radius —over 250 meters— allows a longer horizontal section to be drilled and makes completion easier. Long radii are mainly used for development wells while short ones chiefly apply to recompletion operations.

D. Diameter of the horizontal section

In contrast to vertical wells, the productivity of horizontal wells is not very sensitive to the diameter of the hole. For wells with a long curvature radius, i.e. a long horizontal section (300 m and more), the most commonly used diameter is 8.5 inches. Wells with a short radius require the use of flexible, articulated drillpipe and are drilled with a smaller-diameter bit, 4 inches usually.

10.4.3.4 Drilling the well

When reservoir engineers decide to drain a reservoir with a horizontal well, they usually give the driller an entry point into the reservoir and a minimum horizontal length that needs to be drilled in a given direction. In oilwell terminology, the entry point is called the target and the horizontal section the drain hole.

The driller's first objective will then be to hit the target. To do so, he will have to follow a well course which is determined in one of the essential stages of the preparatory studies. The course must take into account surface installation constraints as well as completion requirements.

The diagram in **Fig. 10.25** shows the different phases of a horizontal well.

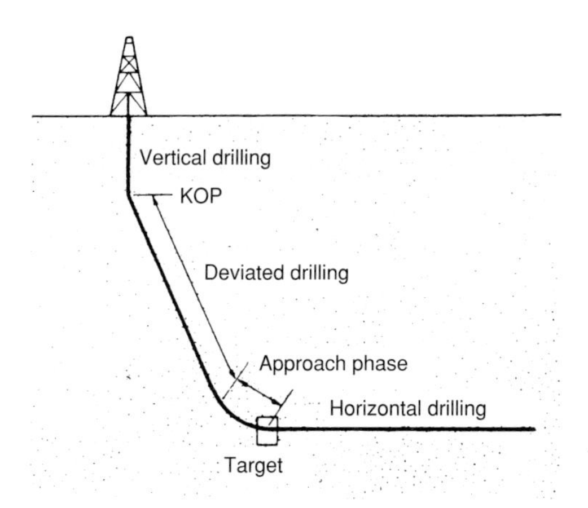

Fig. 10.25
Typical course of a long-radius horizontal well.

A. The vertical section

Except for wells drilled with a slant mast rig, horizontal wells comprise a vertical section of varying length between the surface and the kickoff point.

B. The deviated section

This part of the well goes from the kickoff point to the phase approaching the target. Roughly speaking, this corresponds to drilling a section that goes from vertical to 70° inclination. To achieve this, a choice is made during the drilling engineering phase between an inclination buildup in two separate steps separated by an equal-inclination section, and a steady inclination buildup.

Plotting this part of the course must be optimized in order to:
- reduce friction forces and torque,
- help keep the hole clean and stabilize the walls,
- make completion easier.

The two main difficulties that may come up during this phase are:
- the azimuth may need to be corrected,
- it may be impossible to build up the inclination as quickly as planned.

Simulation models showing the behavior of drill strings are operational and low cost today, they can help avoid big mistakes when drilling is in the planning stage.

During the whole deviation phase, the quality of drilling fluids is crucial for success. All along the course, the borehole must be kept clean and the walls stabilized. Mud properties must be adapted not only to the nature of the formations involved but also to the angle that the borehole follows as it goes through them. Recent studies have pointed out the importance of mud density versus the angle of inclination and the contact time, in particular when shales are being drilled.

It may be possible one day for this phase of the well to be drilled with a remote-controlled drill string. The idea is to connect up the information circuit from the MWD tool to the remote-controlled components of the drill string in a loop, via a surface processing unit. This will enable the drill string's course to be controlled.

The pieces of the puzzle already exist: stabilizers, remote-controlled bent subs and a three-dimensional simulation model are all operational and available. What is missing is the surface processing unit, the expert so to speak that will make the system work.

C. Approaching the target

This is the toughest phase in the operation because the target's exact position is never really known. Geologists and geophysicists can not give the precise position as there are too many geological unknowns that mean the reservoir may be located higher or lower than expected. This may not be of enormous importance for a deviated or vertical well, but accuracy is crucial for a horizontal well.

As shown in **Fig. 10.26**, the driller is like a pilot during this phase. He must land his plane, but does not know the exact altitude of the landing strip. If he pulls up too soon, he will fly over the runway and if he is too late, he will crash into it. Here altimeter indications do not help much and the driller must do what the pilot does. First choose an angle of descent that will enable him to cope with any eventuality, then during his descent look around and deduce his position in relation to the target. He uses the MWD tool and more particularly its capacity to measure stratigraphy (gamma ray and resistivity) to do this. By following the progress of marker beds, he will be able to determine the distance between the target and the tool (**Fig. 10.27**).

This of course means that the markers have been identified beforehand by studying neighboring vertical wells.

Most horizontal wells have "landed" in this way. However, when the marker beds are not reliable or when there is not enough information from neighboring wells, the reservoir may first be identified by an initial penetration at a high angle, 75° for example, and then the wellbore can be sidetracked.

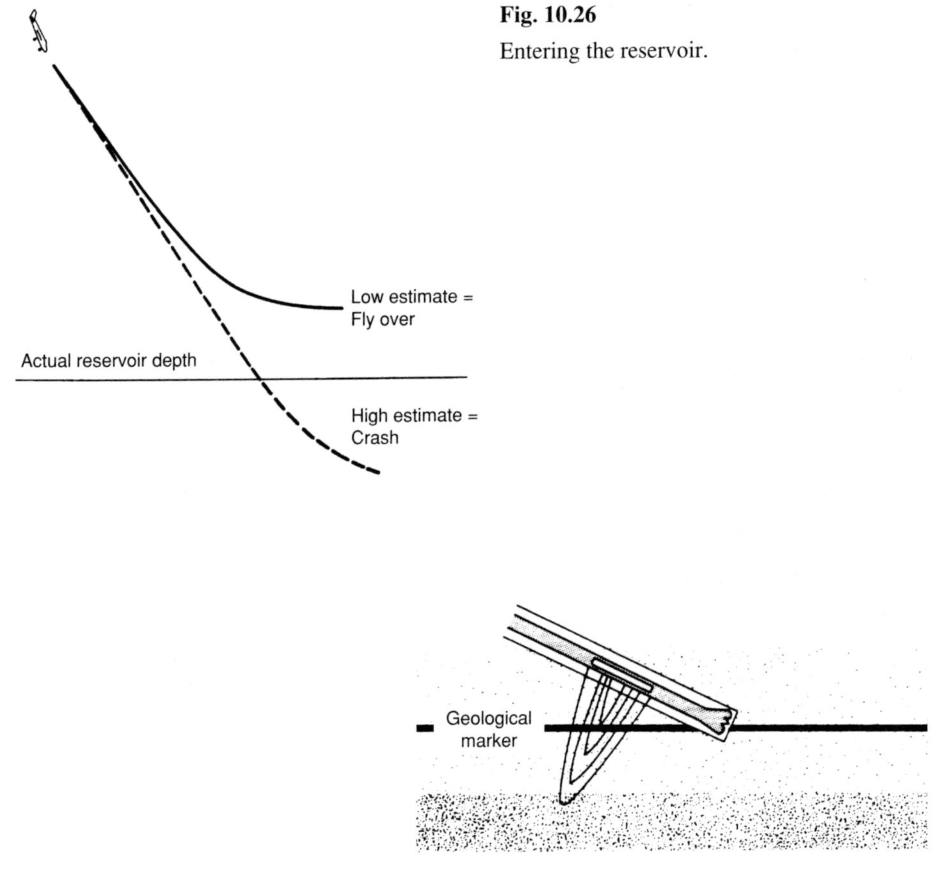

Fig. 10.26
Entering the reservoir.

Fig. 10.27
Locating a position by identifying a marker bed.

This method of approach by stratigraphic monitoring needs improvement. It is risky and its use is limited by the fact that it requires vertical wells in the same area as the horizontal well. This is why a tool is needed for the future to provide relative not absolute positioning. Since the target is a marker bed, two solutions seem possible:

- logging tools with a deep radius of investigation,
- geophysics-type methods.

In both cases, information needs to be rapidly accessible during drilling.

D. Drilling the drain hole

During this phase, the major difficulty is that the drain hole must be kept within very strict altitude limits. The limits may be between the top and the bottom of the reservoir or a minimum distance from a gas/oil or water/oil interface.

Given the scale of the drain hole, the top of a reservoir is seldom flat. A variation of a couple of meters may prove catastrophic.

Take the case of drilling a thin bed with an underlying aquifer. The ideal well course consists in staying as close as possible to the top of the reservoir in order to maximize recovery. But for such a course to be achieved, it is necessary to see and if possible foresee what the bit is coming up against. Most sensors on the market are placed behind the bit at a distance of around ten meters. This means that when the sensor picks up the fact that the bit has gone out of the reservoir into the overlying or underlying bed, it is actually quite a good distance outside. Additionally, the sensors have lateral vision and high vertical resolution because they are designed for vertical wells where the well course is orthogonal to the bedding of the formations. For a course above a water-bearing formation, if the top of the reservoir is flat enough, all that is required is to keep to a given depth. Here a long-range sensor capable of "seeing" the water level would make safe navigation possible.

In the horizontal part of the well, the second difficulty consists in keeping the borehole free of cuttings in order to reduce friction forces, torque and damage to the formation. Here it is hard to get rid of the cuttings because the mud properties must be such that the cuttings move out of the horizontal section as well as the inclined sections. In addition, the rotating drill string regrinds the cuttings.

Keeping the hole stable is also more complex because the distribution of stresses around it is not isotropic. Paradoxically this problem is less crucial in soft formations. In a "liquid" formation, stress distribution is totally isotropic, but it is not possible to maintain this stability with excessively weighted-up mud. Just imagine 300 meters' worth of drill string resting on the bottom of the hole with a high differential pressure due to the mud !

As mentioned earlier, the longer the drain hole, the more it produces. Once the navigation problems have been solved, it is the mechanical difficulties of drilling that curtail the length of the drain hole. These difficulties are generally caused by friction and torque forces.

Additionally, using the drill string under compression stresses may cause buckling, which in turn increases friction forces. Friction and torque depend on the degree to which the drill string is pressed sideways against the borehole wall. This lateral force depends on two main parameters:

- the weight of the drill string,
- the geometry of the well profile (winch effect).

To reduce friction and torque, it is therefore necessary to:

- use lighter-weight drill string components, such as aluminum drillpipe in critical instances,
- lessen the winch effect by keeping well profiles as smooth as possible,
- use drilling mud with high lubricating power, oil-base mud or additives with water-base mud.

The use of a power swivel is becoming increasingly widespread for drilling beyond 2000 meters. It is a fact that the power swivel provides drilling "comfort" that can have quite an impact on the success of operations.

To conclude this part, it can be said that the success of horizontal wells depends on being able to do three things:

- locate the bit in relation to a marker bed or a fluid interface situated several meters above, below or in front of the bit,
- orient the bit's movement so that it stays at the right distance from the chosen reference point,
- reduce friction and torque that restrict the horizontal length of the drain hole.

Chapter 11

FISHING JOBS

While a well is being drilled, several types of accidents can occur and keep drilling from being carried on normally.

The term "fishing job" means an operation in a well using specific implements and tools to restore a situation where the drilling program can most likely be continued.

We will first of all list the main types of accidents possible during drilling:

1. Pieces of metal on the bit-gage surface for a variety of reasons: broken drill bits or rig floor tools falling into the well.
2. Tubulars that fail inside the borehole: drill collars, drillpipe, or a string of casing or tubing may break off. The problem that has to be solved is then to get these lengths of steel cylinders out of the well.
3. Stuck drill string which frequently ends up as the problem described in 2. above. The failure may occur due to attempts to get the drill string unstuck or because the drill string has been unscrewed to leave the stuck part in the borehole temporarily.

11.1 CAUSES OF FISHING JOBS

Three parameters are involved in the causes of accidents: equipment failure, borehole-related problems and human error.

11.1.1 Equipment

Drill bits may be broken due to poor quality or inappropriate implementation. A bit that is unsuited to the formation that is being drilled can become excessively worn and this may cause unexpected failure. Likewise, poorly chosen mechanical parameters may have the same effects. Depending on how much is known about a region, the odds of breaking a bit in an exploratory well may be high when the driller attempts to reach optimum rate of penetration.

The drill string is particularly susceptible to fatigue, wear, poor maintenance and supervision, and inappropriate use. The most common failure occurs at drill collar threads where stresses are the highest.

11.1.2 The borehole

Loose formations cause considerable friction that can keep the whole drill string stuck in extreme cases. Swelling clays will have the same consequences. Highly permeable formations can cause the drill string to get stuck by differential pressure, i.e. the difference between the hydrostatic pressure of the mud and the formation pore pressure. Rotary drilling carried out in crooked boreholes "machines" the formation into a keyseat shape with drillpipe rotation (**Fig. 11.1**). Since the tool joint is wider than the body of the pipe, it catches on the keyseat and may get stuck when the drill string is pulled out.

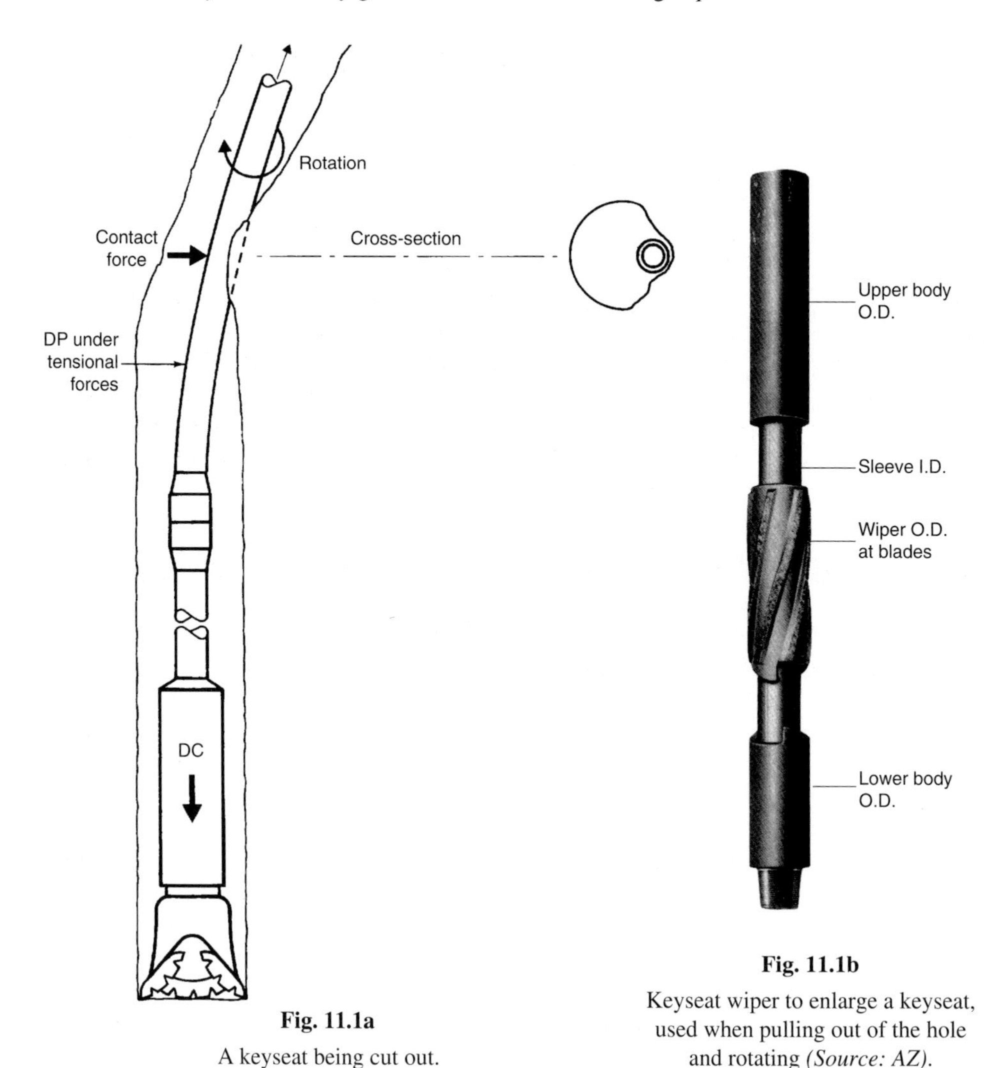

Fig. 11.1a
A keyseat being cut out.

Fig. 11.1b
Keyseat wiper to enlarge a keyseat, used when pulling out of the hole and rotating *(Source: AZ)*.

11.1.3 Human error

There are factors that can not be anticipated, but mainly there are technical mistakes and negligence. This is the case when tools and equipment fall into the borehole. The decision to pull out a bit before it fails is governed by technical circumstances. The toolpusher's experience is crucial in reducing risks, although his job is precisely to take risks to improve efficiency by applying the best performance parameters.

11.2 FISHING JOB PROCEDURES

Before implementing substantial, i.e. costly, equipment and materials, it is indispensable for the toolpusher to evaluate the chances of success and, of course, the cost. Given the prices charged by service companies, the trend today is not to insist too long, but rather to consider a sidetracking operation fairly soon. We will discuss sidetracking later on. If a fishing job is to be a success, the circumstances of the accident must absolutely be known in detail. The various recordings covering the drilling phase under way and the rig geologist's daily report must be available for analysis. Information on the "fish" is of course also a must. It is simpler to describe the fish by figuring out what is missing when the drill string is pulled out. The driller must comply with the rules and regulations in force in the profession which dictate that all downhole material and equipment are to be measured and recorded in a log book. This applies in particular to outside and inside diameters, lengths, thread type, etc.

11.2.1 Equipment, tools and methods used for broken drill bits

Depending on the size of the pieces of metal, the following will be used:

- **The basket sub** (**Fig. 11.2**) to fish out bit rollers, ball bearings and teeth. It is placed above a bit and mud is circulated for a few minutes, then suddenly stopped. This allows the pieces of metal carried along the mud flow to fall back down into the basket. The tool is effective in getting rid of pieces of metal remaining on the bottom of the hole after a fishing job.
- **The junk basket** (**Fig. 11.3**) to fish out bit rollers, for example. By taking a core sample several dozen centimeters long and because of the position of apertures that allow reverse circulation in the bottom of the hole, the tool makes it easier for debris that was above the bottom to be sucked inside the basket. This works only in relatively soft formations.
- **The poor-boy junk basket** (**Fig. 11.4**) is often built on the site. It is a piece of casing with saw teeth on the bottom that can bend and trap junk when a weight of several tons is applied to the tool.

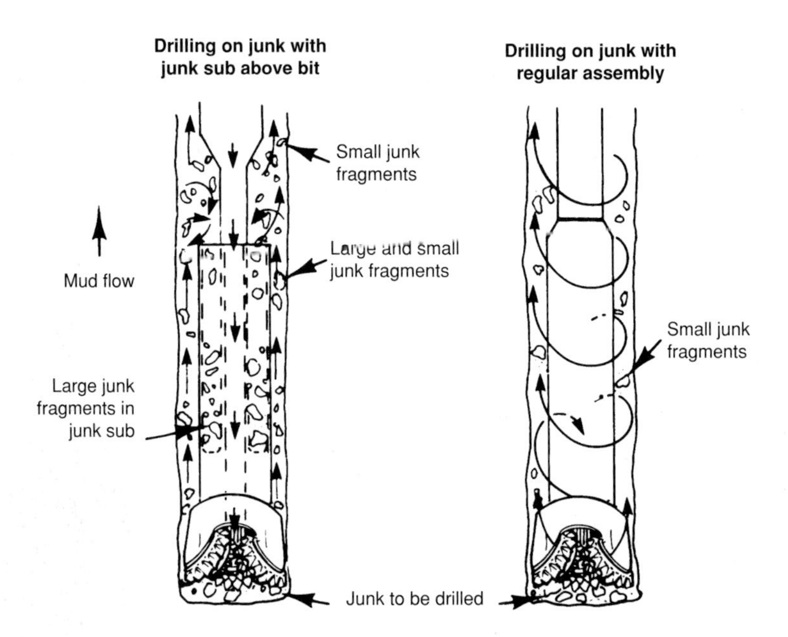

Fig. 11.2
Basket sub.

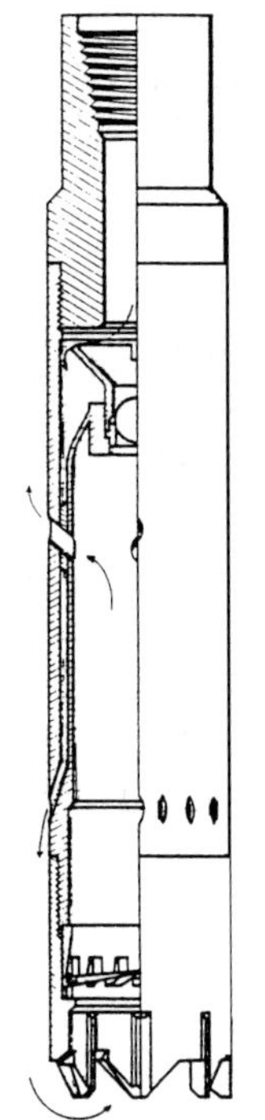

Fig. 11.3
Junk basket with reverse circulation *(Source: Bowen)*.

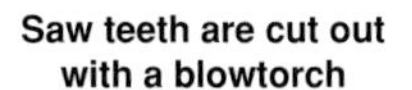

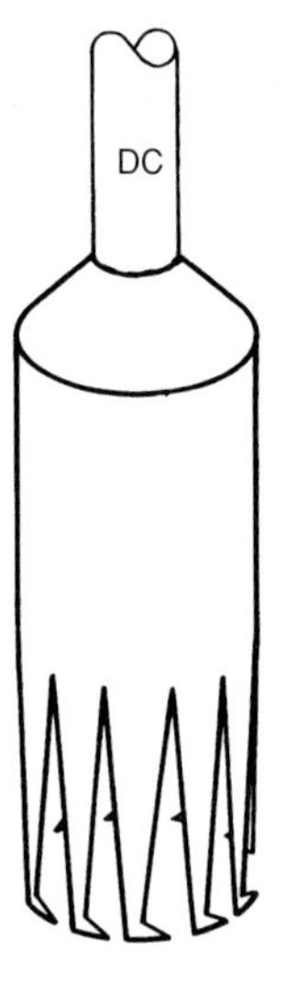

Fig. 11.4
Poor-boy junk basket.

- **The fishing magnet (Fig. 11.5)** is a permanent magnet used to get rid of small-sized metal debris on the bottom of the hole.
- **The junk mill (Fig. 11.6)** is often run in before using a basket sub. It is a drill bit that breaks up pieces of metal on the bottom of the hole.

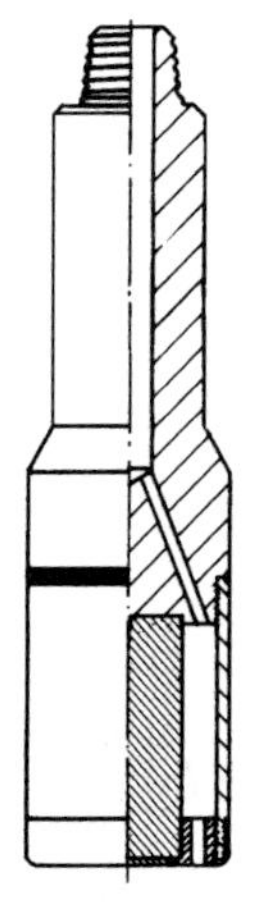

Hole diameter (in)	Outside diameter of fish (in)	Thread API reg (in)
6	5	2 7/8
8 1/2	7	4 1/2
12 1/4	11 1/2	6 5/8
17 1/2	16	7 5/8

Fig. 11.5
Permanent magnet
(Source: Drillstar Industries).

Fig. 11.6
Junk mill
(Source: Tristate Oil Tool).

11.2.2 Equipment, tools and methods used for a broken drill string

The driller realizes that the drill string has failed when there is a variation in three parameters:
- decreased load on the weight indicator,
- drop in mud pump pressure (the mud no longer circulates through the bit jets),
- variation in torque at the rotary table.

Failure is often preceded by a gradual reduction in the mud pump pressure due to a hole in the drill string (wash out). If it is not spotted, it leads to failure of the thread or the pipe.

Most drill string failure occurs:
- at the pin-end thread of drill collars that bend in alternate directions,
- at about 50 centimeters from the box-end tool joints where slip inserts grip causing stress concentration and accelerated corrosion.

As soon as the driller is sure of a drill string failure, he pulls out the upper part of the drill string. He can then see the what the top of the fish looks like, its diameter and depth.

The overshot (**Fig. 11.7**) is the tool that is normally used to attempt to fish it out. This tool also uses the principle of the cone and slips system.

When the overshot has caught hold of the fish, it is pulled up slightly. The cone-shaped parts of the overshot body move upward with respect to the cone-shaped parts of the grapple, thereby placing it between the body and the pipe fish. A packer inside the body secures a seal around the head of the fish and restores circulation through the fish. The packer also makes it easier to clean the well and pull out the tool plus fish together.

If the fish is stuck and the driller wants to pull out the drill string, he releases the overshot by rotating slowly to the right. This allows the spiral grapple to open and disengage from the fish.

The basic problem is to be able to catch hold of the fish when the head is sometimes located where the borehole has caved in. The overshot must then be equipped with a number of accessories that will make it easier to center the head of the fish (**Fig. 11.8**).

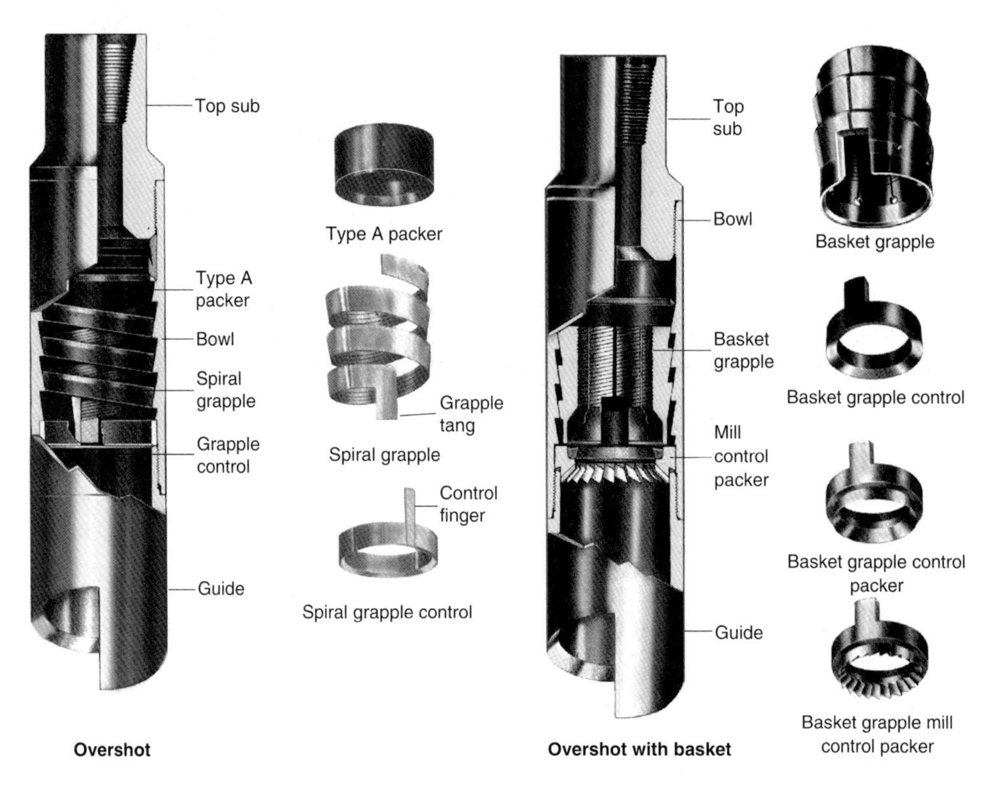

Fig. 11.7
(Source: Bowen).

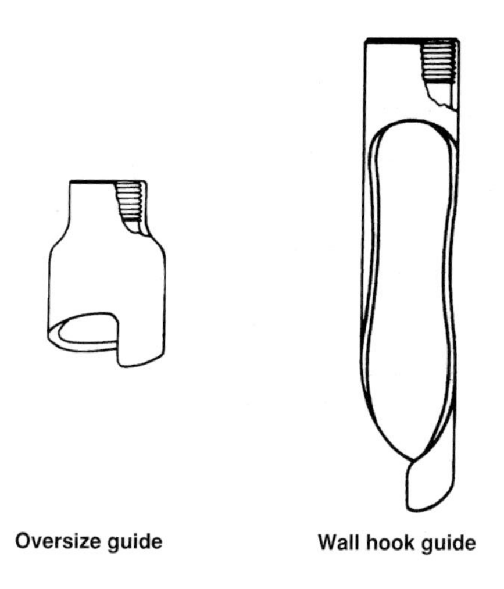

Fig. 11.8
Overshot guides *(Source: Drillstar Industries)*.

11.2.3 Equipment, tools and methods used for stuck drill string

The methods differ according to the cause of sticking:

- caved in walls, borehole narrowed down due to formation instability,
- differential pressure sticking.

In the first instance, the problem is a purely mechanical one which can only be solved by a variety of mechanical operations. In the second, a physical phenomenon due to pressure is involved and in addition to the same type of mechanical operations, an attempt can be made to solve the problem by acting on the differential pressure.

11.2.3.1 Unstable formation

To free the drill string, the driller can alternately pull and compress it using the drawworks or apply torque by means of the kelly or the power swivel. Another possibility is cleaning the annulus between the drill string and the borehole walls by circulating drilling mud or any other fluid that can help get the drill string unstuck.

- Pulling must be restricted to the yield strength tension value of the weakest component of the drill string.
- Only drill collars and heavyweight drillpipe (under certain bit diameter conditions) can be compressed.
- The kelly is very limited from the standpoint of twisting. It can not be rotated and pulled at the same time without pulling it out of the kelly bushing. The best situation is when there is a power swivel.

If these attempts are unsuccessful, fluid slugs can be circulated down to the stuck point. Fluids that help get the drill string unstuck are lubricants, diesel oil and acids that can dissolve carbonate rocks.

For better accuracy in these operations and in further procedure, it is necessary to locate the stuck point. Initially, the driller will use the law of elasticity: relative elongation is proportional to pulling force. He knows the pulling force he is applying and deduces the relative elongation:

$$F/s = E \times \Delta l/L$$

He measures Δl on the kelly after having applied a pulling force F. He can then calculate L which represents the free length of the drill string. (E = Young's modulus, s = pipe cross-section).

To locate the stuck point more accurately, a logging service company can be called in to run a sonde down the drill string. When the sonde, a free point locator (**Fig. 11.9**), is actuated, it transmits drill string stretch under pulling force. When the sonde is located under the stuck point, it no longer transmits elongation, thereby situating the stuck point.

If the drill string is still stuck, the driller can attempt a backoff operation to unscrew the free part of the drill string.

Principle of the backoff: The objective is to unscrew the drill string near the stuck point, in order to pull out the free part and then run in the fishing string best suited to the problem. Accordingly, the first thing to do is make sure that the tightening torque on all threads is uniform by retightening the drill string. This is done by torquing to the right with the rotary table, then to the left (i.e. unscrewing) up to 80% of the make up torque applied previously. The tension on the drill string is regulated to locate the neutral point (zero axial stress) on the joint that is to be unscrewed. Then an explosive run inside the drill string on a wireline is set off at this same depth. The impact of the explosion and the left-handed torque unscrew the coupling that is the least prestressed.

Operations can then carry on in a number of different ways:

- a jarring string can be run in,
- a washover assembly can be run in,
- the fish can be cemented in situ and the well sidetracked.

A. *The jarring string*

The way it catches hold of the fish is discussed in 11.2.2 (overshot), but the string requires an additional safety joint above the overshot which will allow all the string above the joint to be unscrewed and saved. But the real reason for the string is to be able to run a bumper sub into the well and use blows, mainly upward but also downward, to loosen the fish. There are mechanical ones (see **Fig. 11.10**) or hydraulic ones as shown in **Fig. 11.11**.

B. *The washover assembly*

The idea is to redrill the annulus between the tubular and the hole in order to release the fish. The assembly consists of a washover shoe and washover pipe compatible with the

Magna-Tector:
This instrument measures both stretch and torque movement in a drill string. It comprises two electromagnets connected with a telescopic joint and a microcell. Electric current is sent down through the conductor cable pushing the electromagnets against the inside wall of the tubular. The pull or torsional force applied to the drill string above the stuck point is measured by the microcell and read on the surface. The tool can be run in with a collar locator and an explosive, a string shot, to proceed with a backoff as soon as the stuck point is located.

Spring-Tector:
The principle is the same as above, except that instead of electromagnets there are friction springs that perform the same function. This tool is suitable for aluminum or nonmagnetic drill strings.

Figure 11.9 Free point locator
(Source: N.L. MacCullough).

diameter of the fish and of the well, plus a safety joint and drill collars. The washover assembly is always mechanically vulnerable and the maximum practical length is 150 meters (**Fig. 11.12**).

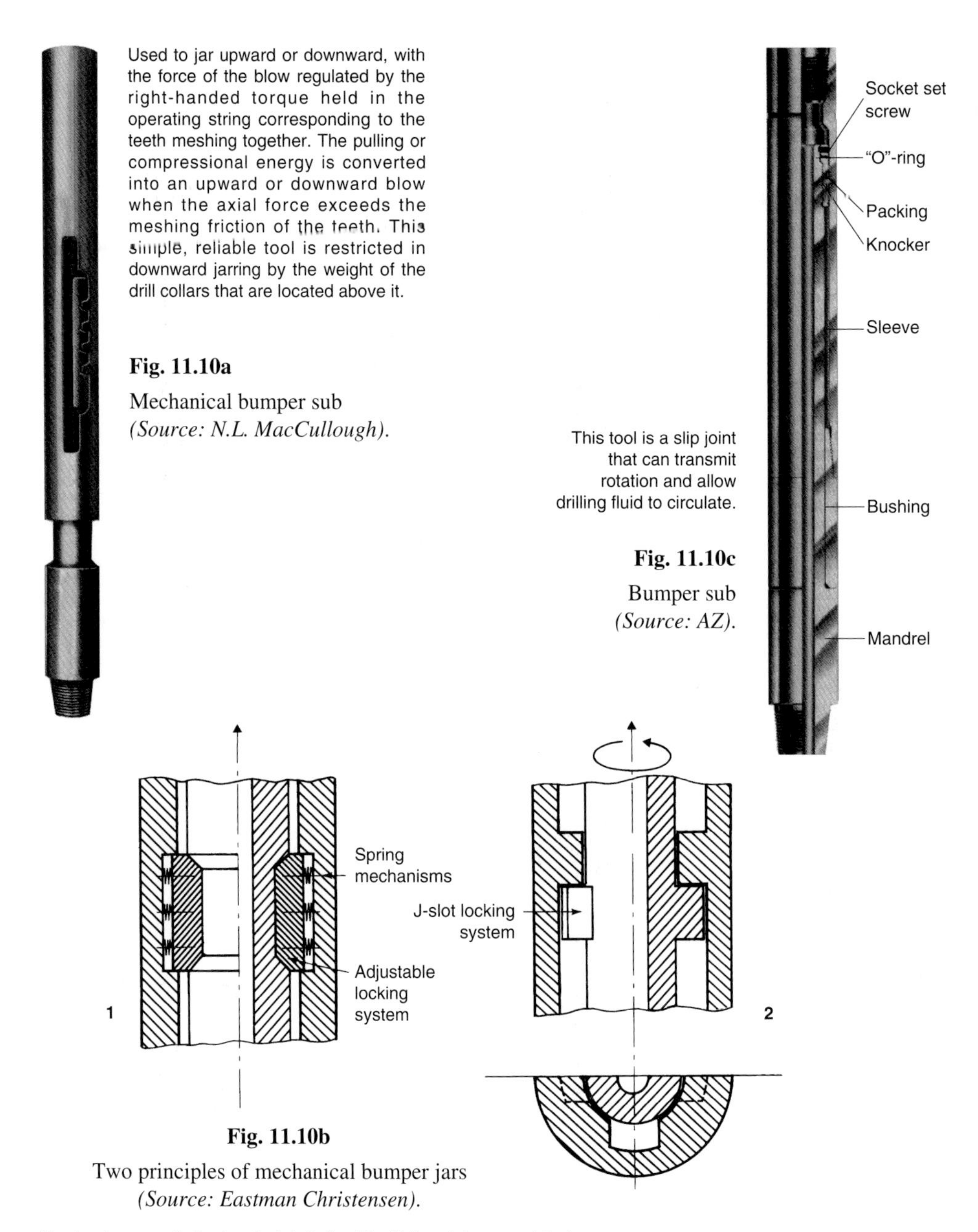

Fig. 11.10a

Mechanical bumper sub *(Source: N.L. MacCullough).*

Fig. 11.10c

Bumper sub *(Source: AZ).*

Fig. 11.10b

Two principles of mechanical bumper jars *(Source: Eastman Christensen).*

The two jars are similar in principle to the MacCullough jar except that:

1. There is no need for torque, the jar is regulated before it is run in by compressing the lock springs to a variable degree.
2. It is a simple bayonet, the pulling energy is applied when the bayonet is locked. The blow is produced when the bayonet is unlocked by rotating 90°.

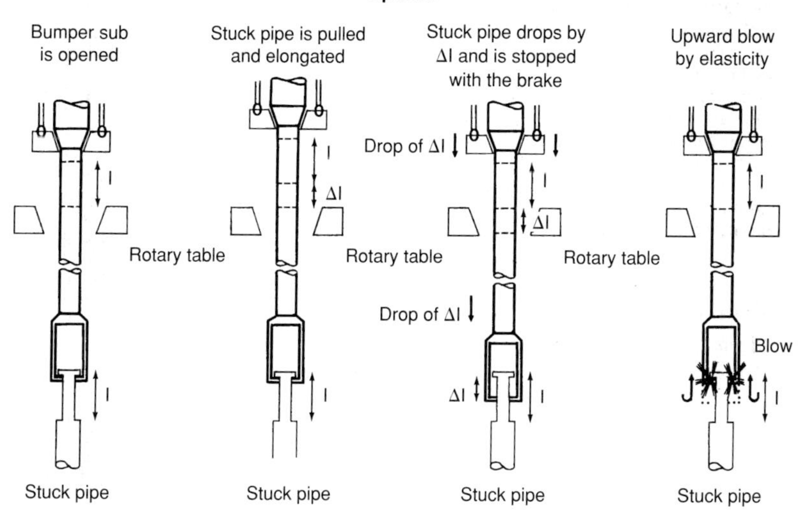

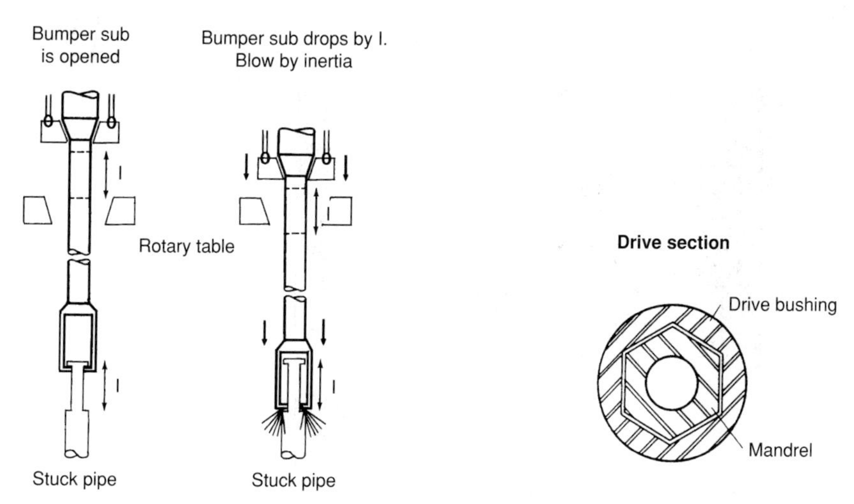

Fig. 11.10d

Principle of jarring upward and downward.

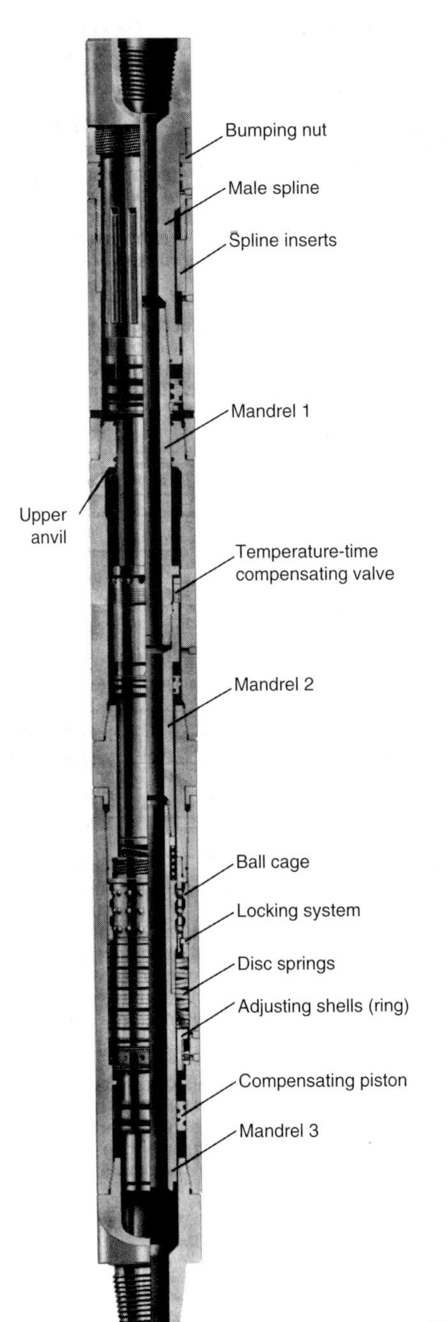

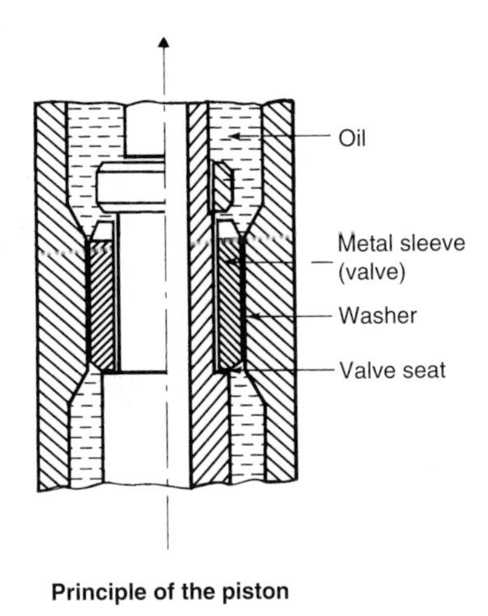

Principle of the piston

These tools can only jar upward hydraulically. Some can be used as downward bumper subs. They are very effective. The principle is based on a hydraulic lock with a controlled leakage piston (valve). When the piston is level with the widened section of the sleeve there is no hold and the hammer strikes the anvil.

Fig. 11.11

Hydraulic bumper sub *(Source: Eastman Christensen).*

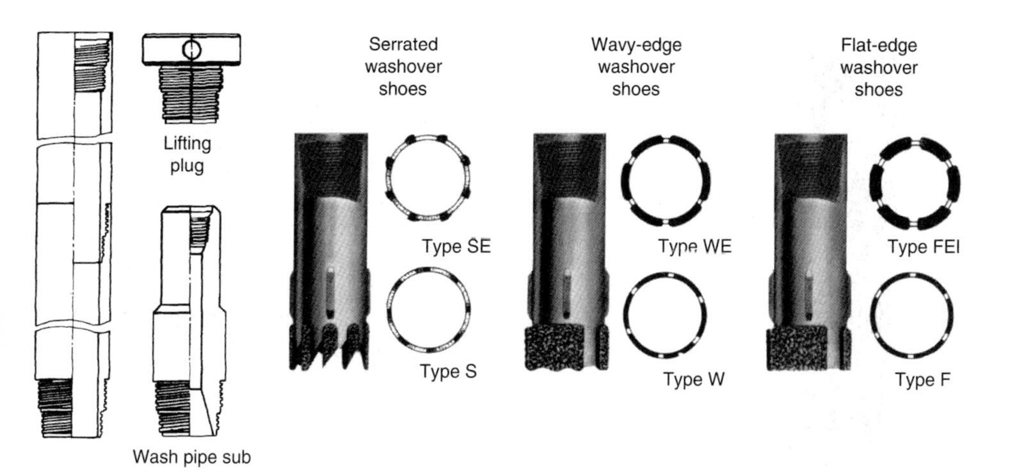

Fig. 11.12

Example of washover assembly components *(Source: Drillstar Industries).*

C. Sidetracking (Fig. 11.13)

This operation is decided when all the possible attempts to bring up the fish have failed. The procedure can also be adopted when retrieving the fish is clearly antieconomical compared to the cost of drilling a slightly deviated well. The aim is to start the deviation above the head of the fish so as to drill toward the target parallel to the abandoned well course at about ten meters' distance. We list a possible sequence of operations below:

- cement slurry is placed at the depth planned for starting the sidetrack and the crew waits for the plug to set,
- when sidetracking is done in a cased borehole, a bladed mill is run in to cut out a window in the casing,
- the sidetracking string is run in: a downhole motor and a bent sub with a drill bit of a smaller diameter than the ongoing phase,
- the buildup is drilled for one drillpipe length, then deviation is measured,
- a free-hanging drill string is run in to ream out the buildup and bring the well course back to vertical,
- doglegs are reamed out with a special string.

Even though this operation is fairly common nowadays and the cost is relatively affordable, the consequences on the well architecture must be considered. There is a dogleg where the sidetrack begins, causing considerable wear on the casing there by the rotating drillpipe. This casing string may have greatly impaired resistance to pressure. In addition, if a development well is sidetracked, sucker-rod pumping may not be feasible.

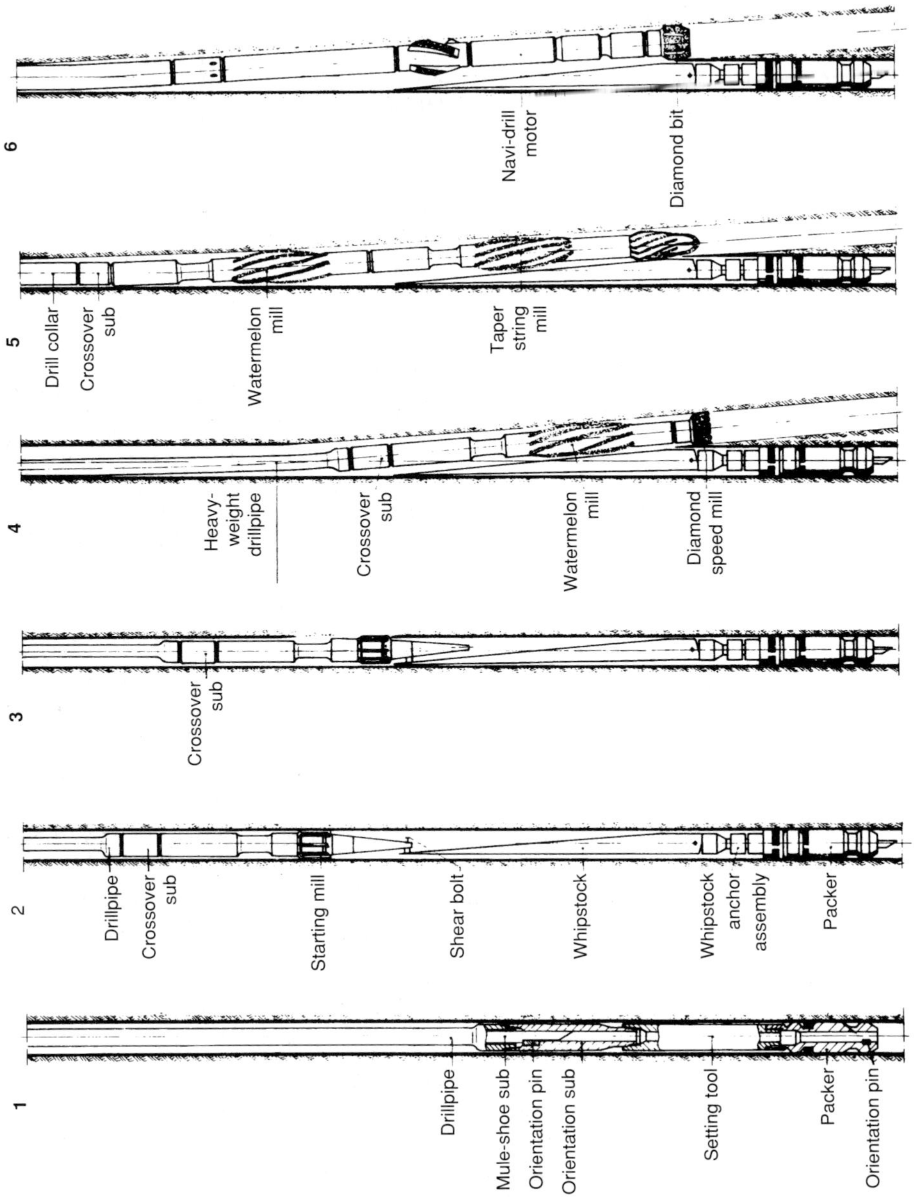

Fig. 11.13a Sidetracking operation in a cased borehole with a packer *(Source: Eastman Christensen).*

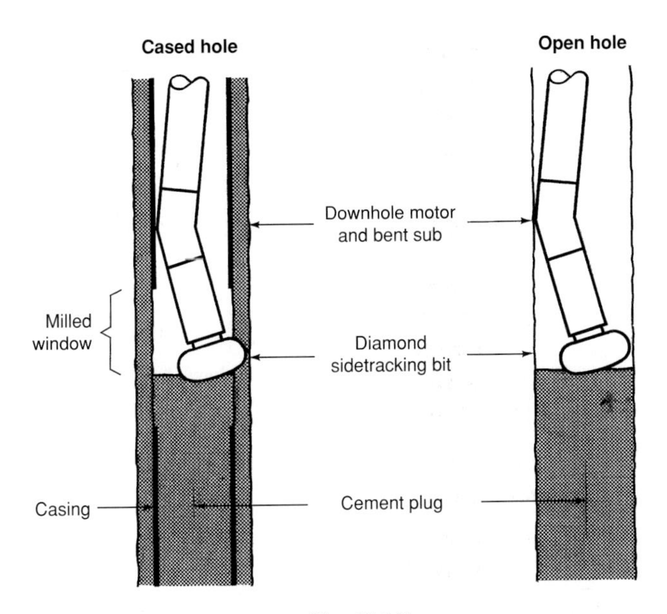

Fig. 11.13b

Open hole and cased hole sidetracking with a cement plug.

11.2.3.2 Differential pressure sticking

This type of sticking is very common, since it occurs due to the well control technique of keeping borehole hydrostatic pressure higher than the formation pore pressure. It is this differential pressure that causes the drill string to stick to the walls of porous and permeable formations. The sticking force is directly proportional to the Δp and to the size of the sealed contact area between the drill collar and the borehole wall. The contact area is large only if the cake is thick as it is very permeable. Differential pressure sticking can therefore occur only after the drill string has been totally stationary, initiating contact and the beginnings of a leak-proof seal (**Fig. 11.14**).

Solving the problem: the basic principle is obviously to act directly on the causes of sticking, i.e. the Δp, the contact area and f, the steel/cake friction coefficient.

- Once the stuck point depth is identified, a slug of lubricating product can be circulated in an attempt to create a greasy film on the surface of the drill string by diffusion in the cake and thereby reduce f. While the slug is acting, the driller should keep on pulling on the drill string within admissible mechanical strength limits.
- Reducing hydrostatic pressure can of course be considered only if it is sure that no kick will be triggered off. There are several possible techniques: lighter-weight mud, injecting light fluid slugs or even the DST method after a packer has been set. The free part of the drill string has to be unscrewed to run in a test assembly of course.

- The conventional method of freeing stuck pipe can also be attempted, i.e. backoff and a jarring string.

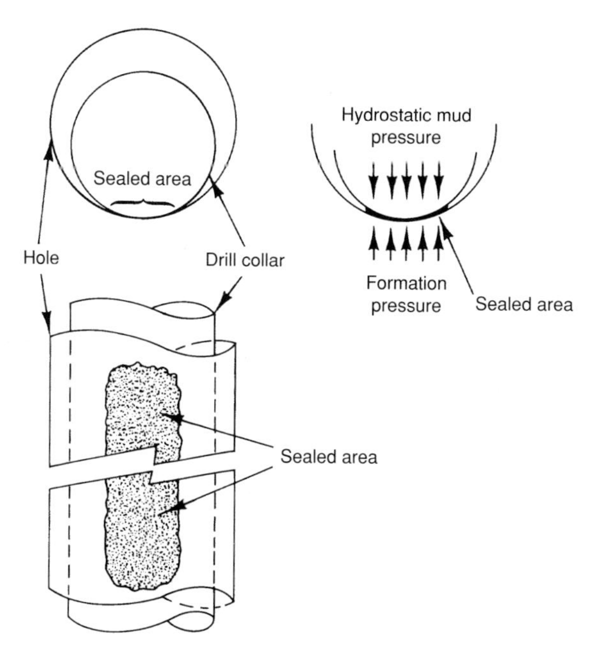

Fig. 11.14
Principle of differential pressure sticking.

11.3 TYPICAL FISHING JOB STRINGS

11.3.1 Jarring strings (Fig. 11.15)

It is recommended to place a safety joint above the fishing tool, whether overshot, pin-tap, fishing tap or bell socket. These subs unscrew to the left with low torque, which guarantees that everything above the joint can be pulled out if the decision to abandon the well is finally taken. It also enables the string to be changed since reconnecting is also easy. Hydraulic jars are enhanced when used with a jar accelerator placed at the top of the drill collars. The accelerator is a pneumatic spring that boosts the speed of the drill collars when the hydraulic jar is tripped.

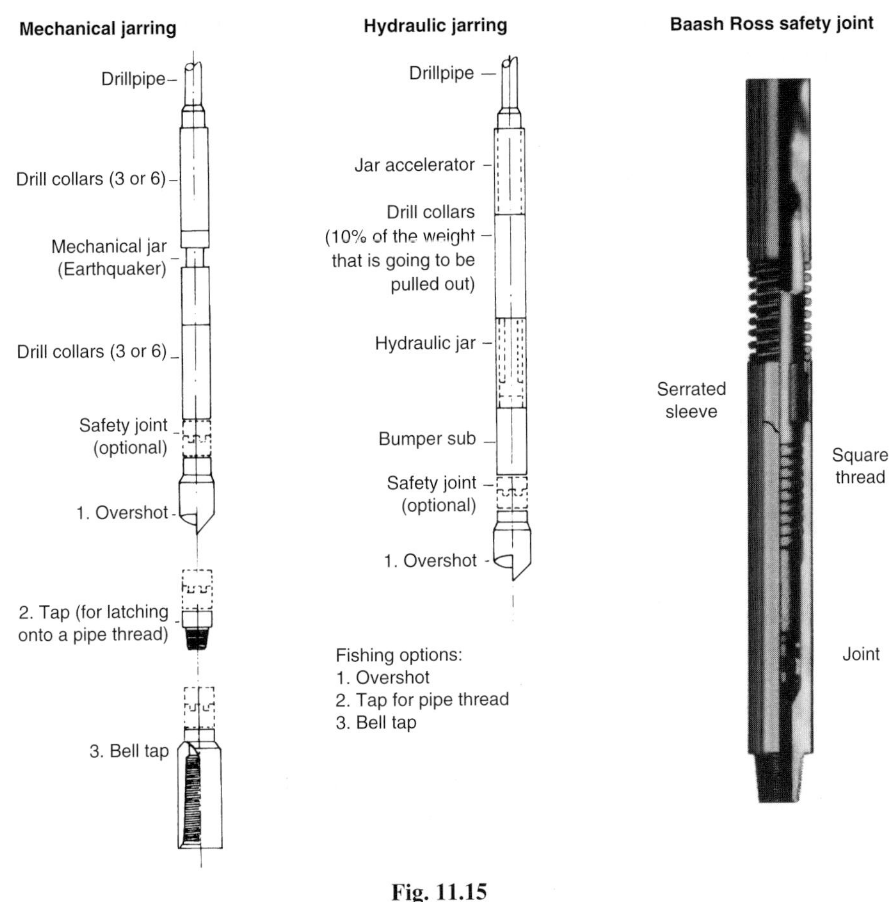

Fig. 11.15
Jarring strings.

11.3.2 The washover assembly (Fig. 11.16)

The string requires a means of jarring upward and downward. The basket sub collects the chips of steel components drilled during the washover operation. The operation is tricky since it generates considerable and highly irregular torque. The relatively low washover string makeup torque makes it all the more dangerous.

11.3.3 The milling string (Fig. 11.17)

In both a cased or an open hole, milling bits must be properly stabilized.

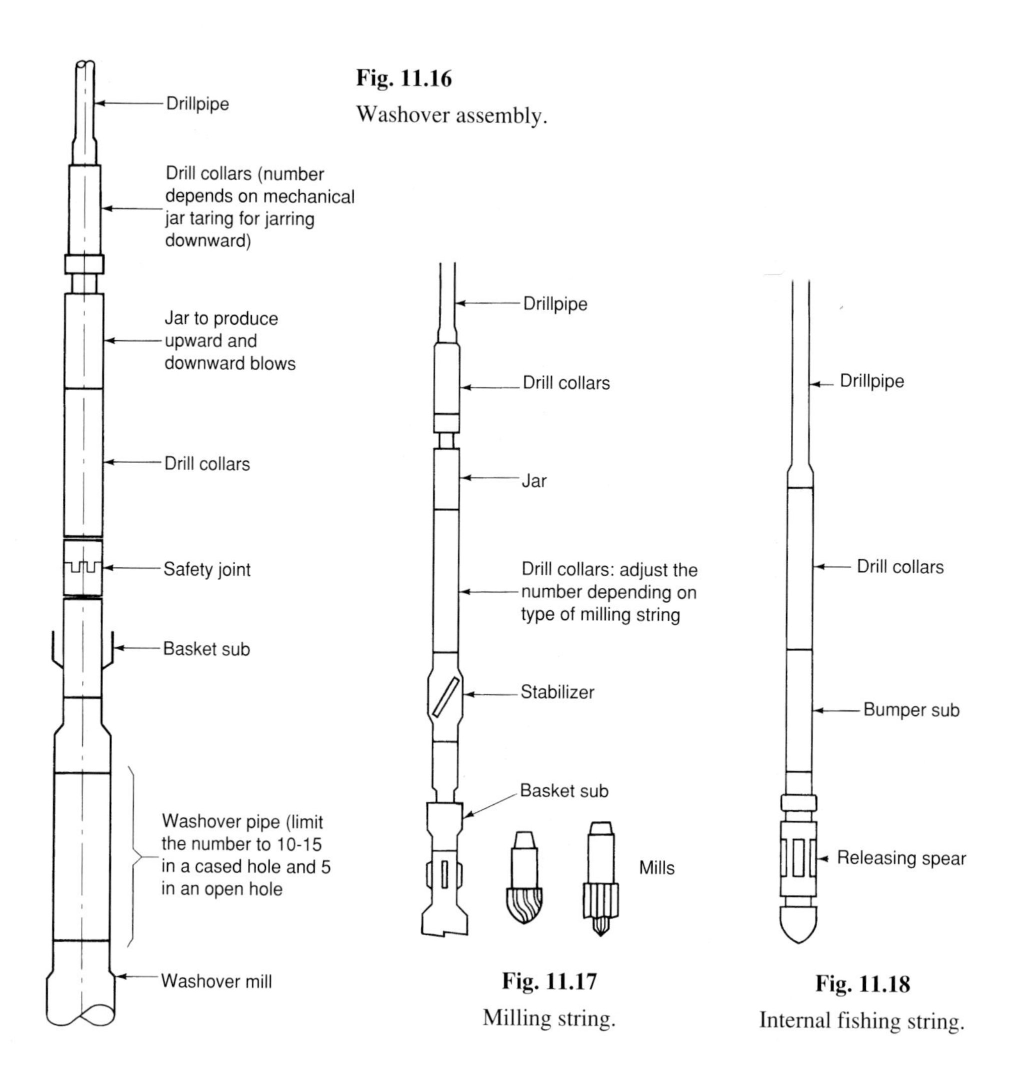

Fig. 11.16
Washover assembly.

Fig. 11.17
Milling string.

Fig. 11.18
Internal fishing string.

11.3.4 The internal fishing string with a releasing spear (Fig. 11.18)

This string is used to fish out a tubular such as a casing or tubing from the inside of the tubular. Since anchoring requires rotation to the left, no safety joint can be used.

11.3.5 General view of typical strings (Fig. 11.19)

This collection is not exhaustive, but shows the variety and complexity of equipment used in the different possible fishing operations.

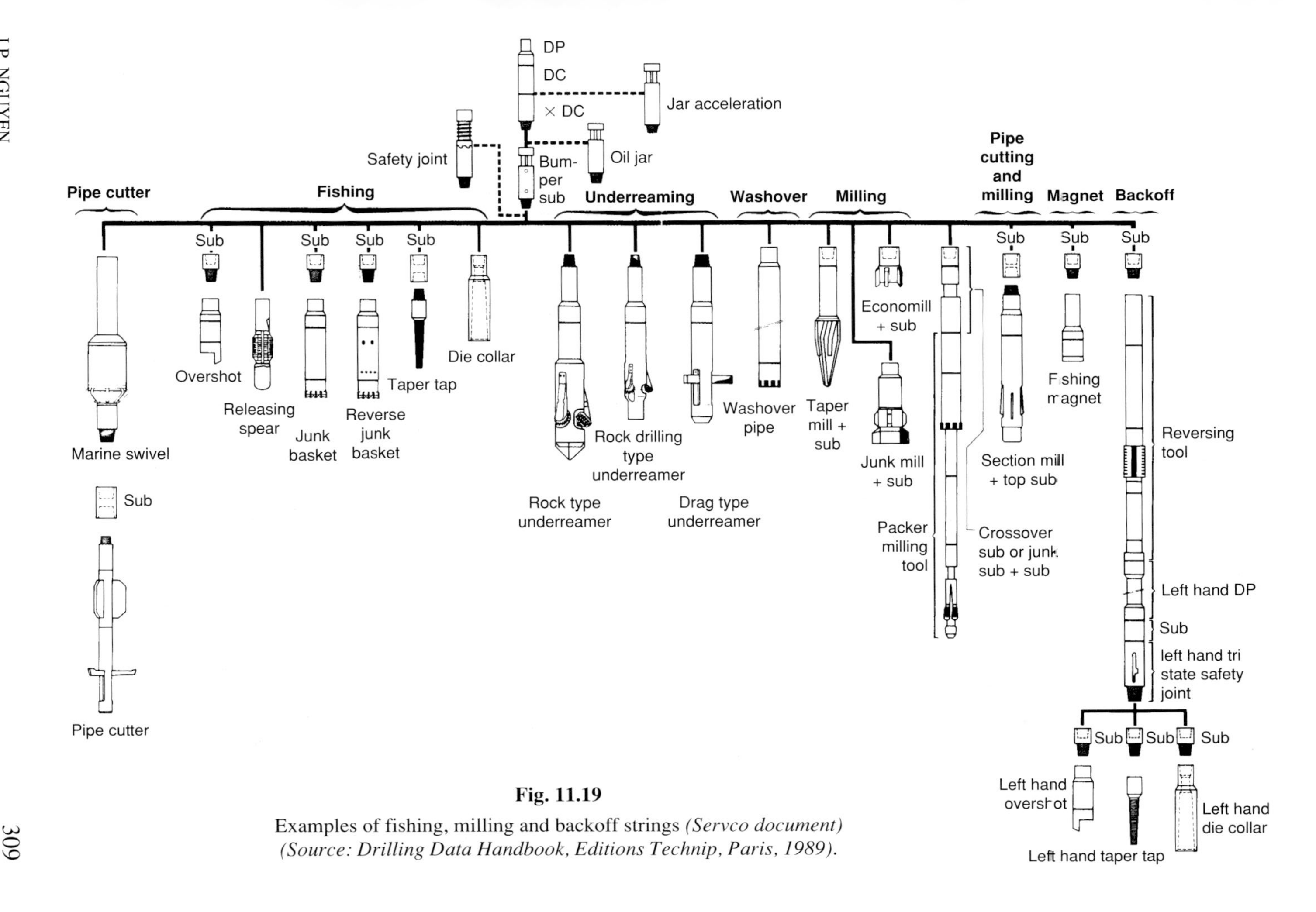

Fig. 11.19

Examples of fishing, milling and backoff strings *(Servco document)*
(Source: Drilling Data Handbook, Editions Technip, Paris, 1989).

Chapter 12

THE DRILL STEM TEST (DST)

Deep-well exploratory techniques (coring, logging and mud logging) can give only assumptions on the type of fluids in porous zones while the formations are being drilled. However, it is important to know for sure whether fluid shows come from formations containing oil, gas or water and to assess the fluid flow rates and the formations' static pressure.

This information needs to be available during drilling and not after testing once drilling has already ended. The reason is that if there is oil or gas in a drilled formation, the rest of the drilling program is often modified. In the event of positive results, the decision will be made to take core samples frequently. Alternatively, even continuous coring can be performed while drilling through the reservoir to determine its porosity, permeability and saturation profile. Meanwhile, safety measures will be reinforced. Then the casing program will provide for a production string to be landed so that longer-lasting tests can be run and the well can be brought on stream if possible.

If the results are negative, either drilling will be pursued toward deeper targets or the well will be abandoned.

Drill stem tests can be defined as temporarily bringing the well on stream, without altering the well's equipment. They serve to collect samples of the fluids contained in the rock, get a rough assessment of the flow rate and measure reservoir pressure.

These data can be gathered:

- during drilling,
- after drilling,
- after cementing a casing string.

12.1 PRINCIPLES

Since formation fluids are usually controlled during drilling by the pressure exerted on the borehole walls by the column of drilling mud, running a test means that:

- the pressure of the mud column on the formation that is going to be tested must be canceled or brought down to lower than the pressure of the formation fluids,

- the fluids must be conveyed up to the surface without any risk of contaminating the mud or triggering a blowout,
- the formations that are not being tested must be kept under the same pressure as during drilling for the whole duration of the test so that the walls do not cave in and the fluids they contain do not intrude into the wellbore,
- it must be possible to stop the fluid flow temporarily without using the hydrostatic pressure of the mud,
- surface facilities must correspond to the different fluids that may be produced.

The principle of the test setup is shown schematically in **Fig. 12.1**. The conditions listed above are met by:

- A packer, or rubber sleeve, that is compressed above the formation that is going to be tested. It presses against the borehole walls thereby sealing off and separating the well into two noncommunicating zones.
- A test assembly which is a series of equipment components, with the test valve being the main one. The test valve is kept closed while the assembly is being run in and the inside of the drillpipe is full of air or a water cushion of a given height.

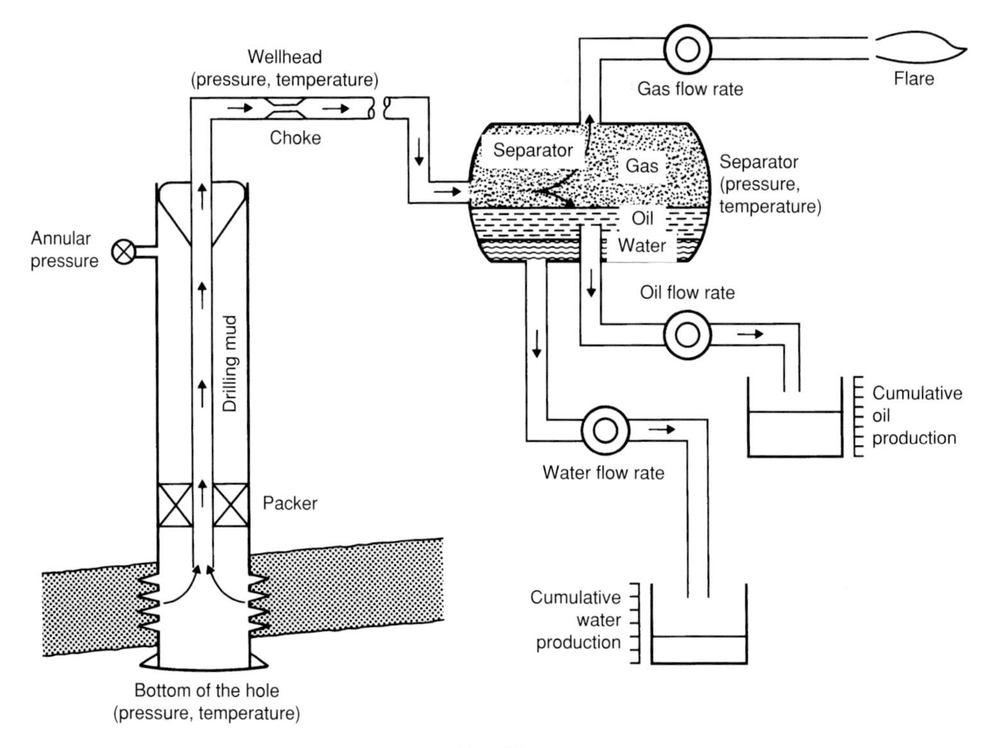

Fig. 12.1

Schematic layout of a test facility.

When the packer is anchored, opening the valve decompresses the fluids below the packer and those contained in the formation. The fluids reach the pressure prevailing above the valve, either atmospheric pressure or the hydrostatic pressure of the water cushion.

Decompression causes the fluids to flow into the drill stem. When the main valve is closed (without unseating the packer), the flow can be stopped without using the hydrostatic pressure of the mud. The pressure gages run into the bottom of the hole can then record the fluid recompression curve.

The measurement instruments used to record the well pressure under the series of bottomhole valves can be:

- read later on, requiring the test string to be pulled out for the recordings to be examined,
- read directly by means of a Schlumberger-type sonde run in on a wireline and connected to the tester.

12.2 THE TEST ASSEMBLY

12.2.1 In an open hole

Figure 12.2 gives the conventional make up of a test assembly for an open hole and a stationary drilling rig.

The series of tools included between the packer and the bottom of the hole are the support structure that is generally less than 30 m long.

We will now discuss the function and operation of each of the various pieces of equipment.

12.2.1.1 The shoe

This is a sub that is used for support on the bottom of the well and to apply compressional force to the test assembly.

12.2.1.2 Perforated liners

These are perforated tubes that the fluids flow through, usually 4 3/4" × 2 1/4" drill collars in sections that are 1, 3 and 5 feet long. They are easy to transport, the perforated holes are 3 to 4 mm in diameter and should keep the various test components from getting plugged up. When a formation over 12 m thick is to be tested, a 9 m drill collar is added to the support section of the assembly.

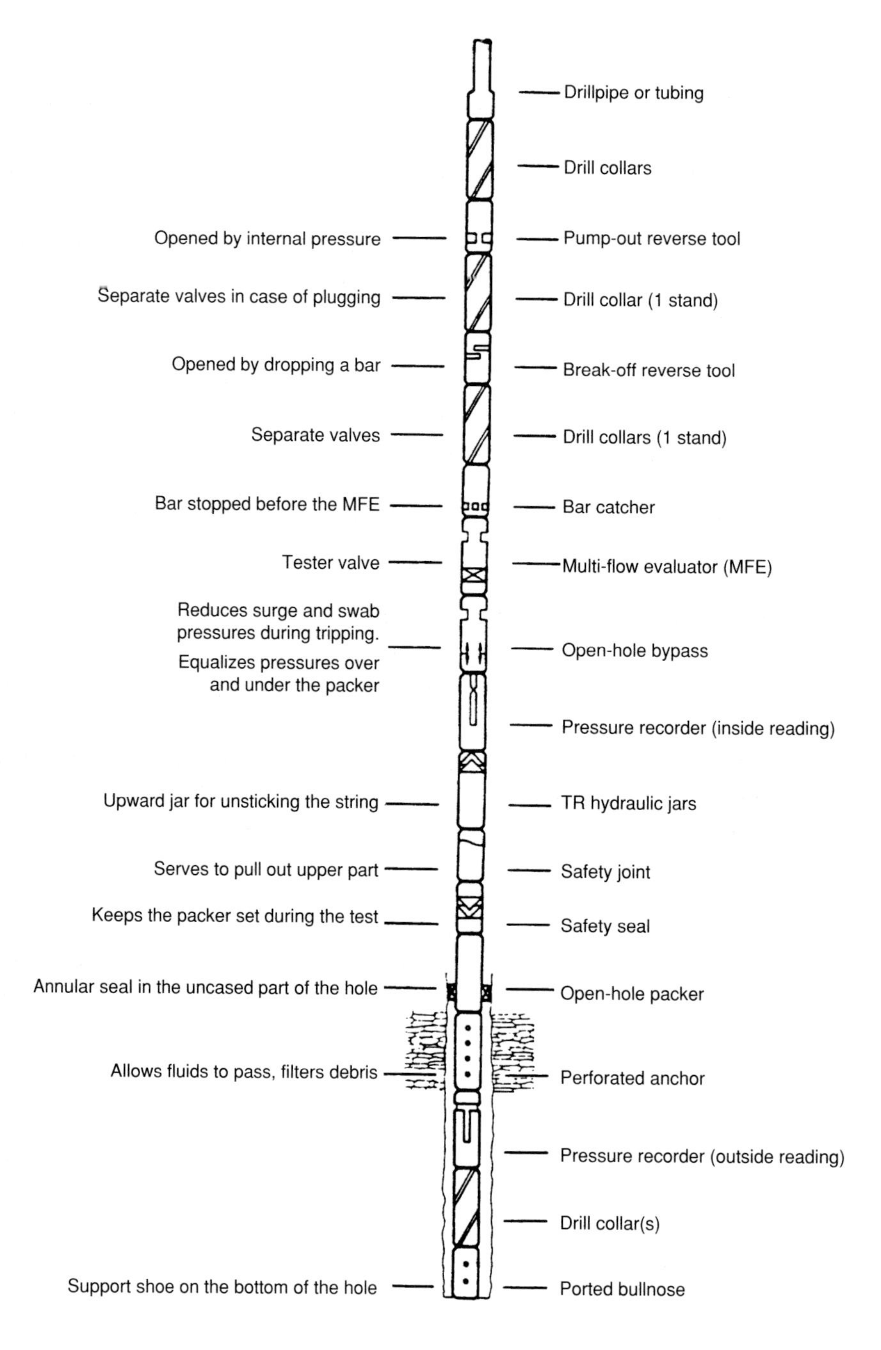

Fig. 12.2

Example of open-hole test assembly *(Source: Flopetrol)*.

12.2.1.3 The packer

The packer closes off the annular space between the walls of the borehole and the drillpipe. It has a rubber sealing element that is compressed between the support section of the assembly resting on the bottom of the hole and the drill stem, thereby causing it to expand in the annulus.

The sealing element must be soft enough to fit snugly against the walls of the hole. But it must also go back to its original shape so that it can be pulled out of the hole without getting stuck at the end of the test.

Conventional, open-hole, full-gage packers with single sealing elements have been replaced by packers with a system to prevent extrusion of the sealing element. The packer sealing element diameters leave one inch of clearance in relation to the diameter of the hole.

The weight required on the packer to compress the sealing element and achieve a good seal is approximately one ton per inch of hole diameter, i.e. around ten tons in an 8 1/2" diameter hole. The packer location will be chosen so that it is in a well consolidated, gage section of the borehole.

12.2.1.4 Setting the packer hydraulically (safety seal)

A seal is achieved with the packer by using weight as a parameter. But the testers used above the packer are also actuated by weight. As a result, the packer has to be set hydraulically.

The hydraulic setting system works by trapping the oil between the outside body of the setting system and an incorporated inside packer mandrel once the packer has been anchored. A differential check valve allows the packer to be unseated as soon as the pressures are identical on either side of it.

12.2.1.5 The safety joint

The odds are high that the packer will get stuck and if it does, the assembly can be unscrewed at the safety joint. The unstuck part can then be retrieved. The safety joints are exactly the same as the ones that are used in fishing strings.

12.2.1.6 The jar

Before the safety joint is unscrewed in the event the packer gets stuck, an attempt is made to free the packer. A pulling force is exerted on it by means of the hydraulic jar that is located above the safety joint.

12.2.1.7 The recorder sheath

The most important information during a DST is the pressure recordings. They are indispensable in interpreting the test results. As a safeguard, a second recorder is therefore added on. Since it is located in a different position, it can also help explain any recording anomalies.

12.2.1.8 The pressure equalizing valve

While the test is being run, the pressure conditions under the packer are different from the hydrostatic pressure of the mud which is exerted above the packer. So that the packer can be decompressed at the end of the test, the pressures above and below the packer are equalized by this valve.

The valve is open while the test assembly is being run in so as to reduce surge pressure caused by the large diameter of the packer sealing element. It is closed during the whole test, then opened to equalize the pressures above and below the packer.

The valve is equipped with a hydraulic system to delay opening so that it is not placed in an open position when weight is taken off the tester valve to close it.

12.2.1.9 The tester valve

The tester valve allows alternate opening and closing of the assembly downhole to trigger periods of flow and recompression of the effluent contained in the formation pore spaces.

The valve in the tester can:

- have baffles where the apertures of a moving mandrel come into position opposite the apertures of a stationary component, or
- be a ball valve, with the ball controlled by the displacement of a piston.

The ball valve system is advantageous in that it provides a continuous passageway inside the test assembly when it is open. A pressure recorder can then be run in to the bottom of the hole with an electric wireline which transmits data to the surface.

The valve is opened and closed by a downward movement due to the weight of the drill collars (MFE) or by pressure in the annulus (PCT). The latter type is more particularly suited to drilling from a floating platform and to highly deviated boreholes where weight is used only to anchor the packer (**Fig. 12.3**).

The tools allow several opening and closing sequences. With the MFE, the opening system is hydraulically delayed. It is the weight of the drill collars placed above the test assembly which allows the tester opening sequence.

The opening sequence must be carried out in a specific order. As soon as the test assembly shoe has reached the bottom of the hole, the driller applies a weight of approximately 10 tons on the bottom. The sequence is then as follows:

- the main equalizing valve is closed,
- the packer is anchored,
- the secondary equalizing valve is closed,
- the tester is opened,
- the differential check valve for hydraulically setting the packer is closed.

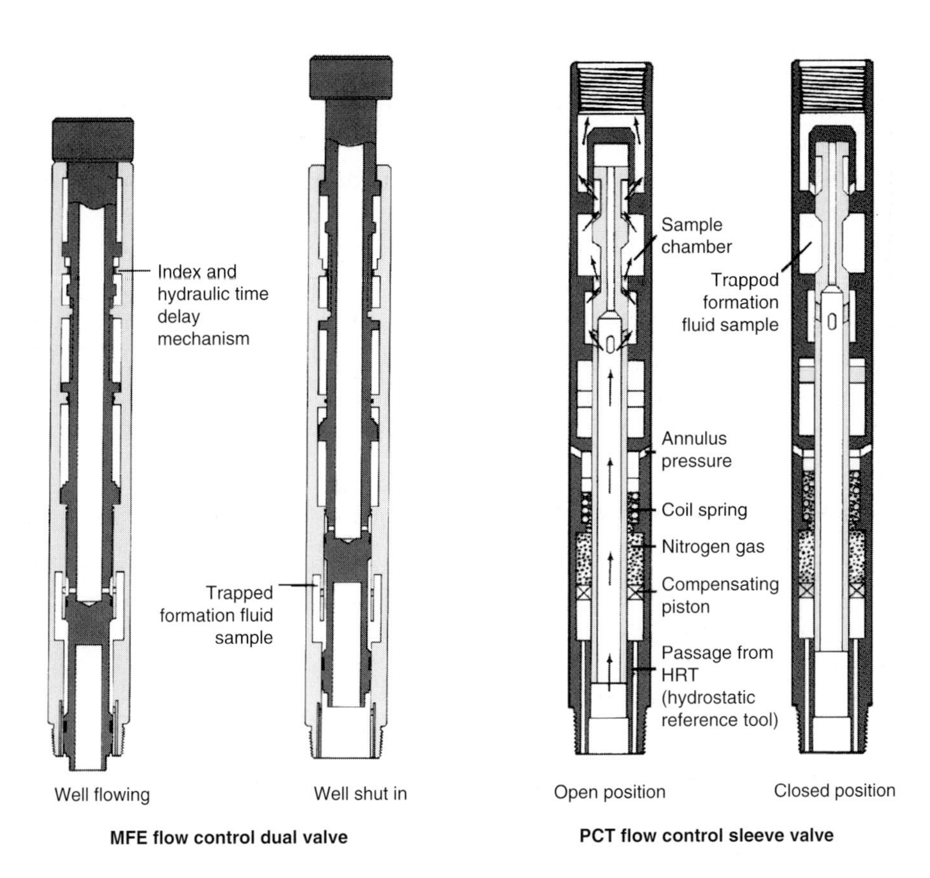

Fig. 12.3

Tester valves *(Source: Flopetrol).*

12.2.1.10 The reverse circulation valve

When the test has been completed and if the well has been productive, the drill stem is partially or completely filled with an effluent. At this time, it is dangerous to start pulling out of the hole with this inflammable fluid inside the drillpipe.

The reverse circulation valve, located from 9 to 30 m above the tester, allows communication with the annulus. It also serves to recover the effluent located in the drillpipe by reverse circulation.

The most common type of reverse circulation valve is the break-off plug. The apertures are plugged up by hollow plugs which are broken off by sending an iron bar down through the inside of the drill stem from the surface.

12.2.1.11 Surface equipment (Fig. 12.1)

The surface facilities must be capable of:
- withstanding wellhead pressures,
- producing back pressures with chokes to modify the bottomhole pressure during the flowing phase,
- recovering samples,
- measuring the flow rate of the air contained in the drill stem at the beginning of the test, which is pushed by the flow rate of the effluent at the bottom of the hole,
- separating the effluent, if it gets to the surface, to measure the volume of oil, gas and water separately,
- storing or flaring the effluent.

12.2.2 In a cased hole (Fig. 12.4)

The test assembly is exactly the same as the one discussed above, except that a particular type of packer is used that can be hung in a casing to produce the seal. The BIAS hydrostatic MFE acts as a safety seal.

12.2.3 The selective or straddle test (Fig. 12.5)

When the formation that is going to be tested must be isolated from lower zones without plugging the well, a selective or straddle test is run. Implementation requires two isolating packers on either side of the section to be tested.

12.2.4 Testing from a floating platform (Fig. 12.6)

The test assembly is required to provide two extra, highly sensitive features:
- disconnection at the subsea BOP level while keeping the well under control by the BOPs and by closing a valve located in the test assembly,
- allow the packer to be anchored and the test assembly to be hung in the wellhead-slip joints must be incorporated.

The most common tester valve today is the type controlled by annular pressure.

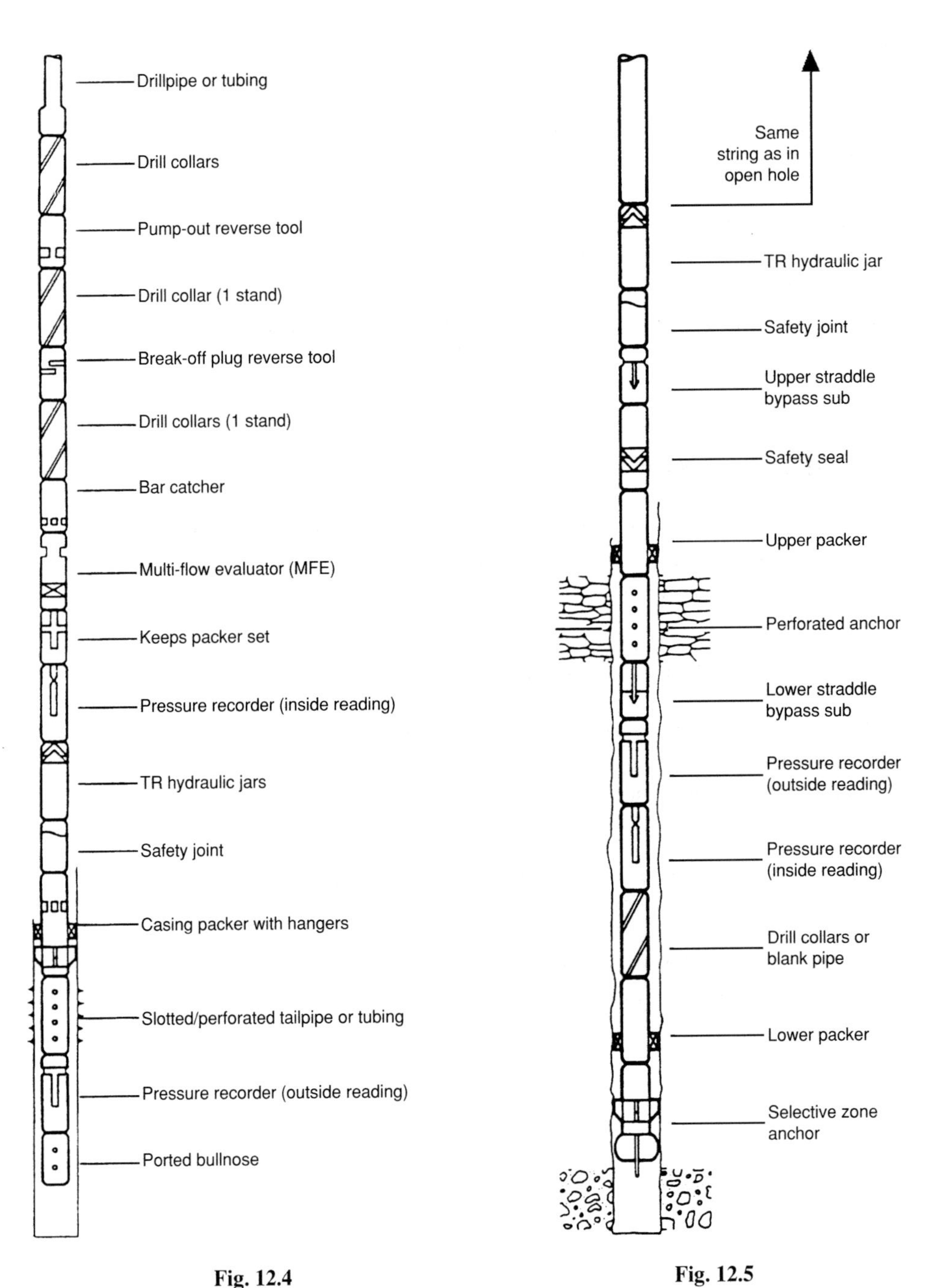

Fig. 12.4

Test assembly for a cased hole *(Source: Flopetrol).*

Fig. 12.5

Straddle test assembly *(Source: Flopetrol).*

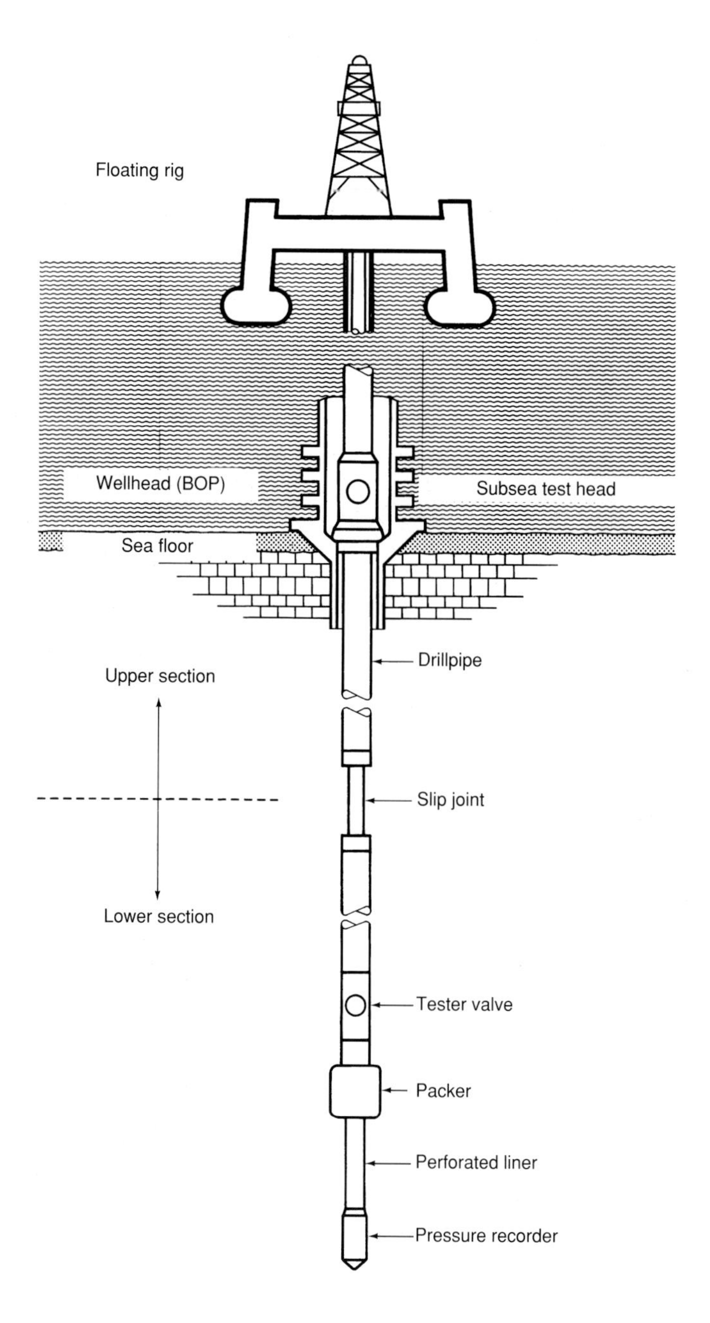

Fig. 12.6

Test assembly in a subsea wellhead *(Source: Dowell Schlumberger)*.

12.3 TEST PROCEDURE

Before going into the operational sequences, we will discuss a number of problems related to this operation.

12.3.1 Operational problems when testing in an uncased hole

- Packers may get stuck when the formation caves in or if the uncased part of the hole is uneven in shape. This may result in the need for a sidetrack or the well may simply be abandoned.
- Tools may get plugged up downhole with rock particles.
- Packers may leak because the wall of the hole is uneven.
- An inadequate amount of data may be collected on the well and reservoir, since testing time is limited for the reasons given above.
- Almost no chance of getting a representative sample of formation fluids.
- Since testing time must be short, priority will be given to the opening phase rather than the pressure buildup phase. The recommended ratio will be for production time to last twice as long as shut-in time.

12.3.2 Sequences in the procedure

The major phases of the operation are as follows:
- preparing the borehole,
- preparing the equipment,
- running in the downhole tools,
- actual testing as such, with the downhole valve open and closed,
- pulling out downhole tools,
- utilizing the data, results, interpreting.

Preparing the borehole consists first in checking that the mud is properly conditioned, or even replacing it with a more suitable fluid. The wellhead (BOPs) is also pressure tested and the casing is perforated if necessary.

The equipment must be checked beforehand to be sure everything will be available at the right time and that it is perfectly operational.

The equipment is prepared in two stages:
- It is checked by the specialized service companies (faulty parts and seals are replaced, etc.).
- Preliminary testing. The tools that are actuated by variations in annular pressure are checked in particular by operating them on the surface where the opening and closing phases can be simulated.

Downhole tools are assembled, then run slowly into the borehole. The mud level in the annulus is carefully and constantly monitored in order to spot any mud losses, to see if the well is flowing and to be sure the downhole valve has not opened too soon.

When the downhole tools have been run in to the depth of the formation that is going to be tested, the test operation as such can begin as described below:

- the test head is installed on the surface,
- surface connections are made,
- the packer is anchored (**Fig. 12.7**),
- the tester valve is opened: initial flow period (**Fig. 12.8**),
- the tester valve is closed: pressure buildup period,
- the tester valve is opened: main flow period,
- the tester valve is closed: pressure buildup period (**Fig. 12.9**),
- reverse circulation (**Fig. 12.9**),
- the packer is unseated (**Fig. 12.10**).

Along with diagrams of test assemblies, we show the variations in pressure recorded versus time. The whole recorded diagram is represented in **Fig. 12.11**.

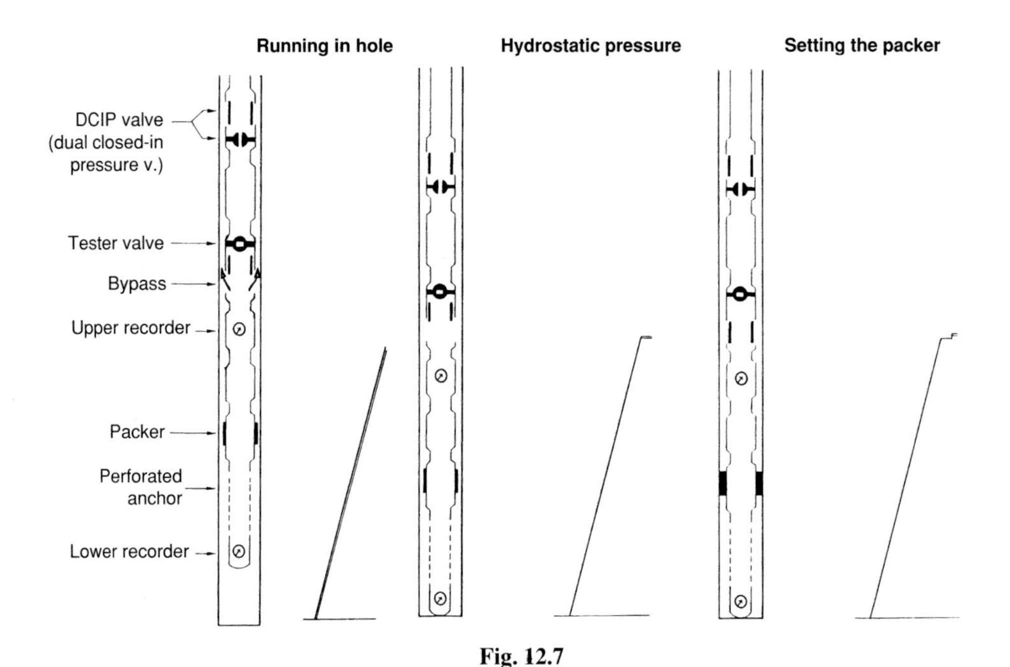

Fig. 12.7

Test procedure: installation phases

(Source: La géologie des fluides en exploitation pétrolière, Geology Division, SNEA(P)).

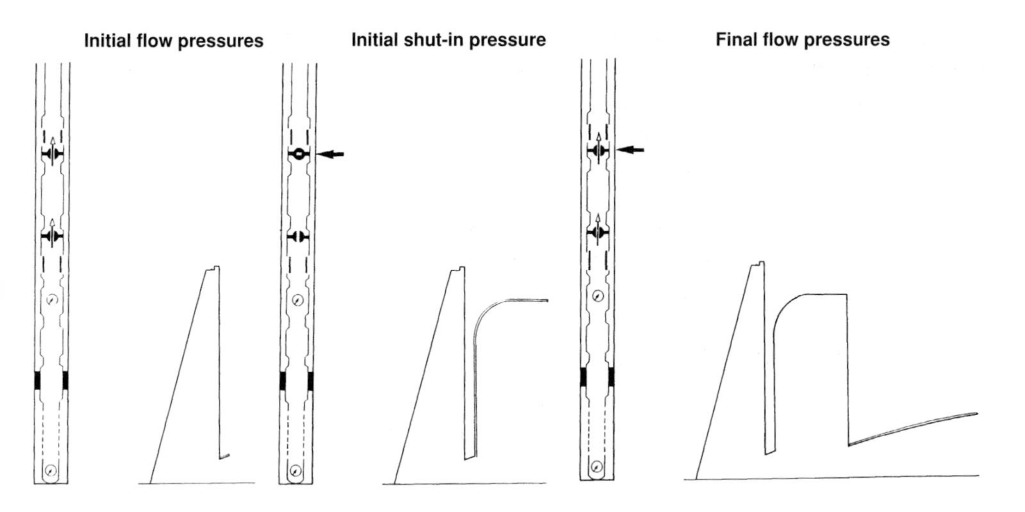

Fig. 12.8

Test procedure: measurement phases
(Source: La géologie des fluides en exploitation pétrolière, Geology Division, SNEA(P)).

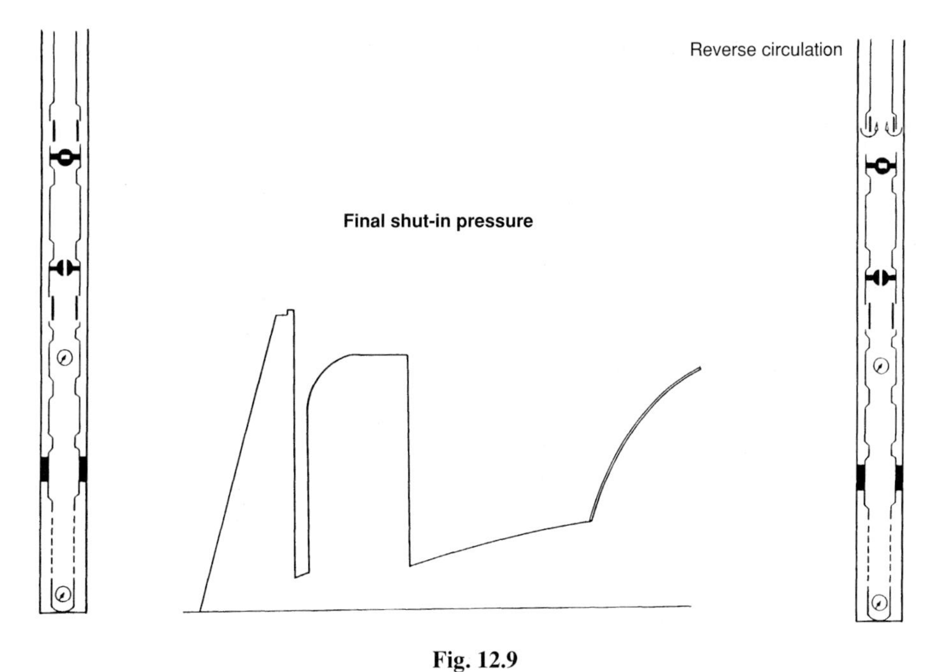

Fig. 12.9

Test procedure: final phases
(Source: La géologie des fluides en exploitation pétrolière, Geology Division, SNEA(P)).

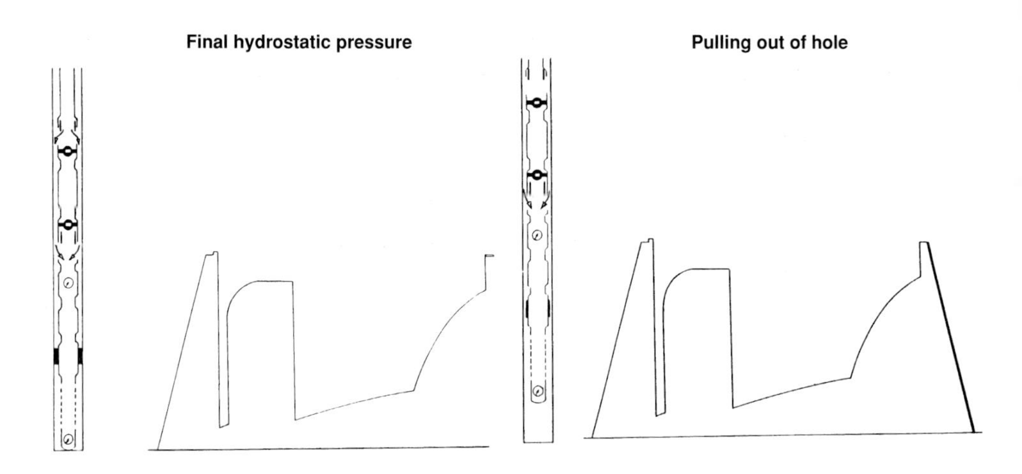

Fig. 12.10

Test procedure: pulling out the test assembly

(Source: La géologie des fluides en exploitation pétrolière, Geology Division, SNEA(P)).

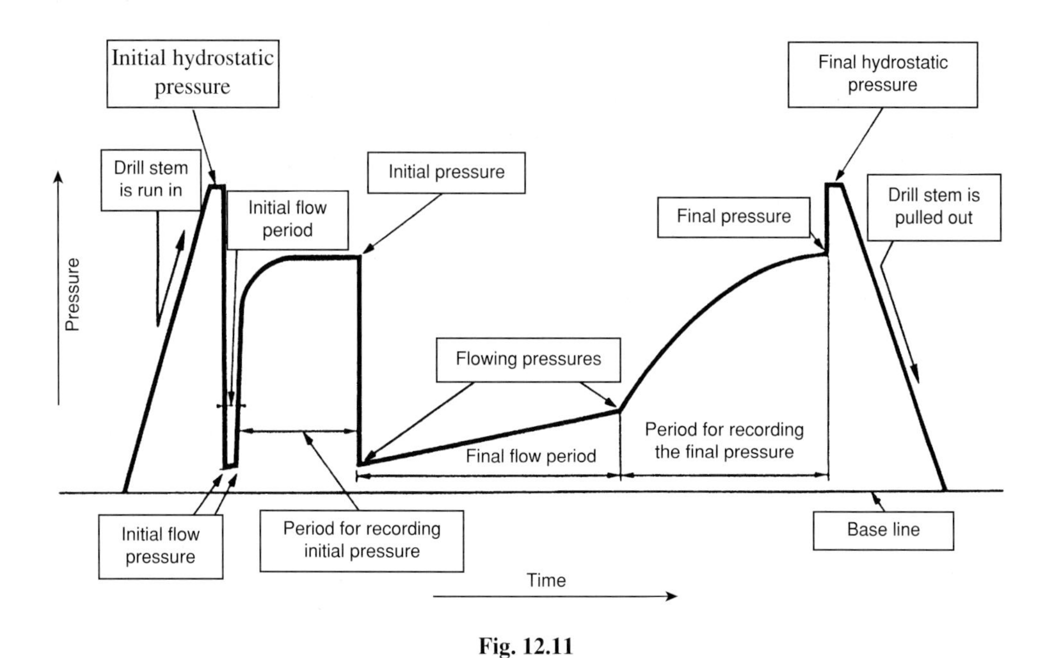

Fig. 12.11

Typical test diagram

(Source: La géologie des fluides en exploitation pétrolière. Geology Division, SNEA(P)).

Generally for a test onshore, the test time will be limited to about three hours because of the risk of getting the packer stuck. Here the operations are as follows:

- an initial flow of 3 to 5 min (open tester),
- initial pressure is measured for 30 min (tester closed),
- a final flow of one hour (tester open),
- a final pressure buildup of one hour (tester closed).

When the tester valve is opened, the mud that was underneath the packer is suddenly decompressed to the pressure prevailing in the drillpipe. The recorder suddenly indicates atmospheric pressure (or approaches it) if there is no water cushion, or shows the hydrostatic pressure due to the water cushion.

If the formation is unconsolidated, a negative ΔP is recommended (formation pressure–water cushion pressure) of less than 10 bar. This is valid for perforations and for putting the well on stream too.

So that the negative pressure is not applied suddenly when the downhole valve is opened, the hydrostatic pressure of the water cushion can be boosted by a certain volume of nitrogen. Nitrogen pressure is adjusted on the surface so that the pressure exerted on the bottom is equal to or slightly greater than the formation pressure.

The nitrogen is then bled off and the ΔP is gradually applied on the formation. This accordingly reduces the risk of fracturing the rock and prevents solid particles from being produced that could plug up downhole tools and damage seals.

If the formation is consolidated, a negative ΔP of approximately 30 to 50 bar maximum is recommended. The trend is to exert a greater ΔP for a very permeable formation than for a relatively impermeable one. The mud filtrate invades a very permeable formation more readily and so the mud cake that is laid down is thicker. Consequently, the force required to "unplaster" the borehole wall must be greater.

In any case however, the recommended ΔP will never exceed 50 bar for the following two reasons:

- too high a ΔP may make the formation plugging effect worse by suddenly accumulating solid particles suspended in the filtrate,
- the cement near the perforations can be damaged by too dramatic a change in stresses.

The purpose of the initial flow period is to unplaster and unplug the borehole walls and produce the filtrate from the invaded zone.

After the valve has been closed, the fluids recompress in the vicinity of the borehole. If a long enough time is allowed to elapse, the pressure indicated by the recorders is then the original reservoir pressure before the formation was drilled. This is also termed the initial pressure.

The valve that communicates with the inside of the drill stem is opened again and the pressure exerted on the recorder is then the pressure that prevailed on shut-in after the initial flow period.

Since the inside of the drill stem is left open, the fluids located in the vicinity of the borehole are once again decompressed and start moving toward the well and up the drillpipe.

The pressure exerted on the recorder then grows gradually until the drill stem is full of oil. If the test is continued, the bottomhole pressure remains constant and is equal to the pressure exerted by the column of oil plus the pressure losses in the drill stem.

When the tester valve is closed again, the reservoir gets recompressed.

The result is a pressure buildup curve that is more or less steep depending on the permeability of the formation. The greater the permeability and the longer the shut-in time, the closer the final pressure at the end of the shut-in period will be to the initial reservoir pressure.

When the tester is closed and the packer is then unseated, particularly when the equalizing valve is opened, the hydrostatic pressure of the mud is immediately restored in the bottom of the hole. The test is accordingly definitely over, the produced fluids are trapped in the drill stem and the formation is controlled by the column of mud.

Reverse circulation is then begun by opening a bypass valve above the tester to empty out all the effluent inside the drillpipe. The assembly can then be pulled out of the hole.

Chapter 13

DRILLING OFFSHORE

13.1 CONSTRAINTS SPECIFIC TO DRILLING OFFSHORE

Drilling an oilwell out at sea means using a floating platform or one that rests on the bottom. The support structure must be capable of fulfilling all the functions that are normally required on a well site onshore. Because of the isolation, it must also have a number of services permanently on board that would only be mobilized when needed onshore (cementing, logging). In addition, there are a number of specific services (divers, meteorological measurements). Accommodation and catering for crews working around the clock require facilities that are sufficiently comfortable and insulated against noise, despite the fact that they must of necessity be close to the drilling machinery.

All these conditions heighten the complexity of an offshore drilling support structure and explain the fact that daily costs are much higher than for an onshore rig with an identical depth rating.

13.1.1 Exploratory wells

The areas are often new prospects with no offshore infrastructures and sometimes even without any port capable of serving as an operations base. The support structure can be chosen according to the following guidelines:

- shallow water: jackup or swamp barge,
- water depth greater than 100 m: drillship or semisubmersible,
- isolated area with icebergs: dynamically-positioned drillship,
- severe sea conditions: semisubmersible,
- water depth greater than 400 m: dynamically-positioned drillship.

Other criteria can be taken into consideration in choosing a support structure for exploratory drilling, in particular availability or mobility for a drilling program that involves several wells.

The capacity, or live load, of the structure is chosen according to the planned drilling program and the distance from the operations base.

13.1.2 Development wells

Development wells are generally drilled from stationary structures in water depths of less than 200 meters. There is always an oil industry environment with an onshore base for storing equipment and materials. Choosing the structure basically depends on meteorological conditions:
- severe sea conditions: compact rig on the platform,
- calm sea: tender,
- shallow water: cantilever jackup.

Development wells that are drilled in water depths of between 300 and 600 meters are a little particular since this is the limit of present-day technology. They are usually drilled from an anchored semisubmersible or from a dynamically-positioned drillship and the production phase uses subsea wellheads.

When the wells are grouped together on stationary structures above the water line or on subsea clusters, the reservoirs are produced by means of directional wells.

13.2 OFFSHORE SUPPORT STRUCTURES

There are two main categories of drilling rig structures used offshore:
- mobile floating platforms and platforms resting on the bottom of the sea,
- those integrated in stationary production structures used exclusively for development wells.

The first category of mobile platforms includes the following types:
- jackups resting on the bottom,
- swamp barges,
- anchored or dynamically-positioned drillships,
- semisubmersibles.

Drilling structures used for developing offshore fields from stationary platforms are of two types:
- compact rigs,
- tenders (barges or jackups).

Compact rigs often belong to the oil company since they have to be built at the same time as the platform. The drilling contractor acts as a supplier of manpower and services. The drilling structure is not necessarily dismantled on completion of the drilling phase when the platform becomes a production structure. If there is any trouble or any maintenance requiring work on the well (workover), mobilizing and bringing a rig to the site may take several weeks when the weather is bad and the corresponding lost production is substantial.

The tender is a compromise solution suited for development wells where only the derrick and the drawworks are installed on the platform, thereby reducing both weight and costs. All the rest, including accommodation, is located on a barge anchored in the immediate vicinity of the platform. This solution is valid only for relatively calm seas, since operations must be shut down in bad weather and the barge must be completely disconnected from the platform.

13.2.1 Drilling support structures resting on the seabed

13.2.1.1 Jackups (Fig. 13.1)

These platforms are used in water depths ranging between 20 and 100 meters and up to a maximum of 120 meters for the highest depth rating. There are two types:

- Latticework platforms usually made of metal, or jackets, with independent legs with mats or spud cans on the ends. The legs penetrate to a greater or lesser extent depending on the composition of the sea bottom and the shape of the feet is chosen according to the soil's penetration resistance.

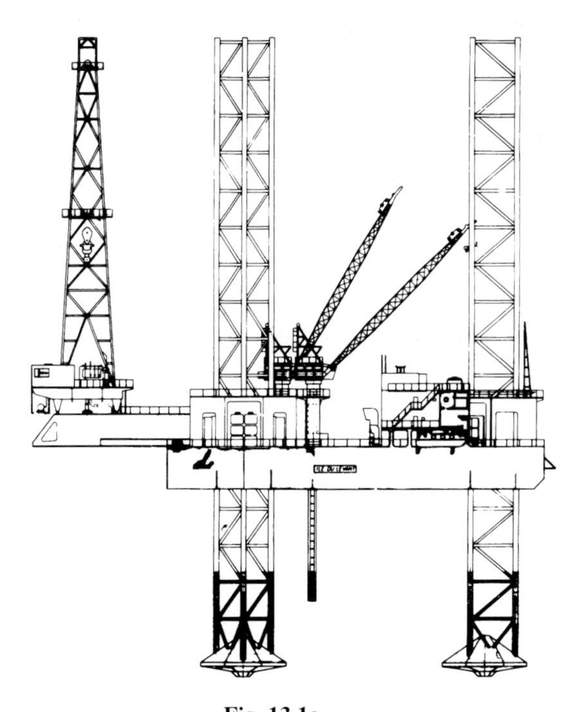

Fig. 13.1a

Jackup *(Source: Forasol)*.

Fig. 13.1b
Jackup drilling a well *(Source: Zapata)*.

- Platforms resting on the sea bottom by means of a large mat connecting up the legs, which are generally tubular and vertical. They are used on flat bottoms in water depths of up to 50 meters. The mat always penetrates the soil only slightly.

Jackups are used both for exploratory and development wells. In both cases, the blowout preventer stack (BOPs) is above the water line, supported by the conductor pipe and held laterally by slings in the rig floor substructure.

In exploratory drilling, the conductor pipe is self-supporting over the whole water depth and its stability is endangered by lateral swell and current forces and by compressional loads. In water depths greater than 50 to 60 meters, compressional loads must be reduced by having the casing strings supported from the mud line, i.e. sea floor. In anticipation of the well resulting in a discovery, the mud-line suspension system can be disconnected so that the jackup can be removed when drilling has been completed (**Fig. 13.2**). Conductor pipes remain self-supporting only in water depths of under 30 meters.

A jackup can be used in development drilling on top of a stationary offshore production structure to drill several development wells. The drilling rig is then of the cantilever type and the drilling mast can move along two axes to position the rotary table over the various conductor pipes. All the wells planned for the production structure can be drilled in this

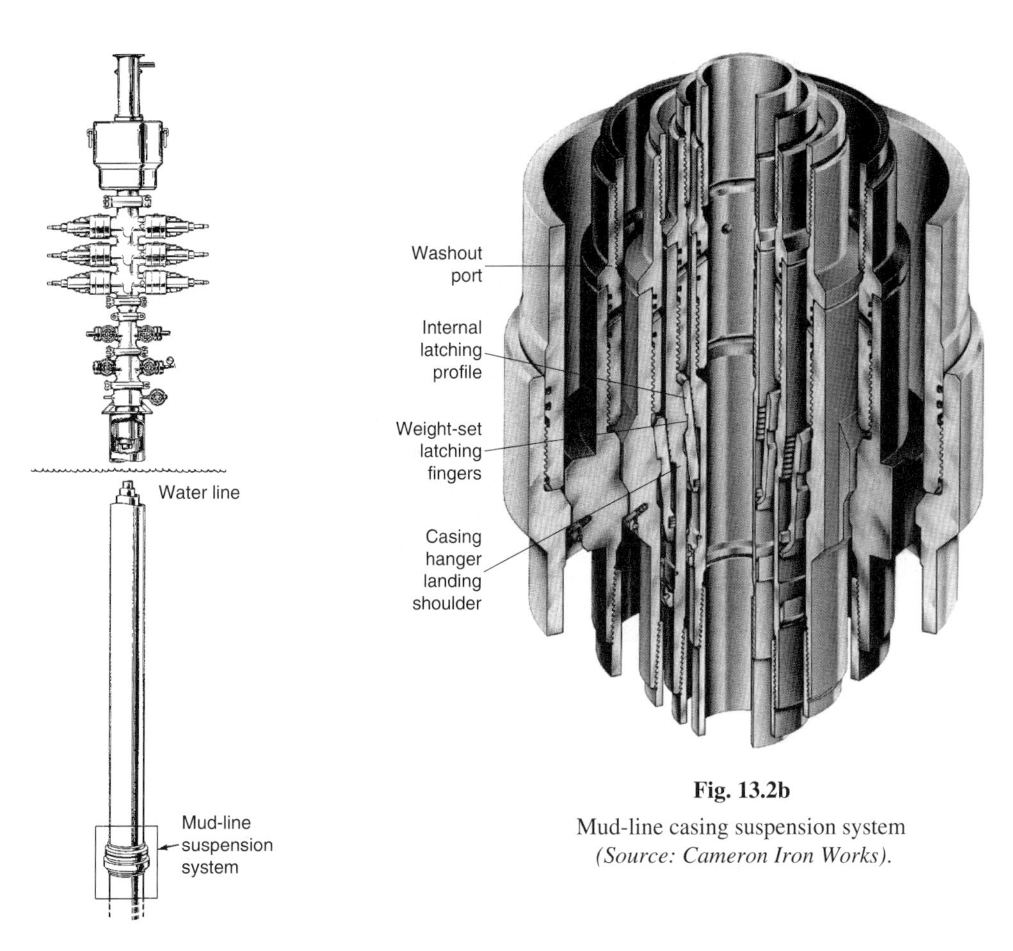

Fig. 13.2a
Wellhead with mud-line suspension
(Source: Cameron Iron Works).

Fig. 13.2b
Mud-line casing suspension system
(Source: Cameron Iron Works).

way one after the other without having to move the jackup. The conductor pipes are guided into the structure and can usually support the other casing strings at the wellhead, the same as for onshore wells.

Getting the jackup platform installed requires the best possible sea and weather conditions. For the operation to succeed, it is obvious that horizontal and vertical movements need to be very limited. Suitable sea conditions are as follows:

- wave height < 1.5 meters,
- wind < 15 knots,
- current < 1.5 knots.

As soon as it has reached the site, the jackup must be immobilized with its anchors and tug boats. Once it is immobilized and oriented, the legs are lowered as quickly as possible (the operating speed of jacks or rack-and-pinion system ranges from 1 to 3 ft/min).

The critical time is when the mat or spud cans touch the sea floor with the jackup undergoing a heaving movement. The jacking up as such is stopped when the waves no longer hit the platform's hull.

The preloading operation is carried out with the hull very close to floating to reduce the risk of accident in the event one of the legs suddenly sinks into the seabed during the test. This operation consists in making the legs penetrate the bottom up to the point where the bearing capacity of the soil is greater than or equal to the force (first of all vertical) exerted by the platform.

The spud cans should provide 30% more operating load according to regulations. Overloading operations by filling the spud cans differ depending on the type of structure:

- for platforms with mats resting on loose soil, it is absolutely necessary to overload the structure evenly to prevent too much differential compaction that could cause a catastrophic accident,
- for platforms with spud cans resting on fairly consolidated soil, the legs can be overloaded one after the other.

Once the preloading tests have been completed, the jacking up process is resumed until the platform sits at a given elevation (air gap) above the water line. The air gap depends on the height of waves and swell in the area.

When the well has been drilled, the platform is then jacked down. Sea conditions must be the same as when it was jacked up, since the process has the same critical points. Once the spud cans have lifted off the sea floor, the legs must be raised quickly to keep them from pounding on the bottom.

In soft soil, the pulling force on spud cans or the suction force on a mat may exceed the installed power of the jacks or be greater than the legs' tensile strength. In this case the forces of interaction between the sea floor and the spud cans or mat must be reduced.

Jetting is often performed to unstick them from the soil. If the current is not too strong (1 knot), it is preferable to do any jetting when the platform is in a floating position. This is because jetting may cause the soil to liquefy under the spud cans. If the platform is high over the water line, it may sink into the sea bottom even more–just the opposite of what is wanted.

13.2.1.2 Swamp barges (Fig. 13.3)

These structures are much smaller than jackups and are made up of two rectangular hulls, one on top of the other and connected to each other by posts.

While the swamp barge is in transit, the bottom hull is deballasted and thereby capable of carrying the whole vessel which is towed along rivers and channels with a draft of at least 2 to 3 meters.

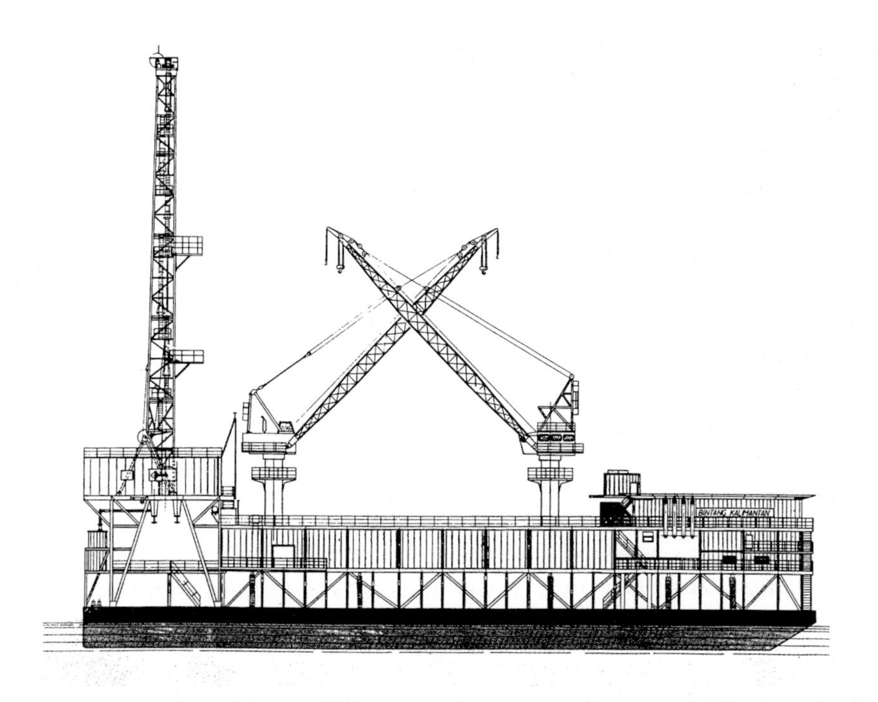

Fig. 13.3
Swamp barge *(Source: Forasol).*

After a place has been dredged out at this depth, the lower hull is ballasted until it rests on the bottom. Given the height of the posts, operations are possible in depths of up to 8 to 10 meters. These barges are often used for exploratory drilling in swampy delta areas (Nigeria, Indonesia, Louisiana).

The drill string passes through a slot in the rear of the hulls with a wellhead above the water line. The barge can always be backed away from the well after deballasting, leaving the self-supporting conductor pipe. Swamp barges are so small that the amount of equipment on board is restricted to a minimum, any further equipment for cementing or other short-duration operations is brought in by flat-bottomed boats.

13.2.2 Rigs incorporated in stationary platforms

These rigs are designed for development drilling on stationary platforms that will later on be converted into production structures.

The technology is similar to onshore drilling, since the conductor pipe going down through the water is guided at several points in the metal jacket structure and can support all of the casing strings at the wellhead. The wellhead stack is completely conventional.

In some cases, compact casingheads are used to reduce the height of the wellhead stack. They include hangers for two casing strings and the production tubing.

13.2.2.1 Compact rigs (Fig. 13.4)

This is the name given to a complete rig that is generally in module form and is set on top of a platform for use under severe sea conditions (North Sea). It can operate under any and all types of weather, provided that supplies can be brought in by boat (casing, consumables).

13.2.2.2 Drilling tender (Fig. 13.5)

In areas with milder weather conditions, a lot of money can be saved by ultra light development platforms that are capable of supporting only the drilling mast and the drawworks. All the remaining drilling equipment, plus accommodation for personnel, is installed on a barge anchored as close as possible to the stationary platform and connected to it by a catwalk leading on to the pipe rack.

Fig. 13.4
Development platform *(Source: Zapata).*

Fig. 13.5a

Development from a platform with a tender
(Source: Monthly bulletin, Sept.–Oct. 1989, Elf Aquitaine).

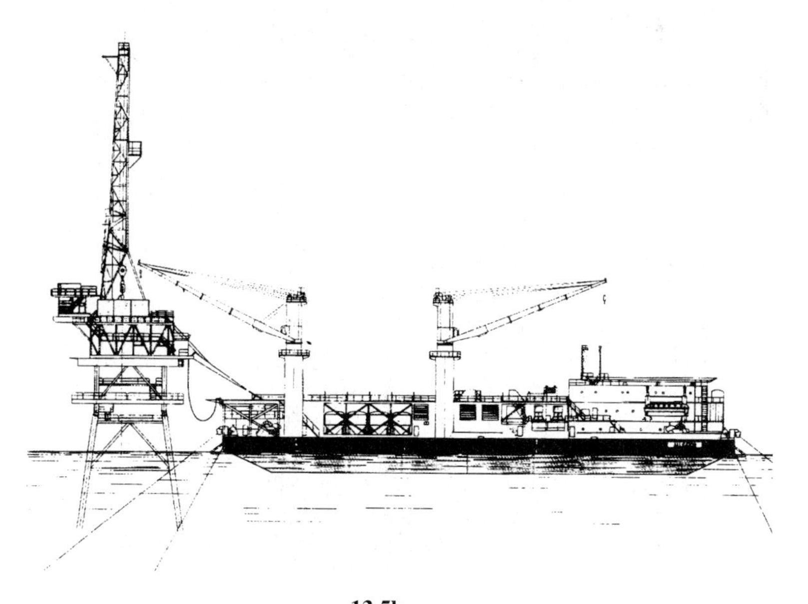

13.5b

Drilling rig and a tender *(Source: Forasol).*

The drawback of this system is that operations must be shut down when the tender barge moves too much due to the swell and wind. It is very commonly used in the calm waters of the Gulf of Guinea and the Persian Gulf where downtime because of bad weather never exceeds 2% of the total time.

13.2.3 Floating support structures

When the water is over 100 meters deep, a floating drilling unit must often be used. It is usually kept on station by a traditional anchoring system (with cables connected to anchors). Because of the movement of the swell and the variation in tide levels, this type of structure can not remain stationary with respect to the sea floor. As a result, there is a device to compensate lengthwise in the drill string.

There are two principles: the bumper sub and the heave compensator.

- **The bumper sub (Fig. 13.6)**
  This piece of equipment consists of a body that a mandrel can slide in. The mandrel is set rotationally in the body by keys or by its specific cross-section (hexagonal or square).
  The bumper sub allows rotation to be transmitted to the drill bit and also has a stroke length of about 5 feet. A sealing system isolates the inside channel. It is generally relatively sophisticated, since bumper subs are preferred in the hydraulically balanced version, i.e. the static and dynamic pressures produce no axial forces.
  Since the available stroke length is only a few feet, several bumper subs will have to be screwed together one on top of the other if a greater heave height is anticipated. Weight on the bit is provided by the series of drill collars placed under the bumper sub. The driller manning the brake must pay attention, once the bit is on the bottom of the hole. He must make sure that the bumper sub is operating according to the heave without extending entirely (taking weight off the bit) and without closing completely (placing extra weight on the bit).
- **Heave compensators (Fig. 13.7)**
  The principle is to keep constant tension on the upper part of the drill string, i.e. where it hangs in the hoisting apparatus. Two solutions are used: a compensator located at the hook and one at the crown block. Both are based on the principle of hydraulic jacks with the body on the traveling block and the rod supporting the hook. Keeping constant weight on the bit, however the drilling mast may move, means keeping constant pressure in the jack, however the rod may move in the body.
  Pressure is regulated conventionally: statically by a hydraulic connection with a large-volume air tank that acts as a buffer when the jack discharges oil or when it takes in oil according to the floater's high or low position. The system is the same when the compensator is located at the crown block, except that the whole pulley system is suspended by means of the jacks. The advantage in this case is that there are no hydraulic hoses in the mast and that it takes up less room in the derrick.

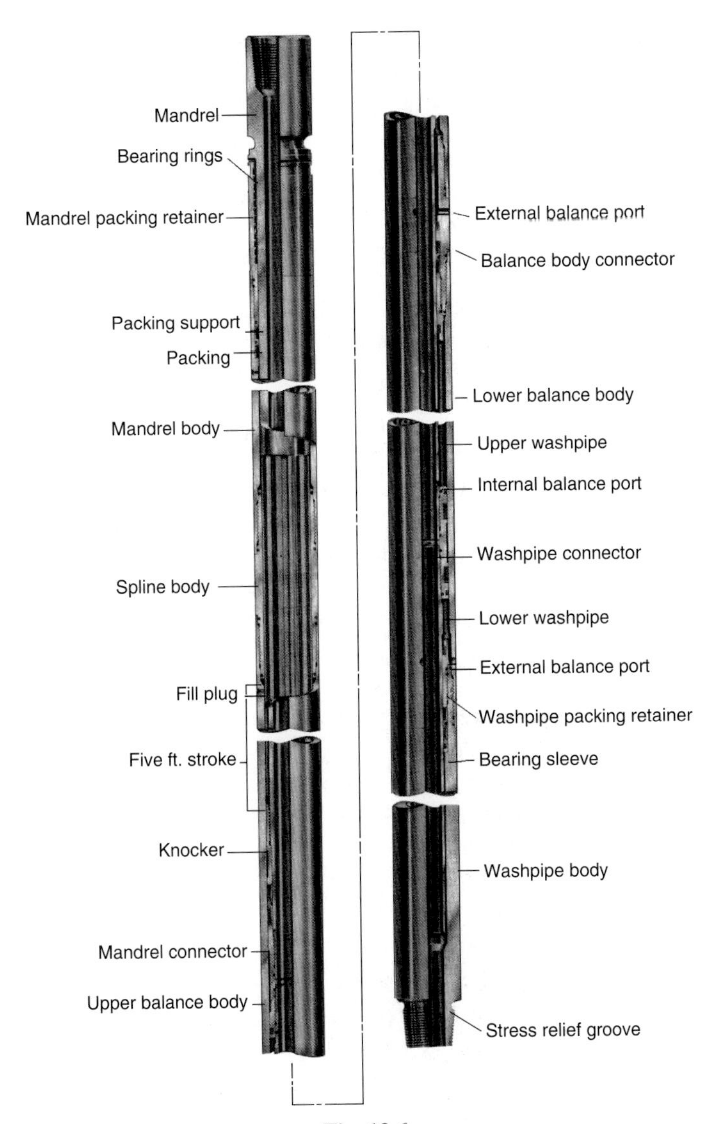

Fig. 13.6

Bumper sub. Example of characteristics *(Source: Bowen).*

Maximum outside diameter (in)	4 3/4	6 1/4	7 3/4
Inside diameter (in)	2	3	3 1/2
Stroke (ft)	5	5	5
Torsional strength (lb/ft)	50,000	100,000	200,000
Allowable tension (lb)	700,000	1,000,000	1,300,000
Length (ft)	34	37	37

However, there is an increase in vertical load.
The stroke of the hydraulic jacks must match the acceptable degree of heave to drill under proper conditions. The double jack Vetco system gives a maximum stroke of 30 ft.

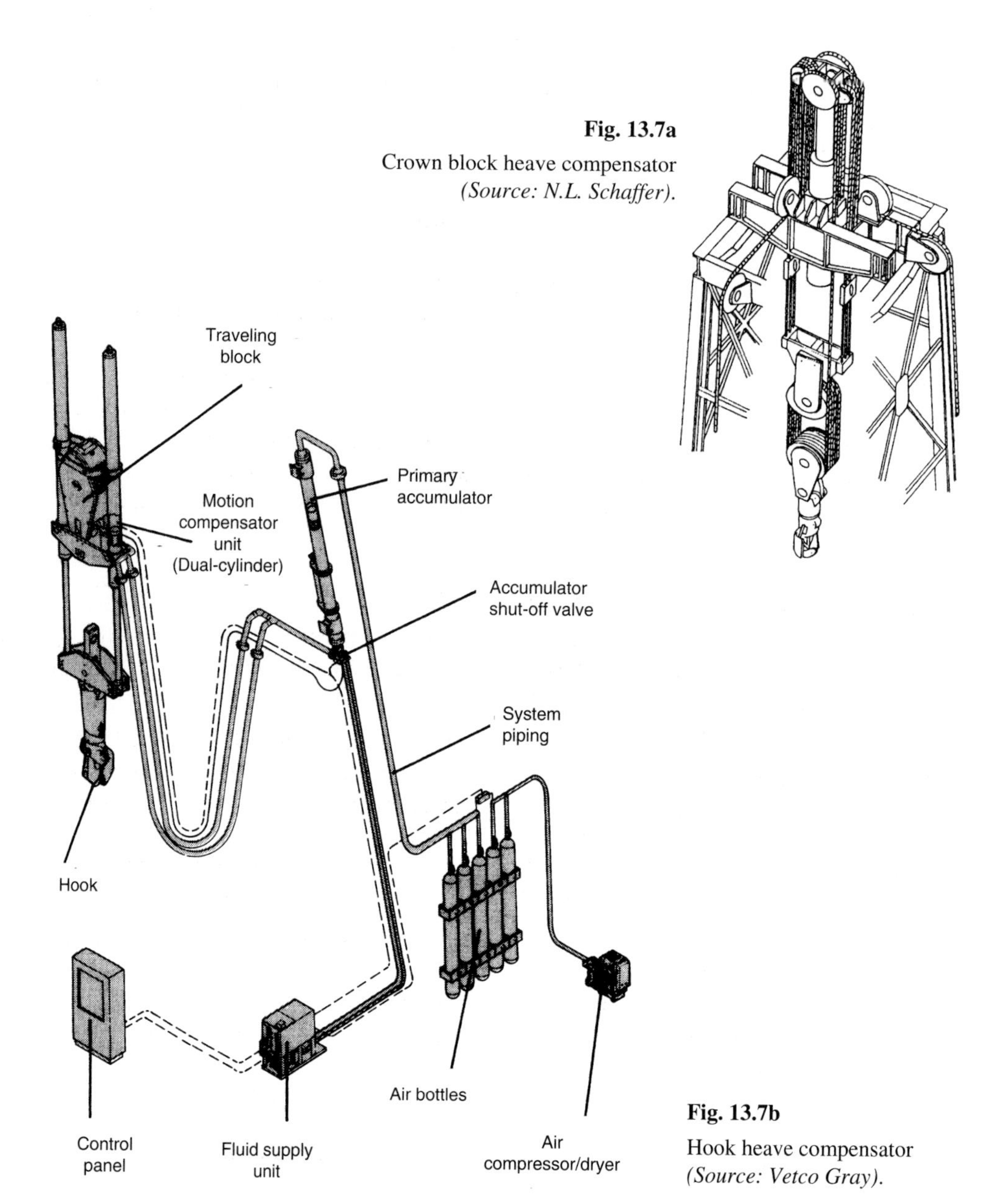

Fig. 13.7a
Crown block heave compensator *(Source: N.L. Schaffer).*

Fig. 13.7b
Hook heave compensator *(Source: Vetco Gray).*

13.2.3.1 Drillships (Fig. 13.8)

In the sixties, drillships accounted for almost all the mobile units for offshore drilling. They were usually fitted out cheaply using the hulls of landing barges retrieved in the fifties. They were characterized by flat bottoms, a 4 or 5 meters draft at the most, and they were seldom over 90 meters long and 15 meters wide. They can be used only in calm waters, such as the Gulf of Mexico or off California.

When they are used for drilling, the well is drilled through an opening, the moon pool, with the rig above it. The deck is converted into a tubular storage area, the pumps and mud conditioning facilities are installed in the hold. They are kept on station over the drilling site by a traditional anchoring system with eight or ten anchors.

They used to be limited to water depths of approximately 200 meters because they had such a small storage capacity, but a new generation of drillships appeared in the early seventies. They have specially-built hulls 150 to 160 meters long, 20 to 22 meters wide, with 7 to 8 meters draft. The variable load capacity of approximately 8000 tons allows storage of 600 to 1000 meters of riser and the equipment required for two wells drilled to a depth of 5000 meters. The conventional anchoring system was replaced by a dynamic positioning system using the main thrusters and transverse tunnel thrusters that keep the ship on station above the well. The installed capacity is in excess of 15,000 hp. Drillships are highly mobile and have made it possible to explore difficult areas full of icebergs in the Arctic seas and in deep water (over 2000 meters) where the reservoirs that have been discovered have as yet not been produced.

13.2.3.2 Semisubmersible platforms (Fig. 13.9)

Since drillships are relatively sensitive to the state of the sea, operational downtime becomes economically unacceptable in bad weather or under severe marine conditions such as in the North Sea.

Sometime around 1966, the first large semisubmersibles came on the market. They have submerged pontoons or buoyancy chambers to keep them afloat that are connected to the deck by columns. Compared with a drillship measuring 160 × 25 meters, the area of the cross-section at the water line is divided by a factor of ten and this ensures good stability even when the swell is strong. The deck is approximately 40 meters high at the base of the pontoons and it has a draft of 22 meters in the submerged drilling position.

It is towed to the well site completely deballasted with a draft of about 7 meters corresponding to the height of the pontoons. The anchoring system is based on 10 to 15-ton anchors and chains or cables. Several types of semisubmersibles have been built:

- three pontoons and three large columns, triangular deck (*Sedco* and *Saipem*),
- five pontoons and five equidistant columns (*Pentagone*),
- two parallel pontoons and six or eight columns (*Aker, Penrod, Pacesetter, Odeco,* etc.),
- four parallel pontoons, two of which are smaller and located laterally (*Odeco*).

Fig. 13.8

Drillship with dynamic positioning *(Source: Forasol)*.

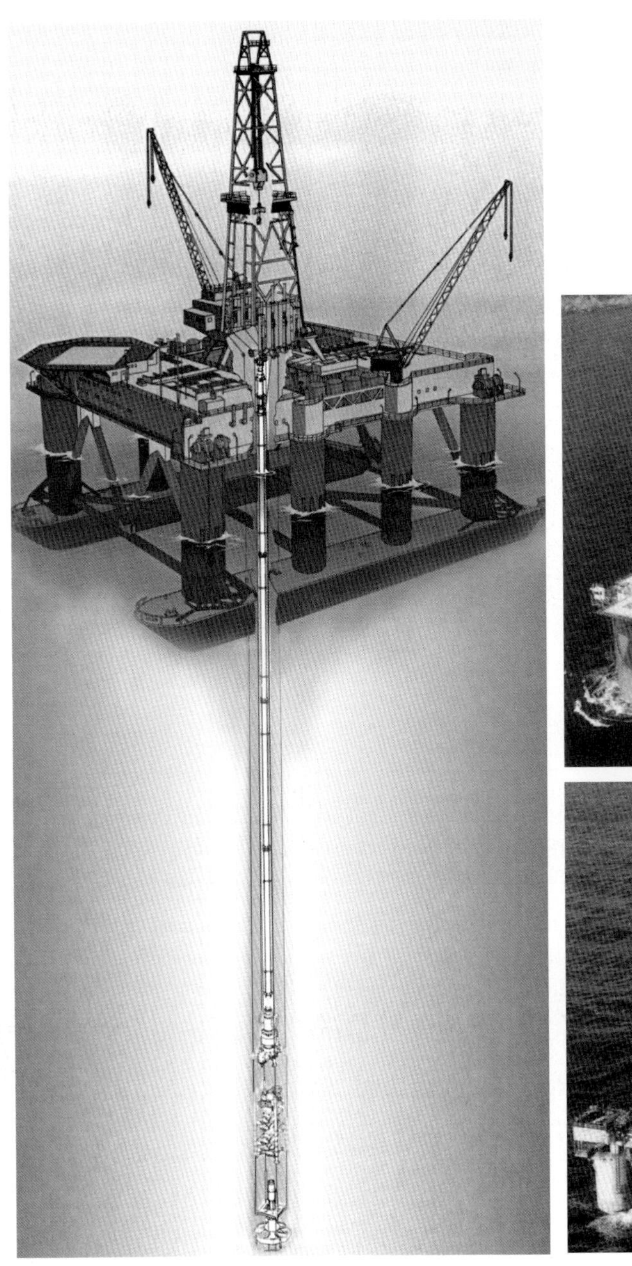

Fig. 13.9a

Semisubmersible platform with subsea equipment *(Source: Vetco Gray)*.

Fig. 13.9b

Semisubmersible platforms.

Toward the end of the seventies, semisubmersibles were able to drill in 500 m of water with a conventional anchoring system at the cost of increased tonnage. Some platforms were equipped with a dynamic positioning system and their displacement reached 50,000 tons *(Henri Goodrich).*

13.2.4 The offshore drilling rig fleet

Most of the mobile offshore drilling units belong to drilling contractors except for those that are the property of national oil companies (*Petrobras*, *Pemex*, China, the CIS, etc.).

The size of the mobile unit fleet has been greatly influenced by the price of the barrel of crude. A look at the statistics on the number of units whose construction was launched versus years, shows two peaks that correspond to the financial fallout of the oil price shocks in 1973 and 1980.

In 1986, the fleet of drilling units capable of being used offshore amounted to some 1150, broken down as follows:

Jackups	447	(94)
Semisubmersibles	168	(33)
Drillships	54	(22)
Swamp barges	25	(12)
Barges	36	(13)
Tenders	57	(24)
Compact rigs (stationary platform)	361	(146)

The figures between parentheses indicate the units that were not under contract at the beginning of 1986. The trend got worse in 1987 with nearly half the units idle, but the market began to recover in 1988.

However, the flow of new building has completely dried up and stagnating crude prices give no hope of an upswing. The number of units on the market is therefore declining slightly because older, less powerful units are being scrapped.

There were 717 mobile units in mid-1988 with a utilization rate of 66%. The breakdown was as follows:

Jackups	432	(293)
Semisubmersibles	177	(117)
Drillships and barges*	70	(49)*
Swamp barges	38	(15)

* 18 of which dynamically positioned.

The cost of building these drilling units increased moderately from 1960 to 1980. For a large-capacity semisubmersible, the cost went from $10 million to $30 million and for a jackup, from $6 million to $25 million. Of course, this increase included the enormous technological progress accomplished in 20 years.

Sometime around 1978, a rule of the thumb was to depreciate the construction cost in 1000 days (less than three years), i.e. a daily rate of approximately $30,000 for a semisubmersible. The dearth of units in 1980-1981 naturally led to a substantial price hike which was compounded by the high cost of building the latest generation of mobile units in overextended shipyards toward 1982. The daily semisubmersible rate reached $85,000 in the North Sea and $47,000 in the Gulf of Mexico at the time.

In the same way, the slowdown in activity due to the drop in crude prices and the surplus of available units in 1986 led to a fall in rates. Rates reached levels lower than those prevailing in 1978, i.e. less than $20,000 for the North Sea and less than $15,000 for the Gulf of Mexico.

13.2.5 Problems specific to floaters

Since the well is a construction that has to withstand high internal pressures, no flexible element can be incorporated in it that would go up through the water head and serve to attenuate the drilling unit's variations in position. This entails the following technical solutions:
- casing hanger equipment rests on the mud line,
- the subsea BOP stack is placed on top of the casingheads,
- the BOP stack is connected to the floater by a pipe called a riser,
- the riser has a flexible joint at the base and a slip joint in the upper part to accommodate the stresses of currents and the movements of the floater,
- the kill line and choke line go up to the choke manifold on the floater, just like the riser (**Fig. 13.10**).

13.2.5.1 Floater anchoring systems

Before the floater is anchored, its orientation, or heading, must be determined to minimize the impact of outside forces on drilling operations.

For a drillship

Any vessel of any type is always vulnerable from the side and the only defense mechanism against the elements is to position it hove to (facing the swell). If it is hit from the side, the following problems arise:
- a larger surface area is exposed to the wind, with consequent strain on the lateral anchor lines, causing the anchors to drag on the sea floor and sometimes breaking the anchor lines,
- too much roll for operations to proceed.

As a result, the sector where the most frequent winds and swells originate must be determined on the basis of statistical tables established for the drilling period. The vessel will then be hove to when atmospheric conditions become unfavorable.

In some regions, the choice of heading can pose an insoluble problem and therefore dictate using a semisubmersible instead of a drillship.

For a semisubmersible

The action of the elements in relation to the main axis is much less important for this type of structure than for a drillship. The choice of heading is especially governed by the position of the helideck, the supply boat mooring stations and the flares.

After the direction and origin of winds and swells have been determined, a heading is chosen so that:

- a helicopter can land and take off facing the prevailing winds without being hindered by the derrick,
- a supply boat can moor facing the prevailing swells to limit its movements during transshipment,
- the flares can be used at all times.

A. *Marking out the site*

The conventional procedure involves placing a number of marker buoys (six should be enough) according to a predetermined layout. This operation is carried out preferably right before the floater gets to the site to decrease the chances of losing buoys.

When the well site is very far from the coast and in relatively deep water, a system of acoustic marking can be used. The system uses a network of submerged transponders and solves the problem of inaccurate positioning due to the buoys swinging on their anchors. Three transponders are enough to give an accurate position. On board the floater there is a receiver that can query the transponders and read their distance. Acoustic marking also is advantageous in that it can be set up several months beforehand, for example when the sea floor is being surveyed.

B. *Anchoring*

As soon as the marking buoys are in sight, the tugboat maneuvers to line up with the heading buoys. During the approach, the better located stern anchor is dropped and left hanging about thirty meters off the bottom. When the floater passes in front of the marker buoy, the anchor is dropped and the cable is allowed to run out under a tension of 20 to 30 tons.

The tugboat gradually slows its hauling speed and stops as soon as the floater gets even with the lateral buoys. At the same time, the stern anchoring winch is stopped. One of the anchor-handling boats helps drop the second stern anchor as soon as possible. Meanwhile the other workboat goes to help drop the anchor that is best located to keep the floater on station depending on the wind and current conditions prevailing at the time.

The other anchors are then dropped in such a way as to hold the floater to the planned heading and counteract any drift. The tugboat can cast off as soon as enough anchors have been dropped to fulfill these conditions.

When the anchoring operations have been concluded, the floater is positioned by maneuvering the anchoring winches as needed.

The anchoring setup must be tested at a tension slightly in excess of the maximum design tension expected for extreme operating conditions. The tension should not go over a value equal to one-third of the breaking load of the line, however.

If the seabed is poorly consolidated, a waiting period of at least 24 hours is recommended after anchoring before proceeding to tests. Testing consists in gradually increasing the tension on two opposing lines in stages of 20 tons. When the test tension has been reached, it is held for 30 minutes, then released with the lines coming back to the original tension.

C. Dynamic positioning

Drillships that are dynamically positioned use lateral thrusters and conventional propellers to remain on station above the subsea wellhead. The major advantage of this technique is that the ship stays mobile since there are no anchors to drop or retrieve. As a result, this positioning system is independent of the water depth.

In oilwell drilling, the record wellhead depth is over 2000 m, with the present-day limit residing in the riser's mechanical resistance and weight.

The position of the rotary table axis in relation to the borehole axis is determined acoustically by a transponder located on the wellhead and hydrophones under the ship's hull. A computer analyzes the relative position of the ship in relation to the wellhead and sends operating commands to the different thrusters depending on the environment (wind, current, waves).

There are two types of thrusters:

- The controlled-pitch propeller which rotates at constant speed and whose action is variable depending on the pitch of the blades. It may be driven by an asynchronous electric motor with the slant of the blades hydraulically actuated. In addition to the motor-driven advantage, the response time is excellent compared to the type of thruster discussed below.
- The fixed-pitch variable speed propeller, which must be driven by a direct current motor with a control loop. But the system's inertia is considerable and the time required to reverse propulsion direction does not promote regulated stabilization. On one hand, the necessary reliability of the dynamic positioning system means complex instrumentation due to redundant safety equipment. On the other hand, power requirements are substantial. Modern dynamically-positioned drillships are equipped with a power plant of over 20,000 hp.

D. The subsea wellhead **(Fig. 13.10)**

Subsea wellhead equipment can be broken down into three main subassemblies:

- the casinghead which acts as a foundation, a casing string hanger and a seal,
- the BOPs (or BOP stack),
- the marine riser and at its base the remote-control platform.

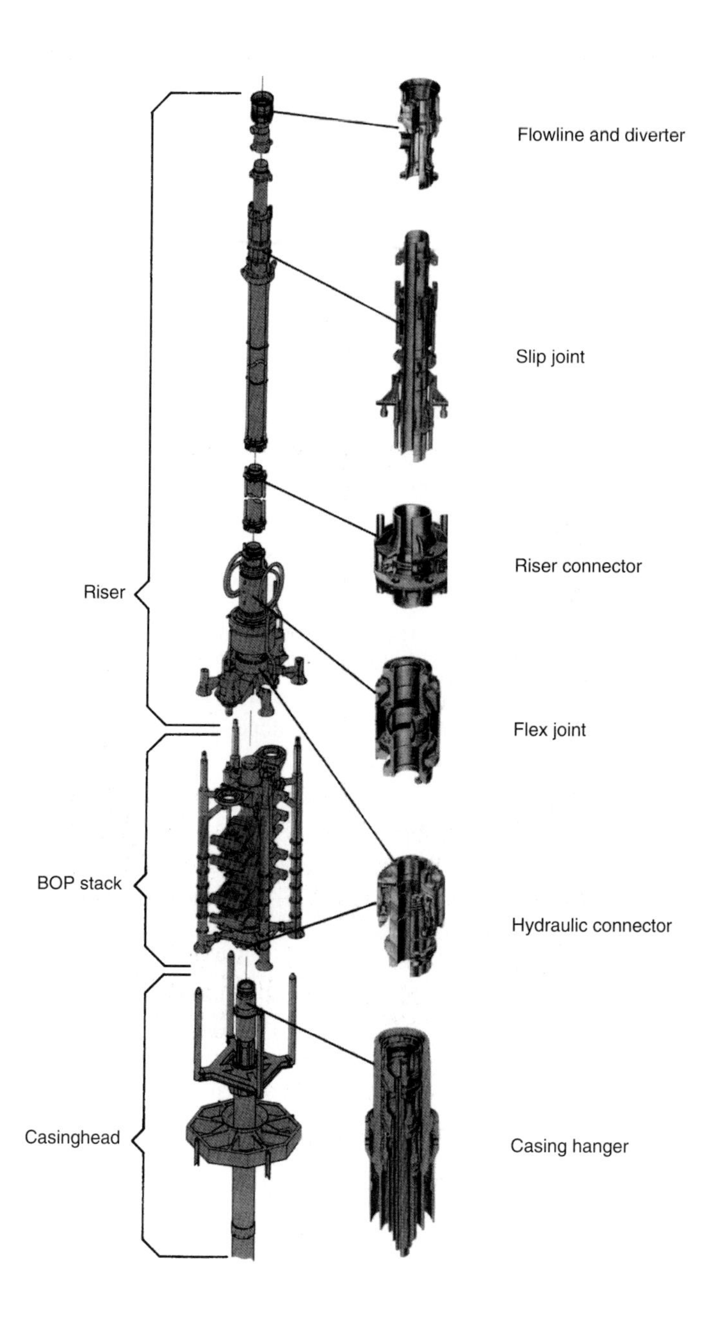

Fig. 13.10

Subsea equipment *(Source: Vetco Gray).*

The subassemblies are interconnected by subs that are operated hydraulically from the floater. In this way, the casinghead can be isolated from the BOPs and the riser or the well can be temporarily abandoned after it has been placed under BOP control by disconnecting at the base of the riser.

The casinghead (Fig. 13.11)

The following drilling and casing program can be considered standard for a floater:

Bit size 36", then 30" casing
Bit size 26", then 20" (or 18 5/8") casing
Bit size 17 1/2", then 13 3/8" casing
Bit size 12 1/4", then 9 5/8" casing
Bit size 8 1/2", then 7" casing.

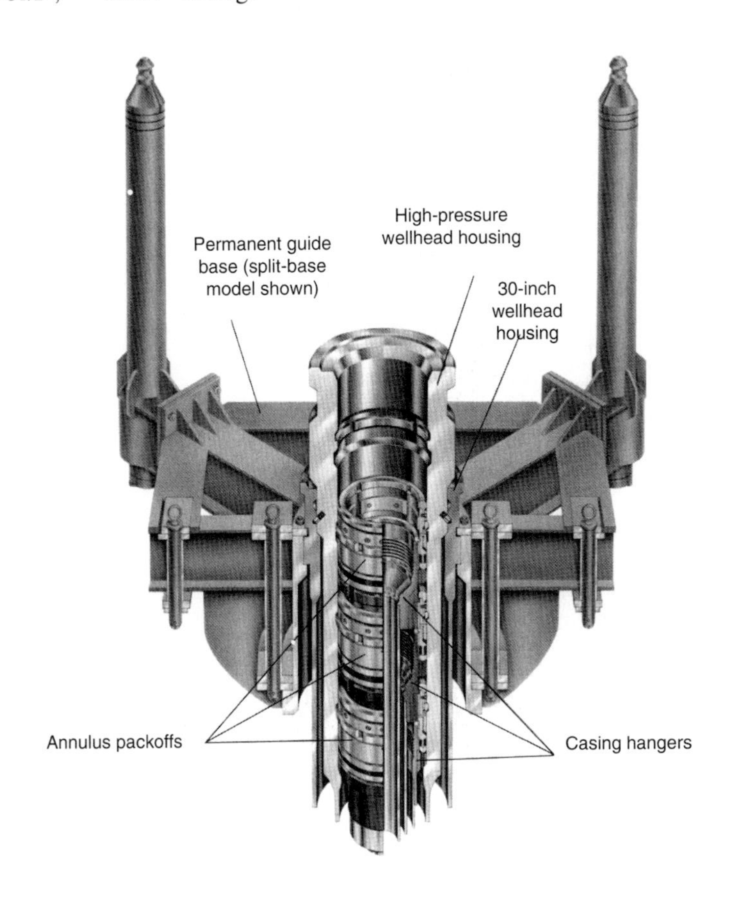

Fig. 13.11

Casing hanger in a subsea wellhead *(Source: FMC)*.

In our discussion of how a well with a subsea wellhead progresses, we will illustrate each operation:

1. Running in the guide base by means of a running tool screwed onto the end of the drillpipe (**Fig. 13.12**).
2. Drilling with a 36" bit without any mud returns coming to the surface (**Fig. 13.13**).
3. Running in the 30" casing assembled by means of squnch joints. Cementing (**Fig. 13.14**).
4. Drilling with a 26" bit with lost circulation or with mud returning to the surface after the riser has been connected (special nonpolluting mud is used) (**Fig. 13.15**).
5. Running in the 20" casing equipped with the casinghead. Cementing (**Fig. 13.16**).
6. Running in blowout preventers at the end of the riser and connection onto the top of the 20" casing (**Fig. 13.17**). Before drilling is resumed, all functions and seals must be thoroughly tested. Except in the event of trouble or if the floater can not stay on station over the well, the BOPs and riser will remain on the casinghead for the following operations:
   - drilling with a 17 1/2" bit, 13 3/8" casing landed and cemented,
   - drilling with a 12 1/4" bit, 9 5/8" casing landed and cemented,
   - drilling with a 8 1/2" bit, 7" casing landed and cemented.

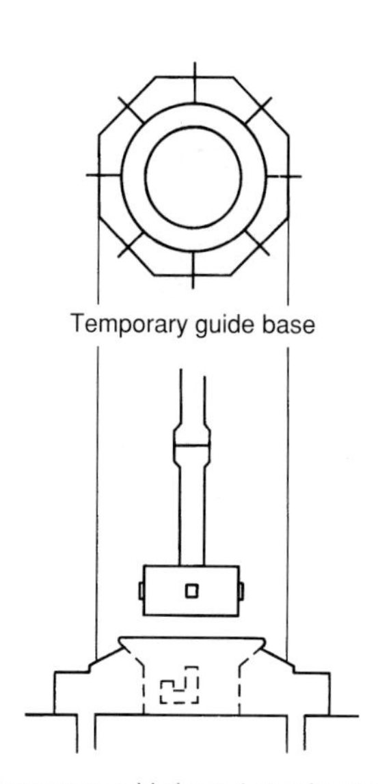

Fig. 13.12
Installing the guide base *(Source: Vetco Gray).*

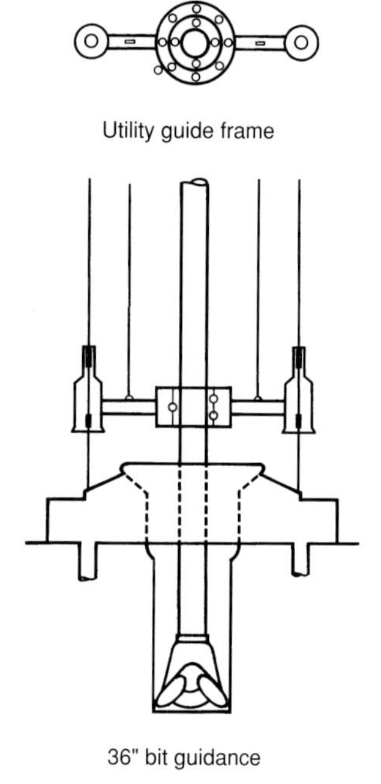

Fig. 13.13
Running in the 36" bit *(Source: Vetco Gray).*

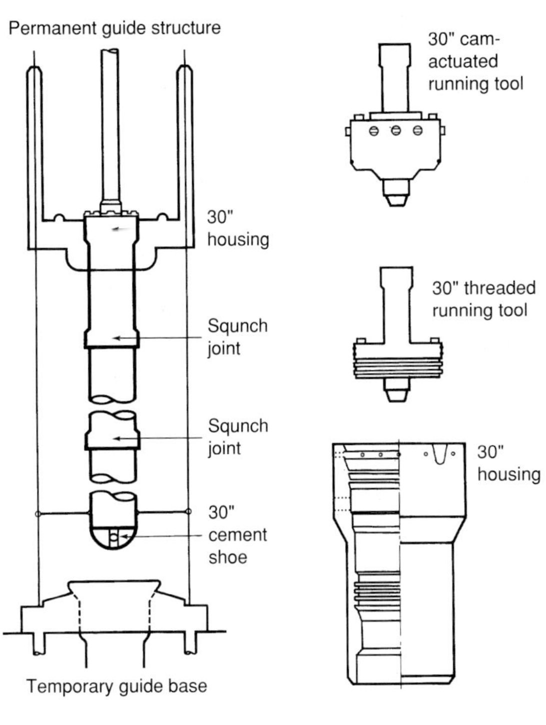

Fig. 13.14

Equipment on the 30" string *(Source: Vetco Gray).*

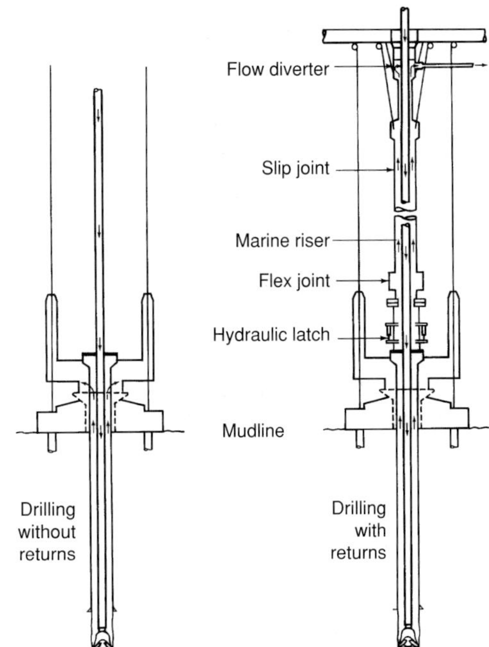

Fig. 13.15

Drilling after 30" casing has been cemented *(Source: Vetco Gray).*

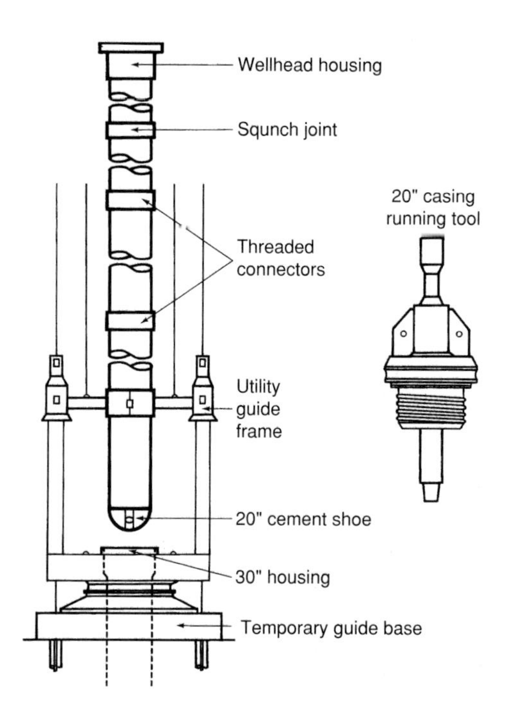

Fig. 13.16

Equipment on the 20" column *(Source: Vetco Gray).*

The operating principle for these wellheads is to:

- run in casing by means of the drillpipe and a running tool screwed onto the casing hanger,
- hang the casing in the 20" wellhead housing,
- proceed to cement as if for a liner,
- actuate (by rotation, weight or hydraulic tool) the annular sealing system, or packoff assembly,
- test the seal,
- install a wear bushing.

The drill bit can then be run in through the riser, the wellhead and the 20" surface casing in order to continue with the program. The sequence of operations for further casing strings is identical. The wear bushings will become shorter as the casing hangers stack up.

The BOP stack

There is absolutely no difference with onshore blowout preventers except for the ram-type BOP hydraulic locking system (**Fig. 13.18**).

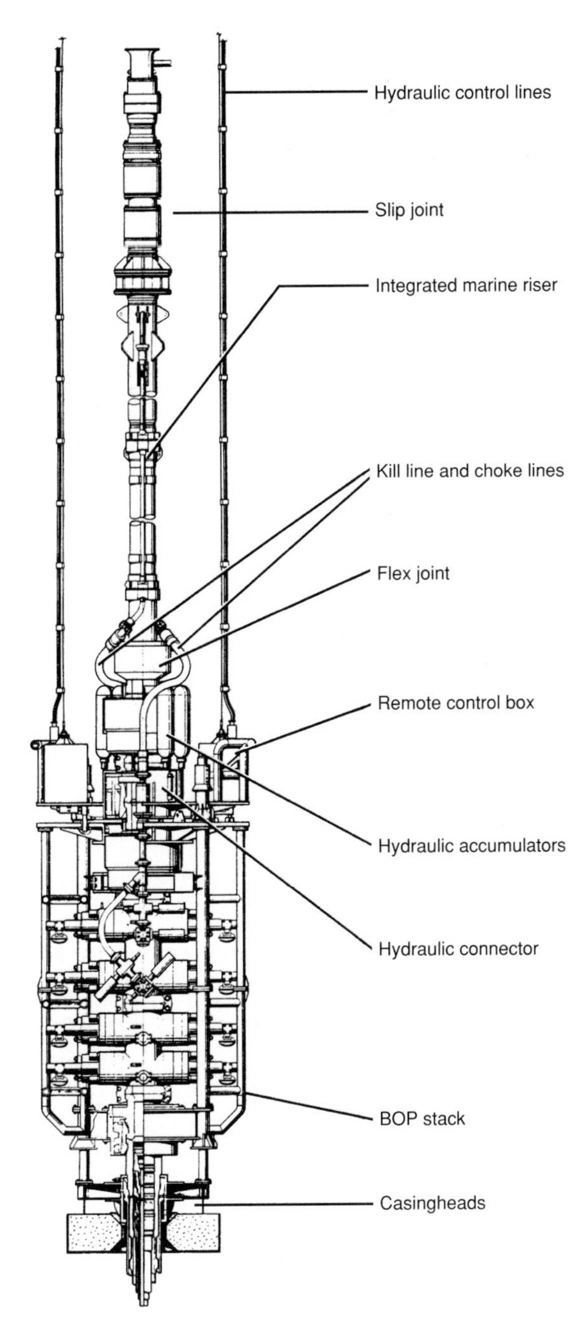

Fig. 13.17

BOP stack and riser *(Source: Cameron Iron Works).*

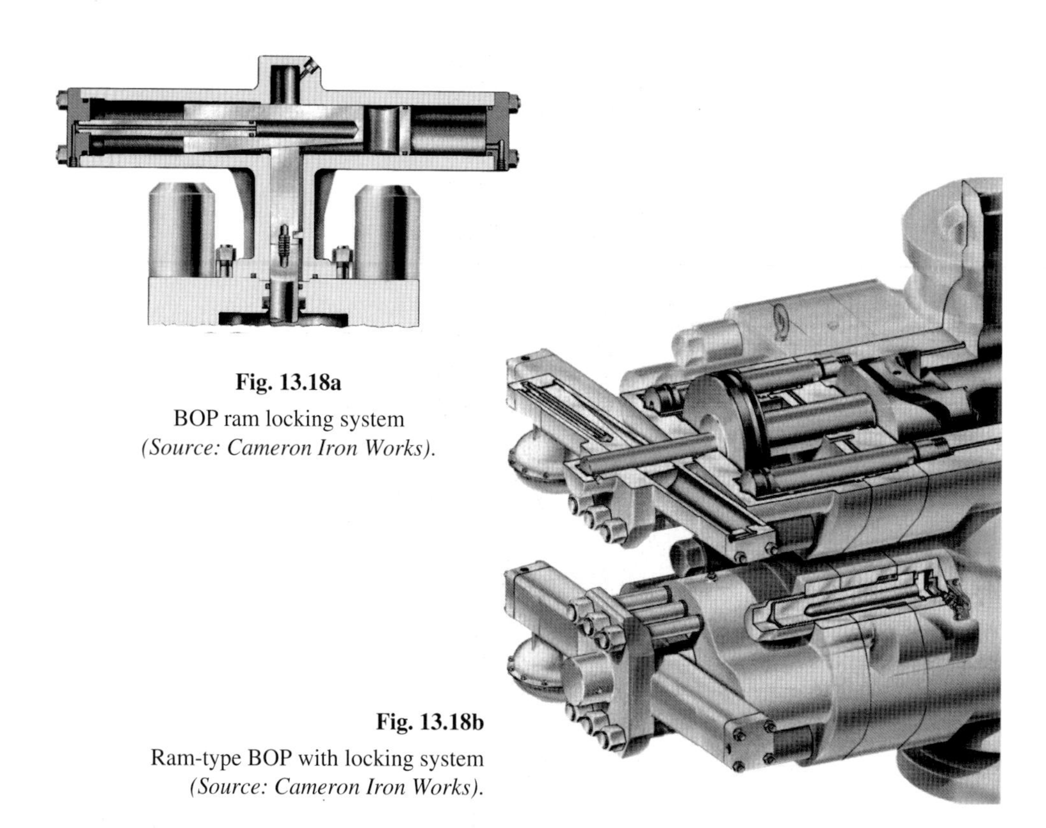

Fig. 13.18a
BOP ram locking system
(Source: Cameron Iron Works).

Fig. 13.18b
Ram-type BOP with locking system
(Source: Cameron Iron Works).

The maximum stack consists of four ram-type BOPs: shear blind rams, two pipe rams and a variable ram. This is not the only setup, it can vary depending on the operator.

The side outlets are generally all connected two by two to the choke line and the kill line. The valves have an installed spare for safety reasons and are termed failsafe, i.e. they are closed mechanically by a spring that opens only with hydraulic pressure (**Fig. 13.19**).

All the possibilities for circulating with the well shut in are therefore ensured owing to a manifold on the surface. The kill and choke lines are connected to the floating support structure by high-pressure conduits that are integrated in each riser element.

Two annular BOPs top off the assembly but are separated by a hydraulic connector. One of them is part of the BOP stack and the other is part of what is termed the lower marine riser package.

The riser and the remote control platform

The connections between the well and the floating support structure are located in the riser:

- the mud returns to the surface through it and it serves as a guide to run into the well,

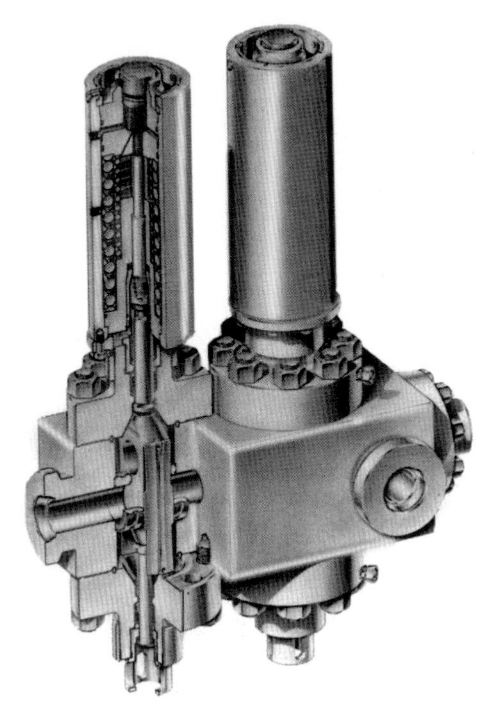

Fig. 13.19
Failsafe valve *(Source: Cameron Iron Works)*.

- it has lateral conduits such as the kill and choke lines, the booster line and the power hydraulics.

The riser needs to be able to withstand the effects of swell, currents, tides and the movements of the floater. So that there is no risk of buckling, an upward pulling force is applied to the riser by means of tensioners located on the sides of the drilling rig. The tensioners are constant tension winches or hydraulic jacks equipped with a pulley system that can absorb heave movements (**Fig. 13.20**).

The main components are, from top to bottom:
- mud return line and diverter,
- slip joint to attenuate heave movements,
- the riser pipes,
- a flexible joint at the top of the BOP stack,
- a hydraulic connector (**Fig. 13.21**).

All of this equipment is extremely heavy, for example a *Vetco* joint 50 feet long made of 21" pipe that is half an inch thick, along with 10,000 psi control lines, weighs approximately 9300 lb (4220 kg). A thousand meters of riser can be estimated to weigh around 275 tons in air (apparent weight 240 t). For deep water drilling, the riser has to be lightened by floats made of syntactic foam with a specific gravity of 0.5.

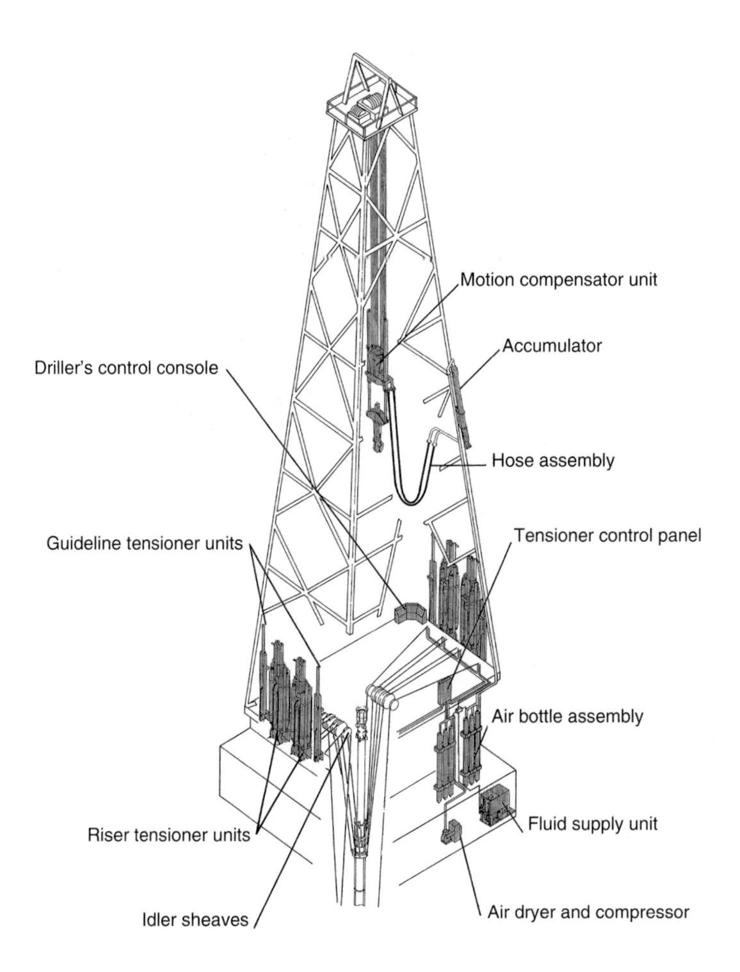

Fig. 13.20

Hydraulic tensioning system *(Source: Vetco Gray).*

The remote control system for the BOPs, the valves and the connectors is usually hydraulic/hydraulic, i.e. the surface control of the four-way valves located on the hydraulic accumulation unit sends a control pressure to the remote control box which is located at the base of the riser (lower marine riser package). The pressure moves a distributer drawer (SPM valve) which drives the pressurized oil from the surface (and bottom) accumulators toward the chosen function.

The remote control system is schematically shown in **Fig. 13.22**. The surface setup is strictly the same principle as for an onshore rig, except that the control panel diagram of the wellhead is more complex (**Fig. 13.23**).

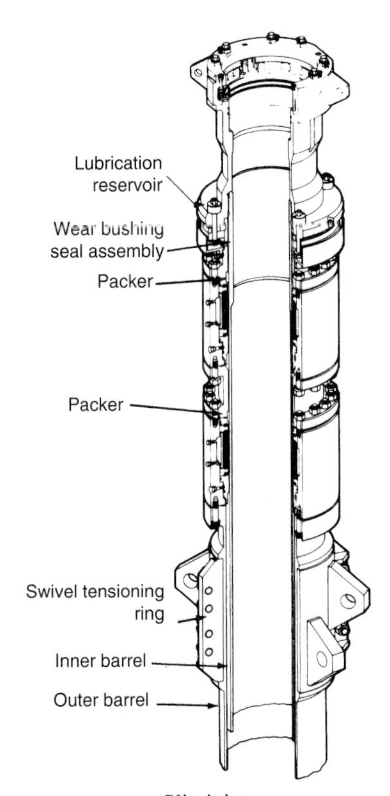

Slip joint
(Source: Cameron Iron Works).

Riser and joint
(Source: Cameron Iron Works).

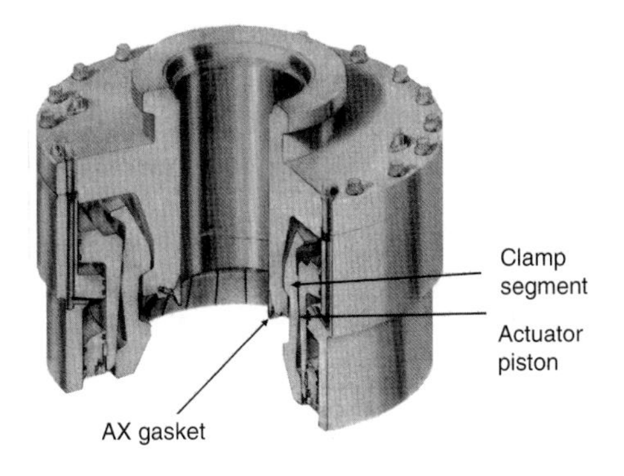

Hydraulic connector
(Source: Cameron Iron Works).

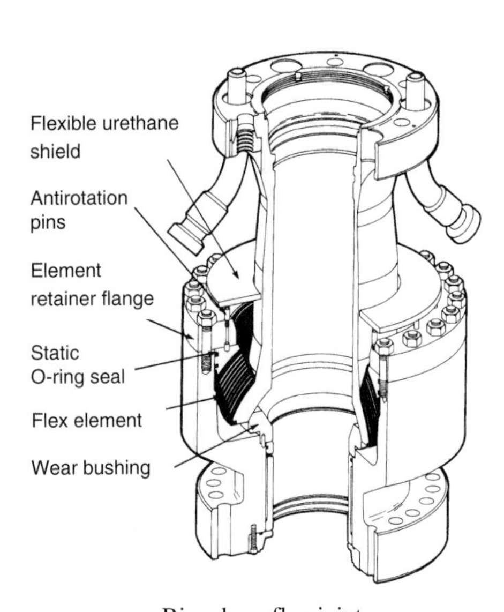

Riser base flex joint
(Source: Vetco Gray).

Fig. 13.21

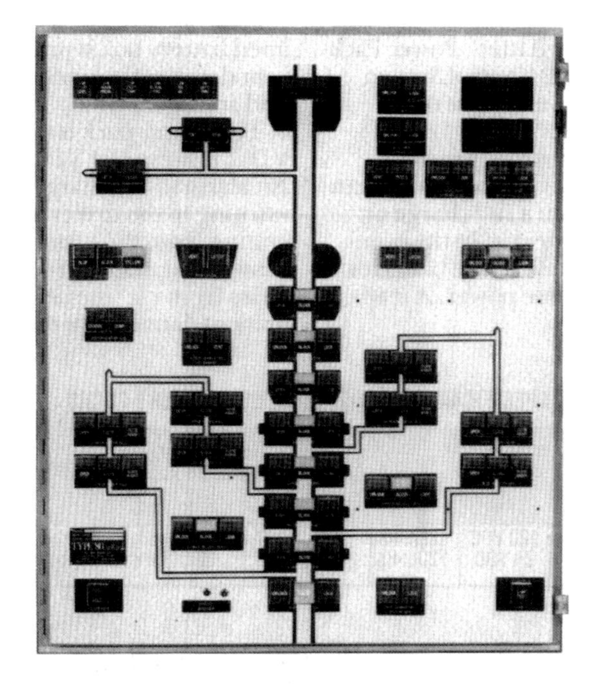

Fig. 13.22
BOP stack control panel *(Source: Koomey Inc.)*.

All the hydrovalves are located in a box called a pod. The pod and its control umbilical can sometimes be brought up and lowered back down alone along the guidelines. All the controls are twinned to enhance safety: there is a yellow pod and a blue pod. A circuit selecting valve, or shuttle valve, is placed on the junction points of the functions so that one pod or the other can be used.

13.3 THE MARINE ENVIRONMENT

The environment offshore has a much greater influence on oil industry activity than the environment onshore. Understanding the factors involved requires a number of studies and surveys before any drilling operation is begun offshore.

The type of prior survey depends largely on the type of drilling support structure:
- anchoring survey for an anchored floater,
- soil and terrain survey for a structure that rests on the seabed (mud line).

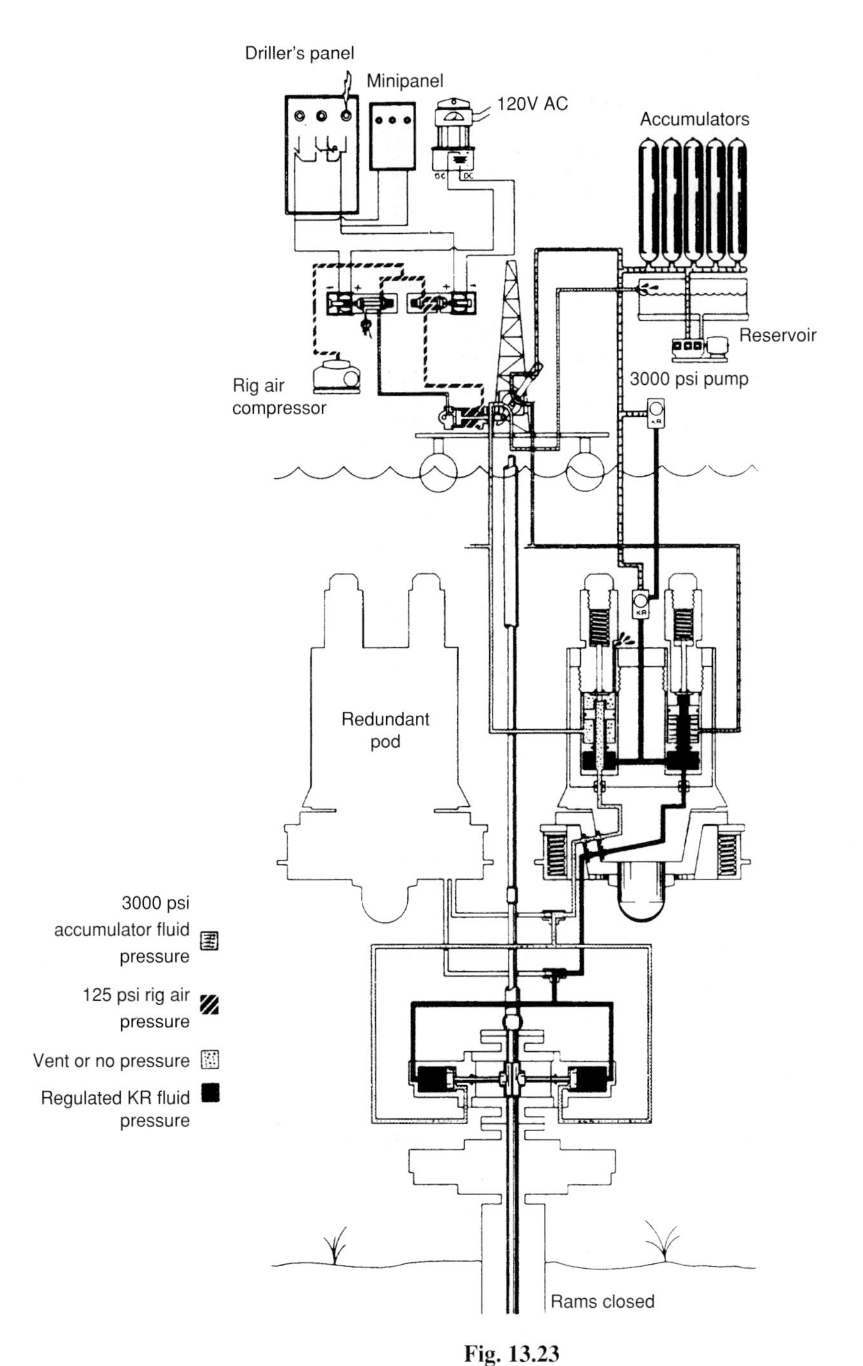

Fig. 13.23

Simplified diagram of BOP stack hydraulic control *(Source: Koomey Inc.).*

In both instances, a program to measure meteorological and oceanographic parameters (winds, currents, waves) in the relevant area must be carried out for a sufficiently long time in order to confirm available statistical records.

13.3.1 For floating support structures

The main idea is to be sure that the equipment on board the unit is suitable for conventional anchoring on the chosen site. Surveys aim to:

- determine the most probable maximum meteorological and oceanographic conditions during the period when the well is to be drilled,
- find out exactly what type of sediments there are for a few meters' depth on the seabed to asses how well the anchors will hold.

This is usually carried out in two phases: installing current meter recorders on the site for several months, preferably during the same season as the planned operation; surveying the sea floor over the area that will be involved in the anchoring layout.

The following measurements will be made:

- depth finder to get an accurate bathymetric reading,
- side-scan sonar to locate obstacles or wrecks,
- shallow sparker to assess the thickness of loose sediments,
- gravity coring to take samples.

A trawler-type vessel 40 to 50 meters long is usually adequate to carry out the program.

13.3.2 For mobile units resting on the bottom (jackups)

The surveyed area can be restricted to 1 km $\times$ 1 km around the theoretical drilling site. The survey aims to assess how far the platform legs will sink into the bottom under the maximum load anticipated during operation and under the most unfavorable meteorological conditions. The meteorological and oceanographic data required are the same as for the floater, however, the data on the type of soil must be more complete. In addition to the surveys mentioned above, the following must be planned:

- deep coring down to 40 meters in loose soil,
- in situ penetrometer test,
- detailed measurements of soil characteristics on core samples,
- pressuremetering in the core sample borehole if need be.

These surveys require the use of a vessel that is specially equipped for coring. Serious accidents have occurred because of inadequate surveying before jackup installation.

REFERENCES

Adams N., *Drilling engineering,* PennWell Publishing Company, Tulsa, Oklahoma, 1985.

Adams N., *Well control problems and solutions,* Petroleum Publishing Company, Tulsa, Oklahoma, 1980.

API Production Division *Standards and Recomended Practices.*

ARTEP, *Techniques et procédures de cimentation,* Ref. IFP 36350, 1988.

Baker R., *A primer of oil-well drilling,* Petroleum Extension Service, Austin, Texas, 1979.

Cendre A., *Les masses-tiges,* Ref. IFP 24147, 1976.

CESFEG, *Énergie et transmissions en forage,* Ref. IFP 28596, 1980.

Chambre Syndicale de la Recherche et de la production du Pétrole et du gaz Naturel.
Comité des Techniciens:

Les mesures en cours de forage, Éditions Technip, Paris, 1982.

Drilling mud and cement slurry rheological manual, Éditions Technip, Paris, 1982.

Directional drilling and deviation control technology, Éditions Technip, Paris, 1990.

Blowout prevention and well control, Éditions Technip, Paris, 1981.

Complétion et reconditionnement des puits, Programmes et modes opératoires, Éditions Technip, Paris, 1985.

Standard agreement for offshore drilling operation, Ref. IFP 27107, 1979.

Choquin A., *Cours de forage, fascicule 1: Paramètres et contrôle du forage,* cours de l'ENSPM, Éditions Technip, Paris, 1982.

Collier S.L., *Mud pump handbook,* Gulf Publishing Company, Houston, Texas, 1983.

Composite Catalog.

Dareing D.W., *Applied drill string mechanics,* Maurer engineering Inc., Houston, Texas, 1982.

Davenport B., *Handbook of drilling practices,* Gulf Publishing Company, Houston, Texas, 1984.

Desbrandes R., *Encyclopedia of well logging,* Éditions Technip, Paris, 1985.

Esso, *Le pétrole, prospection et production,* Esso standard SAF, Département Formation, 1979.

Gabolde G. and Nguyen J.P., *Drilling data handbook,* 6th edition, Éditions Technip, Paris, 1991.

Garcia C. and Parigot P., *Boues de forage,* cours de l'ENSPM, Éditions Technip, Paris, 1968.

Goins W.C., "Better understanding prevent tubular buckling problems", *World oil,* January 1980.

Gutsche W. and Noevig T., "Comparing rig power transmission systems", *World oil,* April 1989.

Henz A., *Mise en œuvre des ciments,* Ref. IFP 23617, 1975.

IADC, *Rotary drilling series, Unit 1: lessons 1 to 12, Unit 2: lessons 1 to 5, Unit 3: lessons 1 to 4,* Petroleum Extension Service, Austin, Texas, 1983.

IADC, *Drilling manual,* tenth edition, 1982.

Jurgens R., "Down-hole motors", *Preprint of a paper given at the 7th International Symposium,* Celle, RFA, 8th of September 1978.

Leblond A., *Cours de forage, Équipement de forage,* Tome 1: Texte et planches, cours de l'ENSPM, Éditions Technip, Paris, 1963.

Lemoigne Y., *Têtes de puits terrestres,* Ref. IFP 23470, 1975.

Lynch P.L., *A primer in drilling and production equipment,* Volumes 1, 2 and 3, Gulf Publishing Company, Houston, Texas, 1981.

Maillet P., *Besoins en puissance,* Ref. IFP 29577, 1981.

McCabe C., "New down-hole motor develops high torque for increased penetration rates", *Ocean Industry,* June 1982.

McNair W.L., *The electric drilling rig handbook,* Petroleum Publishing Company, Tulsa, Oklahoma, 1980.

Menetrier C., *Les opérations de tubage et de cimentation,* Ref. IFP 23617, 1975.

Menetrier C., *Le programme de forage et de tubage,* Ref. IFP 19512, 1971.

Milchem, *Mud facts engineering handbook,* Milchem Incorporated, 1984.

Moore P.L., *Drilling pratices,* Second edition, PennWell Publishing Company, Tulsa, Oklahoma, 1986.

NAM, *L'énergie des profondeur,* Nederlandse Aardolie Maatschappij (NAM), Assen, Pays-bas, 1972.

Ordronneau M., *Manuel du technicien fluides de forage,* Tomes 1 et 2, CECA SA, Département boues de forage, Paris, 1982.

Ormsby G., *Mud Equipment Manual, handbook 6,* Gulf Publishing Company, Houston, Texas, 1984.

Ormsby G. and Young G., *Mud Equipment Manual, handbook 2,* Gulf Publishing Company, Houston, Texas, 1984.

Pasternak G., *Transmissions Diesel-électriques dans le forage pétrolier,* Éditions Technip, Paris, 1970.

Perrin D., *Information complétion,* Ref. IFP 29255, 1981.

Rebeyrol Y., "Le forage... à travers les âges", *Pétrole informations,* Avril 1985.

Robinson L.H., *Mud Equipment Manual, handbook 7,* Gulf Publishing Company, Houston, Texas, 1984.

Servco, *Fishing and milling tool Catalogue,* Servco, division of Smith International, Inc.

Sheffield R., *Floating drilling: equipment and its use,* Gulf Publishing Company, Houston, Texas, 1980.

Short J.H., *Drilling and casing operations,* PennWell Publishing Company, Tulsa, Oklahoma, 1982.

Short J.H., *Fishing and casing repair,* PennWell Publishing Company, Tulsa, Oklahoma, 1981.

Soulié G., *Technologie des têtes de puits sous-marines de forage,* Éditions Technip, Paris, 1978.

Terrien M., *Méthodologie des essais de puits,* Ref. IFP 34191, 1986.

Wilson G.E., Garret W.R., "How to drill a usable hole", *World oil,* October 1976.

Wilson G.E., Garret W.R., "Proper field pratices for drill collar strings", *SPE 5124,* 1976.

INDEX

ACHEVÉ D'IMPRIMER
SUR LES PRESSES DE
L'IMPRIMERIE CHIRAT
42540 ST-JUST-LA-PENDUE
EN DÉCEMBRE 1995
DÉPÔT LÉGAL 1996 N° 1813
N° D'ÉDITEUR : 922

IMPRIMÉ EN FRANCE